2/19/93



# THE GREAT PHYSICISTS *from Galileo to Einstein*

George Gamow

*Illustrations by the Author*

DOVER PUBLICATIONS, INC., NEW YORK

**TO PERKY**



Published in Canada by General Publishing Company, Ltd., 30 Lesmill Road, Don Mills, Toronto, Ontario.
Published in the United Kingdom by Constable and Company, Ltd.

This Dover edition, first published in 1988, is an unabridged and unaltered republication of *Biography of Physics*, originally published by Harper & Brothers, Publishers, New York in 1961 in the "Harper Modern Science Series."

Manufactured in the United States of America
Dover Publications, Inc., 31 East 2nd Street, Mineola, N.Y. 11501

**Library of Congress Cataloging-in-Publication Data**

Gamow, George, 1904–1968.
[Biography of physics]
The great physicists from Galileo to Einstein / George Gamow ; illustrations by the author.
p. cm.
Reprint. Originally published: Biography of physics. 1st ed. New York : Harper & Brothers, c1961. (Harper modern science series)
Bibliography: p.
Includes index.
ISBN 0-486-25767-3 (pbk.)
1. Physics—History. I. Title.
QC7.G27 1988
530'.092'2—dc19
[B] 88-17677
CIP

# Contents

# *Preface*

THERE ARE two types of books on physics. One is the textbook, intended for teaching the reader the facts and theories of physics. Books of this kind usually omit all historical aspects of the development of science; the only information concerning great scientists of the past and present is limited to the year of birth and death (or —) given in brackets after the name. The other type is essentially historical, devoted to biographical data and to character analysis of the great men of science, and simply listing their discoveries under the assumption that the reader, studying the history of a given science, is familiar with that science itself.

In the present book I have tried to keep a midway course, discussing on an equal basis the trial of Galileo and the basic laws of mechanics which he discovered, or giving my personal recollections about Niels Bohr along with detailed discussion of Bohr's atomic model. The discussion in each of the eight chapters is centered around a single great figure, or at most two, with other physicists of that era and their contributions forming more of the background. This accounts for the omission of many names which would be found in most books on the history of physics, and for the omission of many topics that are a "must" in the regular physics textbooks. The aim of this book is to give the reader the feeling of *what* physics *is,* and *what kinds of people* physicists *are,* thus getting him interested enough to pursue his studies by seeking out more systematically written books on the subject.

When one is reading about the great men of the past or present, it is always desirable to know how they looked or look. Because of the limitations in the number of glossy pages, I decided to use them all for reproducing actual photographs of various physical phenomena such as the spectra of light, the diffraction of the electron, and the tracks of nuclear particles in the cloud chamber. Thus, it became necessary to make over the portraits of physicists into pen drawings. Not being an artist, I had to use certain auxiliary devices, such as projection of photographic slides on drawing paper, and the results seemed to carry the likeness well enough to justify their being included.

I hope that this book will give young readers (and maybe some older readers, too) the impulse to study physics; this is its primary aim.

GEORGE GAMOW

*University of Colorado*

# CHAPTER I *The Dawn of Physics*

IT IS VERY difficult to trace the origin of the science of physics, just as it is difficult to trace the origin of many great rivers. A few tiny springs bubbling from under the green foliage of tropical vegetation, or trickling out from beneath the moss-covered rocks in the barren northern country; a few small brooks and creeks running gaily down the mountain slopes and uniting to form rivulets which in turn unite to form streams big enough to deserve the name "river." The rivers grow broader and broader, being strengthened by numerous tributaries, and finally develop into mighty flows—be it the Mississippi or the Volga, the Nile or the Amazon—carrying their waters into the ocean.

The springs which gave birth to the great river of physical science were scattered all over the surface of the earth inhabited by *Homo sapiens,* i.e., thinking man. It seems, however, that most of them were concentrated on the southern tip of the Balkan peninsula inhabited by the people now known as "ancient Greeks"; or, at least, it seems so to us who inherited the culture of these early "intellectuals." It is interesting to notice that whereas other ancient nations, such as Babylonia and Egypt, contributed a great deal to the early development of mathematics and astronomy, they were completely sterile in the development of physics. The possible explanation of that deficiency as compared with Greek science is that Babylonian and Egyptian gods lived high up among the stars, while the gods of the ancient Greeks lived at the elevation of only about 10,000 feet, at the top of Mount Olympus, and so were much closer to down-to-earth problems. According to a legend the term "magnetism" originated from the name of a Greek shepherd, Μάγνης, who was surprised to notice that the tip of his iron-shod staff was attracted by a stone (magnetic iron ore) lying along the roadside. Similarly the term "electricity" comes from the Greek word ἤλεκτρον, for amber, maybe because some other Hellenic shepherd, trying to polish a piece of amber by rubbing it against the coat of one of his sheep, noticed that it acquired a mysterious property to attract loose pieces of wood.

## PYTHAGOREAN LAW OF STRINGS

Whereas these legendary discoveries would hardly stand their ground in any legal priority suit, the discovery of the Greek philosopher Pythagoras, who lived in the middle of the 6th century B.C.,

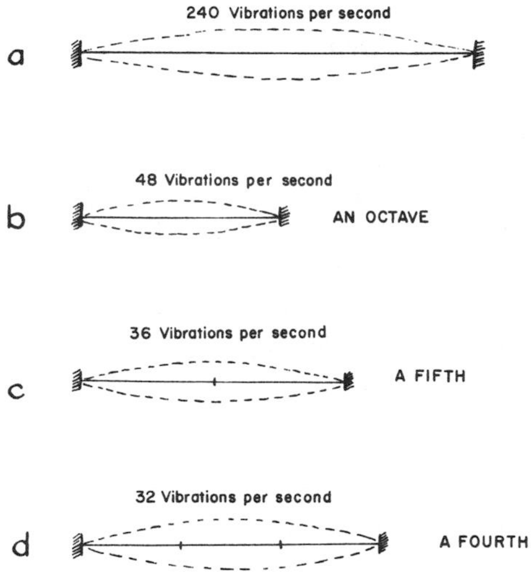


FIG. I–1.
Pythagorean law of strings.

is well documented. Persuaded that the world is governed by number, he was investigating the relation between the lengths of the strings in the musical instruments which produce harmonic combinations of sound. For this purpose he used the so-called monochord, i.e., a single string which can be varied in length and subjected to different tensions caused by a suspended weight. Using the same weight, and varying the length of the string, he found that the pairs of harmonic tones were obtained when the string's lengths stood in simple numerical relations. The length ratio 2 : 1 corresponded to what was known as "octave," the ratio 3 : 2 to a "fifth," and the ratio 4 : 3 to a "fourth." This discovery was probably the first mathematical formulation of a physical law, and can be well considered as the first step in the development of what is now known as theoretical physics. In modern physical terminology we may reformulate Pythagoras' discovery by saying that the *frequency*, i.e., the number of vibrations per second, of a given string subjected to a given tension is inversely proportional to its length. Thus, if the second string (Fig. I-1b) is half as long as the first one (Fig. I-1a), its vibration frequency will be twice as high. If the two string lengths stand in the ratio of 3 : 2 or 4 : 3, their vibration frequencies will stand in the ratios 2 : 3 or 3 : 4 (Fig. I-1c,d). Since the part of the human brain which receives the nerve signals from the ear is built in such a way that simple frequency ratios, as 3 : 4, give "pleasure" whereas the complex ones, as 137 : 171, "displeasure" (it is for the brain physiologists of the future to explain that fact!), the length of strings giving perfect accord must stand in simple numerical ratios.

Pythagoras tried to go one step further by suggesting that, since the motion of planets "must be harmonious," their distances from the earth must stand in the same ratios as the lengths of the strings (under equal tension) which produce the seven basic tones produced by the lyre, the national Greek musical instrument. This proposal was probably the first example of what is now often called "pathological physical theory."

## DEMOCRITUS, THE ATOMIST

Another important physical theory, which in modern terminology could be called "a theory without any experimental foundation" but which turned out to be a "dream that came true," was proposed by another ancient Greek philosopher, Democritus, who lived, thought, and taught around the year 400 B.C. Democritus conceived the idea that all material bodies are aggregates of innumerable particles so small as to be unnoticeable by the human eye. He called these particles *atoms,* or indivisibles (ἄτομος) in Greek, because he believed that they represented the ultimate stage of the division of material bodies into smaller and smaller parts. He believed that there are four different kinds of atoms: the atoms of stone, dry and heavy; the atoms of water, heavy and wet; the atoms of air, cold and light; and the atoms of fire, slippery and hot. By a combination of these four different kinds of atoms all known materials were supposed to be made. The soil was a combination of stone and water atoms. A plant growing from the soil under the influence of sun rays consisted of the stone and water atoms from the soil and the fire atoms from the sun. That is why dry wood logs which have lost all their water atoms would burn, liberating fire atoms (flame) and leaving behind stone atoms (ashes). When certain kinds of stones (metallic ores) were put in the flame, stone atoms united with the fire atoms producing the substances known as metals. Cheap metals like iron contained very small amounts of fire atoms and therefore looked rather dull. Gold had the maximum amount of fire atoms and thus was brilliant and valuable. Consequently, if one could add more fire atoms to the plain iron, one should be able to make precious gold!

A student who would tell all that in his introductory chemistry examination would certainly get an F grade. But, although these particular examples of the nature of chemical transformation were certainly wrong, the fundamental idea of obtaining almost unlimited numbers of different substances by a combination of only a few basic chemical elements was undoubtedly correct and now represents the foundation of chemistry today. It took, however, twenty-two centuries from the time of Democritus to the time of Dalton to make things right.

## ARISTOTELIAN PHILOSOPHY

One of the giants of the ancient Greek world was a man named Aristotle, who became famous on two counts: first, because he was a real genius; second, because he was a tutor and later a protégé of Alexander the Great of Macedonia. He was born in 384 B.C. in the Greek colonial town Stagira on the Aegean Sea to a former court physician of the Macedonian royal family. At the age of 17 he came to Athens and joined the philosophical school of Plato, remaining an ardent student of Plato until his (Plato's) death in 347 B.C. After this, there followed a period of extensive travel until he finally returned to Athens and founded a philosophical school known as Peripatetic which was held at the Lyceum. Most of the works of Aristotle preserved until our time are the "treatises" which probably represent the texts of lectures which he delivered in the Lyceum on various branches of science. There are the treatises on logic and psychology of which he was the inventor, the treatises on political science, and on various biological problems, especially on classification of plants and animals. But, whereas in all these fields Aristotle made tremendous contributions which influenced human thought for two millennia after his death, probably his most important contribution in the field of physics was the invention of the name of that science which was derived by him from the Greek word φύσις which means *nature*. The shortcomings of Aristotelian philosophy in the study of physical phenomena should be ascribed to the fact that Aristotle's great mind was not mathematically inclined as were the minds of many other ancient Greek philosophers. His ideas concerning the motion of terrestrial objects and celestial bodies did probably more harm than service to the progress of science. At the rebirth of scientific thinking during the Renaissance, people like Galileo had to struggle hard to throw off the yoke of Aristotelian philosophy, which was generally considered at that time as "the last word in knowledge," making further inquiries into the nature of things quite unnecessary.

### ARCHIMEDES' LAW OF LEVER

Another great Greek of the ancient period who lived about a century after the time of Aristotle was Archimedes (Fig. I-2), father of the science of mechanics, who lived in Syracuse, the capital of a Greek colony in Sicily. Being the son of an astronomer, he early acquired mathematical interests and skill, and in the course of his life made a number of very important contributions to different branches of mathematics. His most important work in pure mathematics was the discovery of the relation between the surface and volume of a sphere and its circumscribing cylinder; in fact, in accordance with his desire, his tomb was marked with a sphere inscribed into a cylinder. In his book entitled Ψαμμίτης (*Psammites* or Sandreckoner), he develops the method of writing very large numbers by ascribing to each figure in the row different "order" according to its position,* and applying it to the problem of writing down the number of grains of sand contained within a sphere of the size of the earth.

In his famous book *On the Equilibrium of Planes* (in two volumes) Archimedes develops the laws of lever and discusses the problem of finding the center of gravity of any given body. For a modern reader, the style of Archimedes' writings will sound rather heavy and long-winded, resembling in many respects the style of Euclid's books on

* The method we use now in writing numbers in decimal system; i.e., so many units, so many tens, so many hundreds, so many thousands, etc.

FIG. I–2.
Archimedes and the crown.

geometry. In fact, at the time of Archimedes, Greek mathematics was almost entirely limited to geometry, algebra being invented much later by the Arabs. Thus, various proofs in the field of mechanics and other branches of physics were carried out by considering geometrical figures rather than, as we do now, writing algebraic equations. As in Euclid's *Geometry,* over which many a reader had sweated in his or her school days, Archimedes formulates the basic laws of "statics" (i.e., the study of equilibrium) by formulating the "postulates" and deriving from them a number of "propositions." We reproduce here the beginning of the first volume.*

1. Equal weights at equal distances are in equilibrium, and equal weights at unequal distances are not in equilibrium, but incline towards the weight which is at the greater distance.
2. If, when weights at certain distances are in equilibrium, something is added to one of the weights, they are not in equilibrium, but incline towards the weight to which the addition is made.
3. Similarly, if anything be taken away from one of the weights, they are not in equilibrium but incline towards the weight from which nothing was taken.
4. If equal and similar plane figures coincide if applied to one another, their centers of gravity similarly coincide.
5. If figures are unequal but similar, the centers of gravity will be similarly situated. By points similarly situated in relation to similar figures I mean points such that, if straight lines be drawn through them to the equal angles, they make equal angles with the corresponding sides.
6. If two weights at certain distances be in equilibrium, other two weights equal to them will be also in equilibrium at the same distances. [Ain't it clear?]
7. In any figure whose perimeter is concave in the same direction the center of gravity must be within the figure.

These postulates are followed by fifteen propositions derived from them by straightforward logical arguments. We give here the first five propositions, omitting their proof, and quote the exact proofs of the sixth proposition since it involves the fundamental *law of lever.*

*Propositions:*

1. Weights that balance at equal distances are equal . . . .
2. Unequal weights at equal distances will not balance but will incline towards greater weight . . . .
3. Unequal weights will (or rather may) balance at unequal distances, the greater weight being at the lesser distance . . . .
4. If two equal weights have not the same center of gravity, the center of gravity of both taken together is the middle point of the line joining their centers of gravity . . . .

* Quotations in this chapter from Archimedes, Plutarch, Vitruvius, Heron and Ptolemy are reprinted by permission of the publishers from Morris R. Cohen and I. E. Drabkin, *A Source Book in Greek Science,* Cambridge, Mass.: Harvard University Press, copyright 1948 by The President and Fellows of Harvard College.

5. If three equal weights have their centers of gravity on the straight line at equal distances, the center of gravity of the system will coincide with that of the middle weight . . . .

We now turn to the proof of the sixth proposition, modernizing it slightly for the sake of the reader:

6. Two weights balance at distances reciprocally proportional to the weights.

Suppose the weights $A$, $B$ are commeasurable,* and the points represent their centers of gravity (Fig. I-3a):

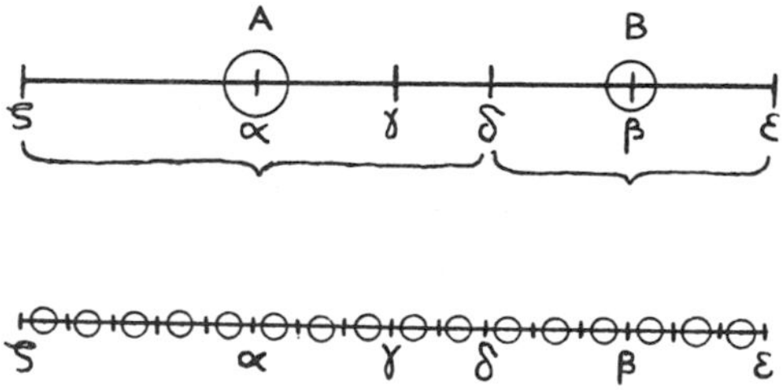


FIG. I–3.
Archimedes' proof of the law of lever.

Draw through $\alpha\beta$ a straight line so divided at $\gamma$ that

$$A : B = \overline{\beta\gamma} : \overline{\gamma\alpha}$$

We have to prove that $\gamma$ is the center of gravity of the two taken together. Since $A$ and $B$ are commeasurable, so are $\overline{\beta\gamma}$ and $\overline{\gamma\alpha}$. Let $\overline{\mu\nu}$ be a common measure of $\overline{\beta\gamma}$ and $\overline{\gamma\alpha}$. Make $\overline{\beta\delta}$ and $\overline{\beta\epsilon}$ each equal to $\overline{\alpha\gamma}$, and $\overline{\alpha\varsigma}$ equal to $\overline{\gamma\beta}$. Then $\overline{\alpha\delta} = \overline{\gamma\beta}$ since $\overline{\beta\delta} = \overline{\gamma\alpha}$. Therefore, $\overline{\varsigma\delta}$ is bisected at $\alpha$ as $\overline{\delta\epsilon}$ is bisected at $\beta$. Thus $\overline{\varsigma\delta}$ and $\overline{\delta\epsilon}$ must each contain $\overline{\mu\nu}$ even number of times.

Take a weight $\Omega$ such that $\Omega$ is contained as many times in $A$ as $\overline{\mu\nu}$ is contained in $\overline{\varsigma\delta}$, whence:

$$A : \Omega = \overline{\varsigma\delta} : \overline{\mu\nu}$$

But,

$$B : A = \overline{\gamma\alpha} : \overline{\beta\gamma} = \overline{\delta\epsilon} : \overline{\varsigma\delta}$$

* That is, that the ratio of these two weights is represented by a rational fraction, as $\frac{5}{3}$, $\frac{117}{32}$, etc.

Hence, *ex aequalis* $B : \Omega = \overline{\delta\epsilon} : \overline{\mu\nu}$ or $\Omega$ is contained in $B$ as many times as $\overline{\mu\nu}$ is contained in $\overline{\delta\epsilon}$. Thus $\Omega$ is a common measure of $A$ and $B$.

Divide $\overline{\varsigma\delta}$ and $\overline{\delta\epsilon}$ into parts each equal to $\mu\nu$ and $A$ and $B$ into parts each equal to $\Omega$. The parts of $A$ will therefore be equal in number to those of $\overline{\varsigma\delta}$ and parts of $B$ equal in number to those of $\overline{\delta\epsilon}$. Place one of the parts of $A$ in the middle point of each of the parts $\overline{\mu\nu}$ of $\overline{\varsigma\delta}$, and one of the parts of $B$ in the middle point of each of the parts $\mu\nu$ of $\overline{\delta\epsilon}$ (Fig. I-4b).

Then the center of gravity of the parts of $A$ placed on equal distances of $\overline{\varsigma\delta}$ will be at $\alpha$, the middle point of $\overline{\varsigma\gamma}$, and the center of gravity of the parts of $B$ placed at equal distances along $\overline{\delta\epsilon}$ will be at $B$ the middle point of $\overline{\delta\epsilon}$. But the system formed by the parts $\Omega$ of $A$ and $B$ together is a system of equal weights even in number at places at equal distances along $\overline{\varsigma\epsilon}$. And, since $\overline{\varsigma\alpha} = \overline{\gamma\beta}$ and $\overline{\alpha\gamma} = \overline{\beta\epsilon}$, $\overline{\varsigma\gamma} = \overline{\gamma\epsilon}$, so that $\gamma$ is the middle point of $\overline{\varsigma\epsilon}$. Therefore $\gamma$ is the center of gravity of the system ranged along $\overline{\varsigma\epsilon}$. Therefore, $A$ acting in $\alpha$ and $B$ acting in $\beta$ balance about the point $\gamma$.

This proposition is followed by proposition seven in which the same statement is proved when the weights $A$ and $B$ are incommeasurable.*

The discovery of the principle of lever and its various applications produced a sensation in the ancient world, as we can see from the description given by Plutarch in his book *Life of Marcellus,* a Roman general who captured Syracuse during the Second Punic War and who was partially responsible for the slaying of Archimedes, who contributed largely to the defense of the city by building ingenious war machines. Writes Plutarch:

> Archimedes, who was kinsman and a friend of King Hiero (of Syracuse) wrote to him that with any given force it was possible to move any given weight, and emboldened, as we were told, by the strength of his demonstration, he declared that, if there were another world, and he could go to it, he could move this. Hiero was astonished, and begged him to put his proposition into execution, and show him some great weight moved by a slight force. Archimedes therefore fixed upon a three-master merchantman of the royal fleet, which had been dragged ashore by the great labors of many men, and after putting on board many passengers and the customary freight, he seated himself at a

* It is the ratio of these two weights that is an irrational number as for example $\sqrt{2}$.

distance from her, and without any great effort, but quietly setting in motion with his hand a system of compound pulleys, drew her towards him smoothly and evenly, as though she were gliding through the water.

The principle of lever plays a very important role in all walks of life from a farmer using a crowbar to move away a heavy boulder, to the intricate machinery used in modern engineering. The law of lever formulated by Archimedes permits us to introduce a very important mechanical notion of the *work* done by an acting force. Suppose we should attempt to lift a heavy stone (Fig. I-4) by using a crowbar with a shoulder ratio $\overline{\alpha\gamma} : \overline{\gamma\beta} = 3 : 1$. We can do it by pressing at the handle of the crowbar with a force which is three times smaller than the gravity force acting on the stone. It is clear from the picture

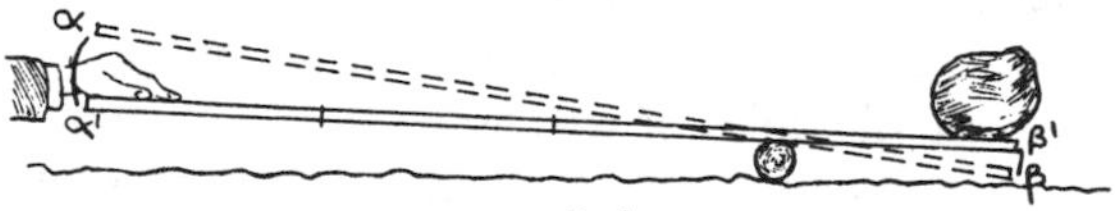


FIG. I–4.

If the left shoulder of the lever is three times longer than the right shoulder, the motion of the left end ($\overline{\alpha\alpha'}$) is three times larger than the motion of the right end ($\overline{\beta\beta'}$).

that when the stone is lifted, say, 1 inch from the ground ($\overline{\beta\beta'}$), the handle of the crowbar goes down by 3 inches ($\overline{\alpha\alpha'}$). Thus we conclude that the product of the force with which we pushed at the handle of the crowbar multiplied by its displacement downward is equal to the weight of the stone multiplied by its displacement upward. The product of the force by the displacement of the point of its application is known as the work done by the force. Thus, according to Archimedes' law of the lever, *the work done by the hand pushing down the long end of the crowbar is equal to the work done by its short end lifting the stone*. This statement can be generalized to any kind of mechanical work; thus, for example, the work done by furniture movers in bringing a grand piano three floors up is equal to the work of bringing three grand pianos only one floor up.*

The principle of equal work done on the two ends of a lever can also be applied to another similar device, *the pulley*, used by Archi-

* Professional furniture movers may dispute this statement, arguing that in the case of three grand pianos one has more trouble in adjusting the straps, etc., but we are speaking here only about the work connected with the actual lifting of the heavy object.

medes for moving a heavy ship, to the great surprise of King Hiero. If, in order to lift a heavy weight, we run a rope attached to it through a wheel fastened to a wooden beam (Fig. I-5a), the weight will be lifted by the distance ($l$) equal to the length ($d$) of the rope pulled, and the force ($Fl$) applied to the rope will be equal to the weight.

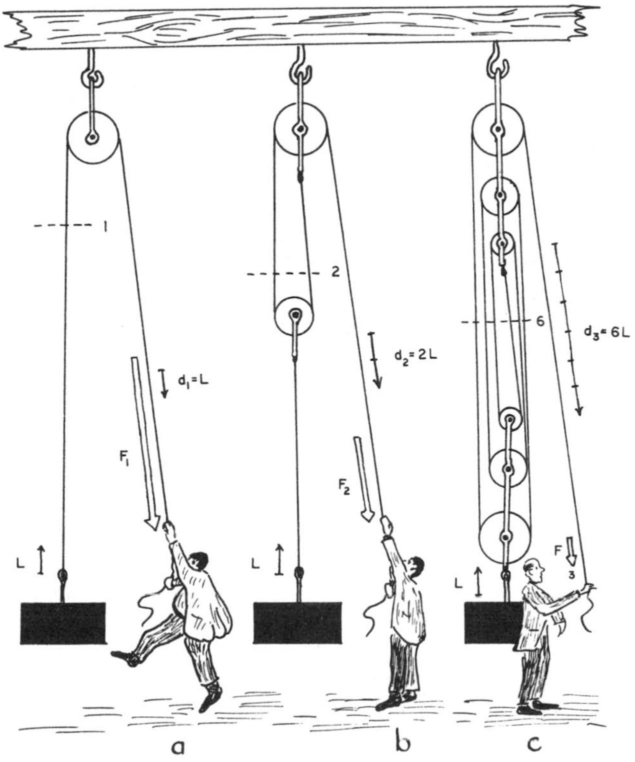


FIG. I–5.
The principle of the pulley.

If, however, we arrange two wheels in a way indicated in Figure I-5b, we will have to pull twice the length of the rope, and the force which has to be applied will be only one-half of the weight. In the arrangement shown in Figure I-5c, the force necessary to lift the weight will be only one-sixth, while the weight will be lifted by only one-sixth of the length of rope pulled.

## ARCHIMEDES' LAW OF FLOATING BODIES

Probably the best-known discovery made by Archimedes is his law pertaining to the loss of weight by bodies submerged into a liquid. The occasion which led to that discovery is described by Vitruvius* in the following words:

In the case of Archimedes, although he made many wonderful discoveries of diverse kinds, yet of them all, the following, which I shall relate, seems to have been the result of a boundless ingenuity. Hiero, after gaining the royal power in Syracuse, resolved as the consequence of his successful exploit, to place in a certain temple a golden crown which he had vowed to the immortal gods. He contracted for its making at a fixed price and weighed out a precise amount of gold to the contractor. At the appointed time the latter delivered to the King's satisfaction an exquisitely finished piece of handiwork, and it appeared that in weight the crown corresponded precisely to what the gold had weighed. But afterwards a charge was made that gold had been abstracted and an equivalent weight of silver had been added in the manufacture of the crown. Hiero, thinking it an outrage that he had been tricked, and yet not knowing how to detect the theft, requested Archimedes to consider the matter. The latter, while the case was still on his mind, happened to go to the bath, and on getting into a tub observed that the more his body sank into it the more water ran out over the tub. As this pointed out the way to explain the case in question, without a moment's delay and transported with joy, he jumped out of the tub and rushed home naked, crying in a loud voice that he had found what he was seeking: for as he ran he shouted repeatedly in Greek, "εὕρηκα εὕρηκα!"

Taking this as the beginning of his discovery, it is said that he made two masses of the same weight as the crown, one of gold and the other of silver. After making them, he filled a large vessel with water to the very brim and dropped the mass of silver into it. As much water ran out as was equal in bulk to that of the silver sunk in the vessel. Then, taking out the mass, he poured back the lost quantity of water, using a pint measure, until it was level with the brim as it had been before. Thus he found the weight of silver corresponding to a definite quantity of water.

After this experiment, he likewise dropped the mass of gold into the full vessel and, on taking it out and measuring as before, found that not so much water was lost, but a smaller quantity: namely, as much less as

* Vitruvius. *On Architecture.*

a mass of gold lacks in bulk compared to a mass of silver of the same weight. Finally, filling the vessel again and dropping the crown itself into the same quantity of water, he found that more water ran over for the crown than for the mass of gold of the same weight. Hence, reasoning from the fact that more water was lost in the case of the crown than in that of the mass, he detected the mixing of silver with the gold and made the theft of the contractor perfectly clear.

The proof of Archimedes' law given by him in his book *On Floating Bodies* is somewhat cumbersome, even though completely correct,

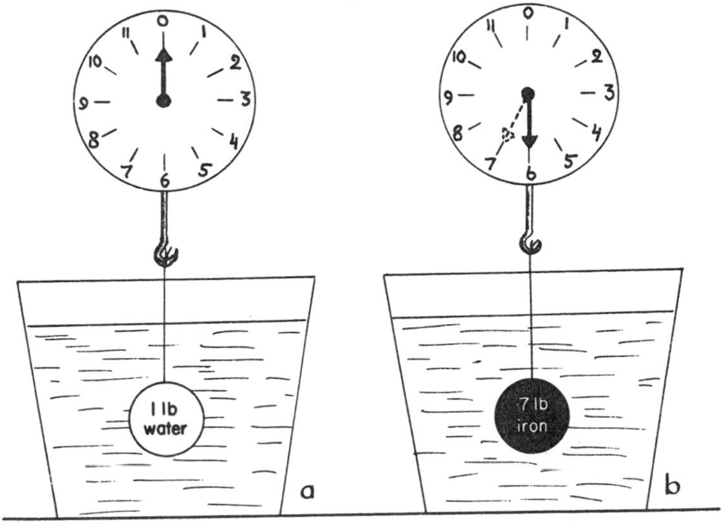


FIG. I–6.
The proof of Archimedes' law of floating bodies.

and we will reproduce it here in more modern language, considering what will happen if we submerge a solid metal ball into a bucket of water (Fig. I-6). Suppose we first take, instead of an iron sphere, a thin plastic sphere of the same diameter filled with water (Fig. I-8a). Since the weight of the plastic shell can be neglected, the situation will be the same as if the water inside the shell would be just a part of water in the bucket, and the scales will show zero. Now replace the water in the shell with iron (Fig. I-8b), which is seven times heavier than the equal volume of water. Since 1 pound of water was supported by the rest of the water in the bucket with the scales showing zero, the change from water to iron will add only $7 - 1 = 6$ additional pounds, and that is what the scales are going to read in this case. Thus, we conclude that the iron sphere weighing (in air) 7 pounds lost 1 pound by being submerged in water, i.e., the weight of water it displaces. This is Archimedes' law, which states that *any solid body submerged in a liquid loses the weight of this liquid displaced by it.*

### ARCHIMEDES, A MILITARY CONSULTANT

Besides being a great mathematician and the founder of the science of mechanics, Archimedes also served, putting it in modern terms, as a "consultant to industry and to the Armed Forces." The best

FIG. I–7.
Archimedes' screw pumping water up by merely turning it. To understand how it works, try to think what happens to the lower parts of the tube when it rotates, and you will find that they move up; not the tube itself, but the positions of its "minima" holding water. It may be helpful to make a spiral, as from a metal wire, and see what happens when it rotates around its axis.

known of his engineering inventions is the so-called "Archimedian screw," shown in Figure I-7, used for raising water. This device, the functioning of which is self-evident, was apparently widely used in irrigation, and for removing subterranean water from the mines.

Archimedes' participation in war work started apparently from his demonstration of the pulley to King Hiero. According to the dramatic description by Plutarch in *Life of Marcellus:*

Hiero, being amazed by this demonstration, and comprehending the power of his art, persuaded Archimedes to prepare for him offensive and defensive engines to be used in every kind of siege warfare. These he had never used himself, because he spent the greater part of his life in freedom from war and amid the festal rites of peace; but at the present time his apparatus stood the Syracusans in good stead, and, with the apparatus, its fabricator.

When, therefore, the Romans assaulted them by sea and land, the Syracusans were stricken dumb with terror; they thought that nothing could withstand so furious an onset by such forces. But Archimedes began to ply his engines, and shot against the land forces of the assailants all sorts of missiles and immense masses of stones, which came down with incredible din and speed; nothing whatever could ward off their weight, but they knocked down in heaps those who stood in their way, and threw their ranks into confusion. At the same time huge beams were suddenly projected over the ships from the walls, which sank some of them with great weights plunging down from on high; others were seized at the prow by iron claws, or beaks like the beaks of cranes, drawn straight up into the air, and then plunged stern foremost into the depths, or were turned round and round by means of enginery within the city, and dashed upon the steep cliffs that jutted out beneath the wall of the city, with great destruction of the fighting men on board, who perished in the wrecks. Frequently, too, a ship would be lifted out of the water into mid-air, whirled hither and thither as it hung there, a dreadful spectacle, until its crew had been thrown out and hurled in all directions, when it would fall empty upon the walls, or slip away from the clutch that had held it. As for the engine which Marcellus was bringing up on the bridge of ships, and which was called "sambuca" from some resemblance it had to the musical instrument of that name, while it was still some distance off in its approach to the wall, a stone of ten talents' weight was discharged at it, then a second and a third; some of these, falling upon it with great din and surge of wave, crushed the foundation of the engine, shattered its frame-work, and dislodged it from the platform, so that Marcellus, in perplexity, ordered his ships to sail back as fast as they could, and his land forces to retire.

Then, in a council of war, it was decided to come up under the walls while it was still night, if they could; for the ropes which Archimedes used in his engines, since they imported great impetus to the missiles cast, would, they thought, send them flying over their heads, but would be ineffective at close quarters, where there was no space for the cast. Archimedes, however, as it seemed, had long before prepared for such an emergency engines with a range adapted to any interval and missiles of short flight, and, through many small and contiguous openings in the wall, short-range engines called "scorpions" could be brought to bear on objects close at hand without being seen by the enemy.

When, therefore, the Romans came up under the walls, thinking themselves unnoticed, once more they encountered a great storm of missiles; huge stones came tumbling down upon them almost perpendicularly, and the wall shot out arrows at them from every point; they therefore retired. And here again, when they were some distance off, missiles darted forth and fell upon them; many of their ships, too, were dashed together, and they could not retaliate in any way upon their foes. For Archimedes had built most of his engines close behind the wall, and the Romans seemed to be fighting against the gods, now that countless mischiefs were poured out upon them from an invisible source.

However, Marcellus made his escape, and jesting with his own artificers and engineers, "Let us stop," said he, "fighting against this geometrical Briareus, who uses our ships like cups to ladle water from the sea, and has whipped and driven off in disgrace our sambuca, and with the many missiles which he shoots against us all at once, outdoes the hundred-handed monsters of mythology." For in reality all the rest of the Syracusans were but a body for the designs of Archimedes, and his the one soul moving and managing everything; for all other weapons lay idle, and his alone were then employed by the city both in offence and defence. At last the Romans became so fearful that, whenever they saw a bit of rope or a stick of timber projecting a little over the wall, "There it is," they cried, "Archimedes is training some engine upon us," and turned their backs and fled. Seeing this, Marcellus desisted from all fighting and assault, and thenceforth depended on a long siege.

When, after two years of siege, in the year 212 B.C., Syracuse was finally captured by Roman legions, a detachment of Roman soldiers broke into the house of Archimedes, who was engaged in his back yard in drawing some complicated geometrical figures on the sand.

"Noli tangere circulos meos" ("Do not touch my drawings!") exclaimed Archimedes in his poor Latin when one of the soldiers trod upon them. In response, the soldier drove his spear through the body of the old philosopher.

When Cicero was quaestor, visiting Sicily in 137 B.C. he found the tomb of Archimedes, near the Agrigentine gate, overgrown with thorns and briers. "Thus," writes Cicero, "would this most famous and once most learned city of Greece have remained a stranger to the tomb of its most ingenious citizen, had it not been discovered by a man of Arpinum."

## THE ALEXANDRIAN SCHOOL

With the decline of the political and economic power of Athens, the center of Greek culture shifted to Alexandria, founded in 332 B.C. on the Egyptian shore of the Mediterranean by Alexander the Great as the key port for trade between Europe and the Orient. By that time Alexandria developed into a beautiful city with ". . . 4,000 palaces, 4,000 baths, 12,000 gardeners, 40,000 Jews who paid tribute, and 400 theatres and other places of amusement." It also boasted a leading university, and a great library which was later unfortunately destroyed by fire as a result of a general conflagration of the city caused by Julius Caesar's order to burn the Egyptian fleet in Alexandria's harbor. Here Euclid wrote his *Elements of Geometry,* and Archimedes acquired his knowledge of the sciences as a young student from Syracuse.

In the field of astronomy, Alexandria was represented by Hipparchus, who lived in the middle of the 2nd century B.C. Hipparchus developed to the highest precision possible at that time the observation of the position of stars, and compiled a catalogue of 1,080 stars, which is still used by modern astronomers as the reference source for the ancient data on stellar positions. He also discovered the phenomenon of precession of equinoxes, which are the points on the celestial sphere at which the sun crosses the celestial equator in its annual motion between the stars. This phenomenon is due to the fact that the axis of the earth's rotation, being inclined to the plane of its orbit, describes a cone in space around the line perpendicular to the orbit, with the period of 26,000 years. The cause of that motion was found almost a thousand years later by Sir Isaac Newton.

As far as physics goes, the Alexandrian school was represented by Heron (or Hero), who was less of a physicist and more of an engineer-inventor. His book *Mechanics* contains many true statements, but many bad mathematical mistakes.

In spite of its defects in the mathematical treatment of the basic problems, Heron's book on mechanics contains the description of a large number of useful gadgets, such as composite pulleys, various types of gears, and cogwheel mechanisms, etc. In his book on "pneumatics" he describes the principle of siphon (Fig. I-8a), and a steam jet engine (Fig. I-8b) which, being similar to an ordinary lawn sprinkler, can be considered, however, as the precursor of modern jet propulsion motors.

Heron had also written a book entitled *Catoptrics,* containing the theory of mirrors and their practical applications. We read:

Catoptrics is clearly a science worthy of study and at the same time produces spectacles which excite wonder in the observer. For with the aid of this science, mirrors are constructed which show the right side as right side, and similarly the left side, whereas ordinary mirrors by their nature have the contrary property and show the opposite sides.

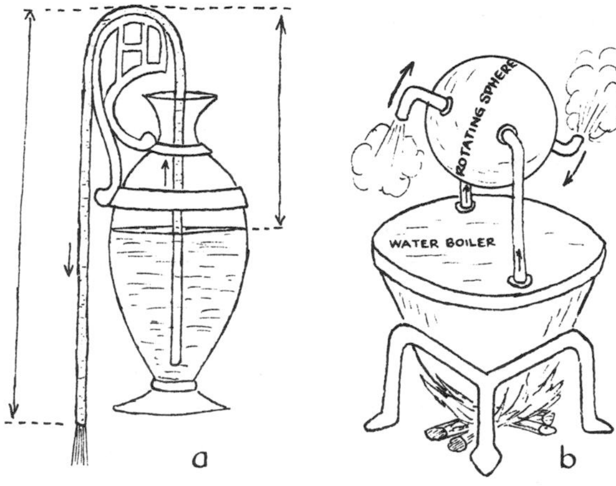


FIG. I–8.

Two devices invented by Heron: (a) The siphon permits the water from a container to flow out all by itself through a bent tube. The reason for the motion of water through the tube is that the weight of water in the left long part of the tube is larger than the weight of water in its right part, which extends essentially only from the water surface in the container to the top of the tube. (b) Heron's steam jet engine in which the rotation of the sphere is caused by two jets of steam coming from the nozzles.

This is done by placing two mirrors, without frames, edge to edge and at a right angle to one another (Fig. I-9).

It is possible with the aid of mirrors to see our own backs [the way the barber shows you the haircut on the back of your neck] and to see ourselves inverted, standing on our heads, with three eyes and two noses, the features distorted as if in intense grief [as in a mirror pavilion in an amusement park].

For who will not deem it very useful that we should be able to observe, on occasion, while remaining inside our own house, how many people there are on the street and what they are doing?

Heron's views on the nature of light are evident from the following quotation:

Practically all who have written of dioptrics have been in doubt as to why rays proceeding from our eyes are reflected by mirrors and why the

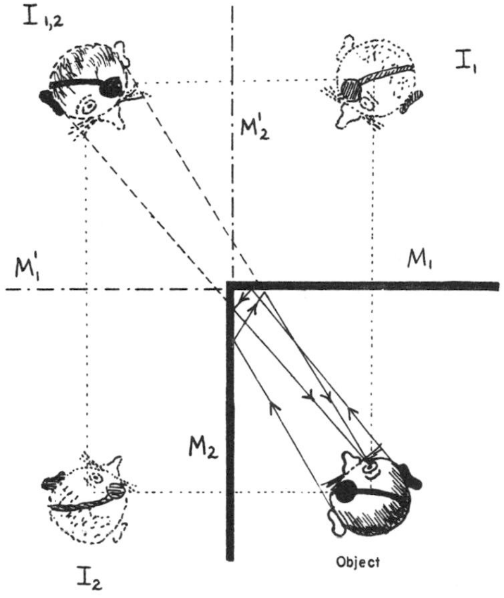


FIG. I–9.
Looking into a composite mirror formed by two plane mirrors $M_1$ and $M_2$ placed edge to edge at a right angle, one sees one's image doubly reflected: first, in mirror $M_1$ and then in the imaginary continuation $M'_2$ of the mirror $M_2$, or first in the mirror $M_2$ and then in the imaginary continuation $M'_1$ of the mirror $M'_1$. Because of the double reflection, the right side remains right and the left side remains left. The actual light rays are shown by solid lines.

reflections are at equal angles. Now the proposition that our sight is directed in straight lines proceeding from the organ of vision may be substantiated as follows. For whatever moves with unchanging velocity moves in a straight line. The arrows we see shot from bows may serve as an example. For, because of the impelling force, the object in motion strives to move over the shortest possible distance, since it has not the time for slower motion, that is, for the motion over a longer trajectory. The impelling force does not permit such retardation. And so, by reason of its speed, the object tends to move over the shortest path. But the shortest of all lines having the same endpoints is the straight line. That the rays proceeding from our eyes move with infinite velocity may be gathered from the following consideration. For when, after our eyes have been closed, we open them and look up at the sky, no interval of time is required for the visual rays to reach the sky. Indeed, we see the stars as soon as we look up, though the distance is, as we may say, infinite. Again, if the distance were greater the result would be the same, so that, clearly, the rays are emitted with infinite velocity. Therefore they will suffer neither interruption, nor curvature, nor breaking, but will move along the shortest path, a straight line.

This passage reveals an amusing fact that Heron, and apparently all his contemporaries, believed that vision is due to some rays emitted from the eye and reflected back by the object, thus being based on the same principle as today's radar.

Another great Alexandrian was the astronomer Claudius Ptolemy (not to be confused with the members of Ptolemaic dynasty who reigned in Egypt many years before the Christian era), who lived and worked during the first half of the 2nd century A.D. Ptolemy's observations of stars and planets, collected in his book known as *Almagest,* represented a considerable addition to the data obtained by Hipparchus two and a half centuries earlier. His important contributions to physics are contained in his book *Optics* which came to us as the Latin translation of the lost Arabic version of the original Greek manuscript. In this book Ptolemy discusses, among other things, the important subject of the refraction of light in passing from one medium into another. He writes:

Visual rays may be altered in two ways: by *reflection,* i.e. the rebound from objects, called mirrors, which do not permit the penetration, and by bending (i.e. *refraction*) in the case of media which permit the penetration and have a common designation ("transparent materials") for the reason that visual ray penetrates them.

He illustrates the phenomenon of refraction by the following simple experiment with a coin placed at the bottom of a vessel filled with water, called a "baptistir"* (Fig. I-10a).

* Presumably used in churches for baptizing children.

Suppose that the position of the eye is such that the visual ray emanating from it and just passing over the edge of the baptistir reaches a point higher than the coin. Then, allowing the coin to remain in its position, pour water gently into the baptistir until the ray that just passes over the edge is bent downward and falls upon the coin. The result is that the object not previously seen is then seen along the straight line passing from the eye to the point above the true position of the object. Now the observer will not suppose that the visual ray has been bent toward the object, but that the object is itself afloat and is raised towards the ray. The object, therefore, will appear on the perpendicular drawn from it to the surface of the water.

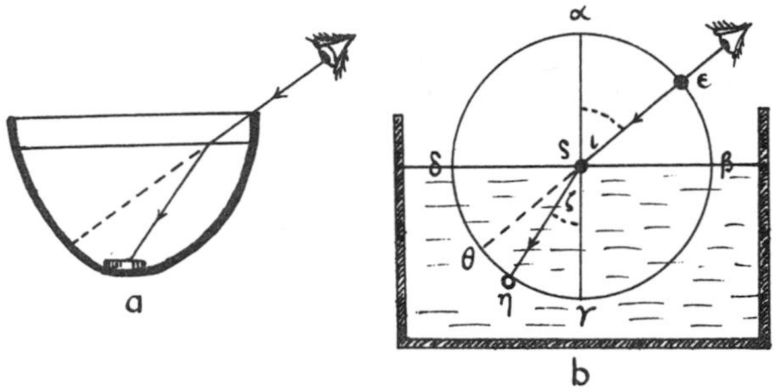


FIG. I–10.

Ptolemy's experiments with the refraction of light: (a) The coin at the bottom of a water-filled container seems to be *higher* than it actually is. (b) The apparatus for studying refraction of light. Ptolemy measured the relation between the angle $\delta\varsigma\eta$ in water and angle $\alpha\varsigma\epsilon$ in air and tabulated the dependence between them.

Later in the text, Ptolemy describes an experiment designed to study in detail the laws of refraction of light.

The amount of refraction which takes place in water, and which may be observed, is determined by an experiment like that which we performed with the aid of a copper disc, in examining the laws of mirrors. On this disc draw a circle $\alpha\beta\gamma\delta$ [Fig. I-10b] with the center in $\varsigma$ and two diameters $\alpha\varsigma\gamma$ and $\delta\varsigma\beta$ intersecting at right angles. Divide each quadrant into ninety equal parts and place over the center a very small colored marker. Then set the disc upright in a small basin and pour into the basin clear water in a moderate amount so that the view is not obstructed. Let the surface of the disc, standing perpendicular to the surface of the water, be bisected by the latter, half the circle, and only half, that is $\beta\gamma\delta$, being entirely below the water. Let diameter $\alpha\varsigma\gamma$ be perpendicular to the surface of the water.

Now take a measured arc, say $\alpha\epsilon$, from the point $\alpha$, in one of the two quadrants of the disc which are above the water level. Place over $\epsilon$ a small colored marker. With one eye, take sighting until the markers at $\epsilon$ and $\varsigma$ both appear on a straight line proceeding from the eye. At the same time move a small, thin rod along the arc $\gamma\delta$ of the opposite quadrant, which is under water, until the extremity of the rod appears at that point of the arc which is the prolongation of the line joining $\epsilon$ and $\varsigma$. Now if we measure the arc between the point $\gamma$ and the point $\eta$, at which the rod appears on the aforesaid line, we shall find that this arc, $\gamma\eta$, will always be smaller than arc $\alpha\epsilon$.

If we place the eye along the perpendicular $\alpha\varsigma$ the visual ray will not be bent, but will fall upon $\gamma$, opposite $\alpha$ and in the same straight line as $\alpha\varsigma$. In all other positions, however, as arc $\alpha\epsilon$ is increased, arc $\gamma\eta$ is also increased, but the amount of the bending of the ray will be progressively greater.

| | | | | *Bending* |
|---|---|---|---|---|
| When $\alpha\epsilon$ is: | 10° | $\gamma\eta$ will be | 8 ° | 2 ° |
| | 20° | | 15½° | 4½° |
| | 30° | | 22½° | 7½° |
| | 40° | | 29 ° | 11 ° |
| | 50° | | 35 ° | 15 ° |
| | 60° | | 40½° | 19½° |
| | 70° | | 45½° | 24½° |
| | 80° | | 50 ° | 30 ° |

This is the method by which we have discovered the amount of refraction in the case of water.

Ptolemy also studied by a similar method the refraction of light rays on the boundary between air and glass, and found that in that case the bending of the ray is larger. He did not try, however (or at least did not succeed if he did), to express the results of his observations by means of a mathematical formula, and the mathematical formulation of the law of refraction of light was not found until the 17th century. It is ironical enough that he could have easily done so, since the mathematical apparatus involved in the formulation of that law was the relation between arcs and chords discussed by Plutarchus one and a half centuries before him and developed by himself at great length in *Almagest* in connection with astronomical observations.

The problem was to find the length of chord *ADB* corresponding to the arc *ACB* of the unit radius circle (Fig. I-11). Using ingenious mathematical methods, Ptolemy built a table, part of which is reproduced below:

| *Arcs* | *Chords* | *Arcs* | *Chords* | *Arcs* | *Chords* |
|---|---|---|---|---|---|
| 116 ° | 1.014557 | 117½° | 1.023522 | 119 ° | 1.032344 |
| 116½° | 1.020233 | 118 ° | 1.025137 | 119½° | 1.033937 |
| 117 ° | 1.021901 | 118½° | 1.030741 | 120 ° | 1.035523 |

This table corresponds to what is now known as trigonometrical tables of *sines,* with the only difference that one used half-arcs (the angle *AOC*) and half-chords *AD*. The length *AD* for the unit radius is

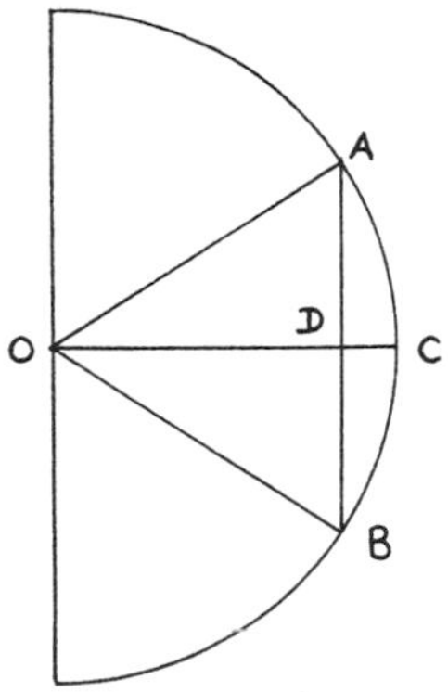


FIG. I–11.

The relation between Plutarch's "chord-tables" and the modern trigonometrical tables. Plutarch tabulated the lengths of the chords *ADB* for various lengths of the arcs *ACB*. In modern trigonometry one tabulates the length *AD* (half-chord) against the arc *AC*. The length *AD* is known as *sine*, or *sin,* of the angle *AOC*, while the length *OD* is known as *cosine,* or *cos,* of that angle.

known as sine *AOC,* while the distance *OD* is cosine *AOC*. Trigonometric functions are extremely useful in solving various geometrical problems involving both lengths and angles.

If Ptolemy would have compared the results of his experiment on refraction of light with his table of sines, he would have found that *the ratio of the sine of the angle of incidence to the sine of the angle of refraction is a constant for any given pair of substances*. He did not, and the above-stated law of refraction was not discovered until fourteen centuries later by the Dutch astronomer and mathematician Willebrord Snell. As we shall see later, Snell's law is of paramount importance for understanding the nature of light.

Ptolemy's work was the last great contribution of ancient Greek culture to the development of science, and after his death research in Alexandria rapidly began to decay. Probably the last name which can be mentioned in connection with the Alexandrian school is that of Hypatia, a daughter of the mathematician Theon, and herself a teacher of science and philosophy. She lived during the reign of the Roman Emperor Julian the Apostate, who tried to protect Greek learning and Greek gods against the ever-increasing power of the Christian church. The reaction which took place after his death resulted, in A.D. 415, in a big anti-Greek revolt organized by Bishop Kyrillos of Alexandria. Hypatia was torn to pieces by the Christian mobs, and the remainder of the city's libraries were destroyed.

# CHAPTER II *The Dark Ages and the Renaissance*

WITH THE EXTINCTION of the Greek culture the development of science in general and of physics in particular came to a virtual standstill. Romans, who were the potentates of the world during this period of human history, cared little about abstract thinking. Theirs was a "businessman's civilization" and, although they encouraged learning, they were mostly interested in the practical applications. After the fall of the Roman Empire the situation went from bad to worse, and the feudal states which grew on its ruins certainly did not represent a fertile soil for any kind of scientific development. The only unifying stimulus during this period, which extended well over a thousand years, was the Christian religion, and the abbeys and monasteries became intellectual centers. Consequently, the main interests became centered around theological problems, and whatever scientific ideas were left over after the fall of the ancient Greek culture were subjugated to religious dictatorship. The Ptolemaic system of the world, with the earth resting in the center and the sun, planets, and stars rotating around it, was accepted as unshakable dogma, since it was best fitted to the concept of the central position of the Vatican as the abode of God's chosen emissary on earth. "Scientific" discussions were mostly limited to such problems as to how many angels can dance on the end of a needle and whether Almighty God could make a stone which is so heavy that He cannot lift it Himself. Primitive "Lysenkoism" was flourishing all over Europe, and the Holy Inquisition took care of stamping out any deviation from the general line of religious belief.

Fortunately for us, Greek science found a refuge in the newly born Arabian Empire which, in the course of the 7th century, engulfed all the lands south of the Mediterranean and splashed over into Spain, over the narrow Straits of Gibraltar. The benevolent potentate Haroun Al-Raschid of the story of "A Thousand and One Nights" founded, in A.D. 800, a school of science in Bagdad, while the city of Cordoba in Spain became a cultural center of the Arabian Empire on European soil. Arabian scholars studied and translated Greek manuscripts salvaged from the partially destroyed Hellenic libraries, and carried the banner of science while Europe was suffocating in the clutches of medieval scholasticism. The Arabian era in the history of science is witnessed by such scientific terms in use today, as: *Al*gebra, *Al*cohol, *Al*kali, *Am*algam, *Al*manac, *An*tares, etc. The Arabs made considerable progress in mathematics, developing algebra, unknown to the Greeks, and introducing arabic numerals which make computation much easier than it was with the Roman system. But, perhaps as the result of Scheherazade's fairy tales, their work in astronomy and chemistry was limited mostly to the pursuit of fantastic aims in predicting man's life on the basis of the configuration of the stars under which he was born (astrology), and finding the methods of turning common metals into precious gold (alchemy). They did not seem to do anything in the field of physics, except, of course, for the fact that alchemistry can be considered as a precursor of modern techniques of transforming one chemical element into another. But, "when the Moor accomplished his task, the Moor must go away," and in the 12th century the Arabian Empire rapidly succumbed as the result of Genghis Khan's invasion and the persistent Christian crusades to the Holy Land.

By this time, European states were slowly emerging from the chaos of the dark Middle Ages and learning was on the upgrade again. In A.D. 784, Charlemagne, the ruler of the French Empire, decreed that all abbeys in his vast domain should have schools attached to them, and in A.D. 1100 the University of Paris was founded. The universities of Bologna, Oxford, and Cambridge were founded soon thereafter and rapidly became the recognized centers of scholastic activity. The customary course of study consisted of the "trivium," which included Latin grammar, rhetoric, and logic, and the "quadrivium," which included arithmetic, geometry, music, and astronomy. However, education was still under the vigilant supervision of the Church, and universities in all the Christian countries had to obtain the Pope's sanction in order to continue their existence. Studies were based mostly on the writings of Aristotle, which reached Europe in Arabian translation. As we stated earlier, the fact that Aristotle, although excelling in many other respects, was not too good in the field of physical sciences certainly did not help the rejuvenation of physics in Europe, which was just awakening from her thousand-year sleep.

One of the important factors in the spread of knowledge was the invention of the printing press in the middle of the 15th century in the shop of a man called Fust in Mainz, Germany, and one of the most important books which came off these early presses was undoubtedly *De Revolutionibus Orbitum Coelestium* (Nuremberg Press, A.D. 1543) by Nicolaus Copernicus, in which he established a new system of the world with the sun in the center of it. But, in order to avoid banishment by the Church, it was found necessary to supply this book with a preface (written, presumably, without Copernicus' knowledge by his editor Andreas Osiander) which stated that all the ideas expressed in it were of purely hypothetical nature and represented more of a mathematical exercise than a description of real things.

## KEPLER'S ORATORY AND LAWS

The mixture of theology and true science during this era is probably best illustrated by the following passages from *Mysterium Cosmographicum* (1596) by Johannes Kepler, the discoverer of the basic laws of planetary motion. Being dedicated to an assortment of German nobility who supported Kepler in his research, the book begins with the following words:

To their Illustrious, Noble and Righteous Lords, Sigismund Friedrich, Baron of Herberstein, . . . to the Most Noble Lords of the Illustrious Estates of Styria, the Honorable Council of Five, my gentle and gracious Lords,

Greetings and Humble Respects!

What I have promised seven months ago, to wit a work that according to the judgment of the learned will be elegant, impressive, and far superior to all annual calendars, I now present to your gracious company, my noble Lords, a work that though it be small in compass and but the fruit of my own modest efforts, yet treats of a wondrous subject. If you desire maturity—Pythagoras has already treated of it some 2000 years ago. If you desire novelty—it is the first time that this subject is being presented to all mankind by myself. If you desire scope—nothing is greater or wider than the Universe. If you desire venerability—nothing is more precious, nothing more beautiful than our magnificent temple of God. If you wish to know the mysteries—nothing in Nature is, or ever has been, more recondite. It is but for one reason that my object will not satisfy everybody, for its usefulness will not be apparent to the thoughtless. I am speaking of the Book of Nature, which is so highly esteemed in the Holy Scriptures. St. Paul admonished the Heathens to reflect on God within themselves as they would on the Sun in the water or in a mirror. Why then should be Christians delight the less in this reflection, seeing that it is our proper task to honour, to revere and to admire God in the true way? Our piety in this is the deeper the greater is our awareness of creation and of its grandeur. Truly, how many hymns of praise did not David, His faithful servant, sing to the Creator, who is none but God alone! In this his mind dwelled reverently on the contemplation of the Heavens. The Heavens, he sings, declare the glory of God. I will consider Thy heavens, the work of Thy hands, the moon and the stars which Thou has ordained. God is our Lord, and great is His might; He counteth the multitude of the Stars, and knoweth them by their names. Elsewhere, inspired by the Holy Ghost and full of joyousness, he exclaims to the Universe: Praise ye the Lord, praise Him, Sun and Moon, etc.

Later on we read:

The fact that the whole world is circumscribed by a sphere has already been discussed exhaustively by Aristotle (in his book on the Heavens), who based his proof particularly on the special significance of the spherical surface. It is for this very reason that even now the outermost sphere of fixed stars has preserved this form, although no motion can be ascribed to it. It holds the Sun as its centre in its innermost womb, as it were. The fact that the remaining orbits are round can be seen from the circular motions of the stars. Thus we need no further proof that the curve was used for adorning the world. While, however, we have three kinds of quantity in the world, *viz.*, form, number, and content of bodies, the curved is found in form alone. In this, content is not important since one structure inscribed concentrically into a similar one (for instance, sphere into sphere, or circle into circle) either touches everywhere or not at all. The spherical, since it represents an absolutely unique quantity, can only be governed by the number Three.

While Kepler was writing these flowery passages, he was working hard on a more prosaic problem: the exact law of planetary motion.

The Copernican system, as presented in the *Revolutionibus,* assumed planetary orbits to be circles, in accordance with the old tradition of Greek philosophy to consider the circle to be a perfect curve and a sphere to be a perfect body. But this assumption did not fit in too well with the detailed measurements of planetary motions carried out by a Danish astronomer, Tycho Brahe, in his private observatory, Uraniborg, on a little island not far from Copenhagen. Being Tycho's student and assistant and having at his disposal a considerable knowledge of mathematics acquired from reading Euclid and other classical Greek works, Kepler set himself the task of finding out what is the exact shape of planetary orbits, and what are the laws governing their motion. After years of work he came to his first important discovery. He found that, while in their motion around the sun the planets do not follow exactly circular orbits, they describe another class of curves just about as famous as the circle in the old Euclidean geometry. This class of curves is known as *conical sections* and can be defined as the intersection of a cone with differently oriented planes (Fig. II-1). If the plane is perpendicular to the axis of the cone we get, of course, a circle in the cross section. If, however, the plane is inclined to the cone's axis we get the elongated closed curves known as *ellipses.* When the plane becomes parallel to the side of the cone, one extremity of the ellipse disappears into infinity, and we have an open curve known as a *parabola.* At still larger tilts the original parabola becomes more "open" and turns into what is known as a *hyperbola.* It should be noticed that, in the case of the hyperbola, we have actually two disconnected branches, the second branch being provided by the intersection of the plane with the second upside-down part of the cone. An ellipse can be alternatively defined as a set of points selected in such a way that the *sum* of distances of each of them from the two fixed points known as *foci* is always the same. Thus one can draw an ellipse by attaching a string to two thumbtacks driven into a cardboard, and moving the pencil in such a way that the string is always stretched. Similarly a hyperbola is a set of points for which the *difference* of distances from the two foci remains constant (Fig. II-2a), the fact which does not provide any convenient practical way to draw that curve.

Analyzing Tycho Brahe's data concerning the positions of planets between the stars, Kepler came to the conclusion that everything would fit very nicely if one assumed that *all planets follow elliptical orbits with the sun located in one of the foci.* He also found that in their motion around the sun, planets move faster when they are closer to the sun (in *aphelion*) and more slowly when they are farther away (*perihelion*). The correlation between the velocities of a planet and its distance from the sun at different parts of its orbit is such that *the imaginary line connecting the sun and the planet sweeps over equal*

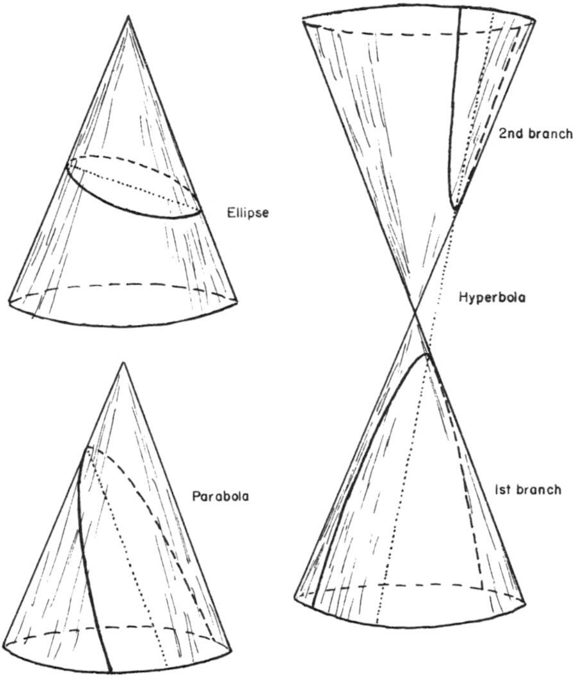


FIG. II–1.
Conical sections obtained by cutting a cone by a plane at different angles.

*areas of the planetary orbit in equal intervals of time* (Fig. II-2a). These two basic laws of planetary motion were announced by Kepler in 1609 and are known as the first and the second Kepler laws.

Having found the laws of motion of individual planets, Kepler began to look for the correlation between different planets, and it took him nine years to find it. He was trying all kinds of possibilities such as, for example, the correlation between planetary orbits and the regular polyhedra of solid geometry but nothing seemed to fit. Finally Kepler made a brilliant discovery that is known today as Kepler's third law. It states that *the squares of the periods of revolution of different planets around the sun stand in the same ratio as the cubes*

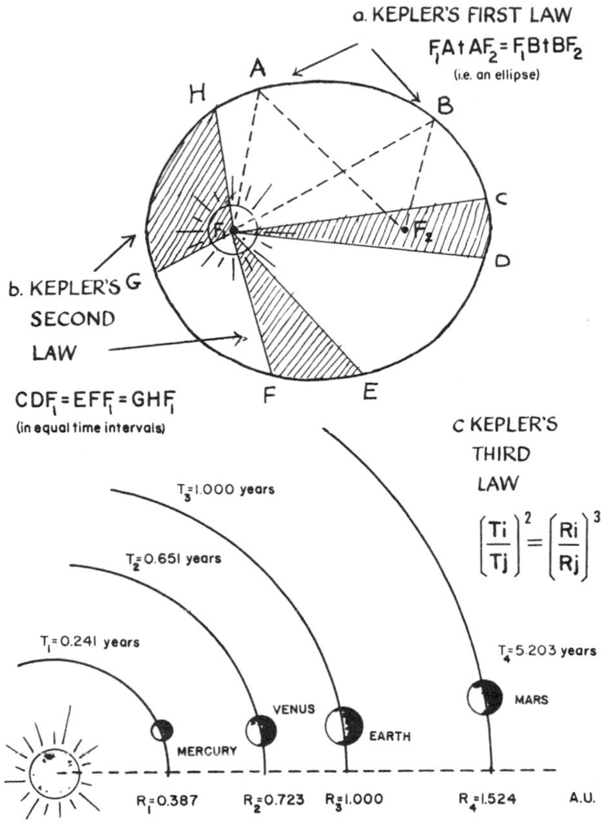


FIG. II–2.
The three Kepler laws of planetary motion.

*of their mean distances from the sun.* In Figure II-2b we give a scheme of the orbits of the so-called inner planets—Mercury, Venus, earth, and Mars—with their distances expressed in terms of the radius of the earth orbit (the so-called *A*stronomical *U*nit), and their revolution periods in years.

Taking the *squares of revolution periods,* we obtain the sequence:

0.058, 0.378, 1.000, 3.540.

On the other hand, taking the *cubes of distances* we get:

0.058, 0.378, 1.000, 3.540.

The identity of the two sequences proves the correctness of Kepler's third law.

Thus early in the 17th century scientists learned *how* the planets move around the sun, but it took more than half a century before they could answer the question of *why* they do so.

### STEVINUS' CHAIN

While Kepler was mostly interested in celestial spheres, his contemporary, a Flemish engineer, Simon Stevinus, had more down-to-earth interests and was extending the works of Archimedes on mechanical equilibrium commonly known as "statics." His main contribution was the solution of the problem of the equilibrium on an inclined plane, which was apparently not attacked by Archimedes, and, as we have seen before, was treated in an erroneous way by Heron. On the cover of Stevinus' book on statics appeared a diagram shown in Figure II-3, which signals a great progress in the understanding of equilibrium problems. A chain formed by a large number of metal balls (ball bearings, we would call them today) is placed on a prism-shaped support with very smooth ("frictionless") sides. What will happen? Since there are more balls on the left (longer) side of the prism than on its right (shorter) side, one would think that, because of the difference of weights, the chain should begin to move from right to left. But, since the chain is continuous, this motion would never stop, and the chain would roll round and round forever. If this were true, we could add to this gadget some cogwheels and gears and run all kinds of machinery without any cost for an indefinite period of time. We would be getting work done for nothing, and humanity would be benefited in a much higher degree than by all the promises of the "atoms for peace" program.

But, being a practical and sober man, Stevinus discarded that possibility, and postulated that the chain should remain in equilibrium. But this means that the pull of a ball located on an inclined plane decreases with the angle between it and the horizontal plane, which is, indeed, in full accord with the fact that no force is acting on a ball placed on the horizontal surface. Since the number of balls located on the left and on the right slopes is apparently proportional to the lengths of these slopes, we can write, denoting by $F_l$ and $F_r$ the forces acting on a single ball on each side:

$$F_l \times \overline{AC} = F_r \times \overline{CB}$$

or

$$F_l/F_r = \overline{CB}/\overline{AC}$$

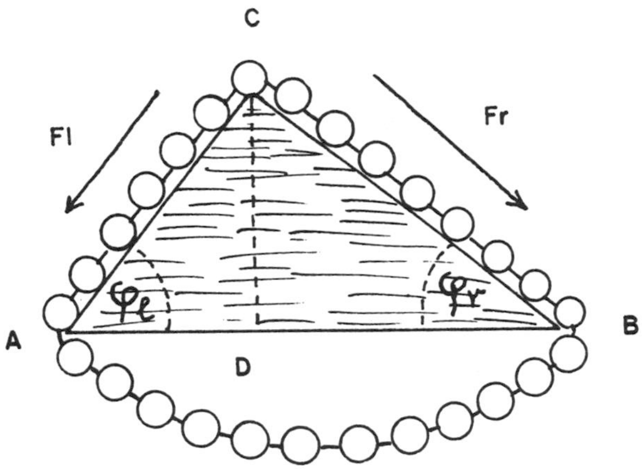


FIG. II–3.
Stevinus' endless chain, demonstrating the law of equilibrium on an inclined plane.

Introducing the *sines* of the angles $\varphi_l$ and $\varphi_r$ which characterize the two slopes, we have:

$$\sin \varphi_l = \overline{CD}/\overline{AC}; \sin \varphi_r = \overline{CD}/\overline{CB}$$

so that the above relation can be rewritten in the form:

$$F_l/F_r = \sin \varphi_l / \sin \varphi_r$$

Expressed in words, this means that *the force of gravity acting on an object located on an inclined plane in the direction of that plane is directly proportional to the sine of the angle of inclination.*

### THE PENDULUM

While Stevinus made considerable advances in his studies of statics, the honor of making the first steps in the science of dynamics, i.e., the study of the motion of material bodies, goes to the son of an impoverished Florentine nobleman by the name of Vincenzio Galilei. Although Signor Vincenzio was himself much interested in mathematics, he planned for his young son Galileo a medical career, as the more profitable profession. Thus, in 1581, at the age of 17, Galileo started the study of medicine at the University of Pisa. But apparently he did not find dissecting dead bodies to be a very exciting occupation and his restless mind was looking for other kinds of problems.

Once, attending a mass in the Cathedral of Pisa, he was absent-mindedly watching a candelabra which was set in motion by the attendant who lighted the candles. The consecutive swings were becoming smaller and smaller as the candelabra was slowly coming to rest. "Does the time of each swing also become shorter?" Galileo asked himself. Having no stop watch—for this had not been invented at that time—Galileo decided to measure the time of the consecutive swings by counting his own pulse. And, probably to his surprise, he discovered that, although the swings were becoming shorter and shorter, their duration in time remained exactly the same. Coming home, he repeated this experiment with a stone tied to the end of a string and found the same result. He also discovered that, for a given length of the string, the oscillation period remained the same no matter whether he used a heavy or a light stone in his experiment. Thus, the now familiar device known as a pendulum came into being. Still having one foot in the medical profession, Galileo reversed the course of his discovery and suggested the use of a pendulum of standard length for measuring the pulse beat of patients. This device, known as the "pulsometer," became very popular in contemporary medicine and was the precursor of the modern white-clad nurse holding the patient's hand and looking at her elegant wrist watch. But this was also Galileo's last contribution to medical science, since the study of the pendulum and other mechanical devices completely changed the direction of his interests. After some argument with his father, he changed his academic plans and began to study mathematics and the sciences.

For a number of years his interests were centered in the field of what is now known as *dynamics*—that is, the study of the laws of motion. Why is the period of the pendulum independent of the "amplitude," i.e., the size of the swing? Why do a light and a heavy stone attached at the end of the same string swing with the same period? Galileo never solved the first problem since its solution requires the knowledge of calculus which was invented by Newton almost a century later. He never solved the second problem, which had to wait until Einstein's work on the general theory of relativity. But he certainly contributed a lot to the formulation of both, if not to their solution! The motion of a pendulum is a special case of the fall caused by the force of gravity. If we release a stone unattached to anything, it will fall straight down to the ground. If, however, the stone is tied to a string attached to a hook in the ceiling, it is forced to fall along an arc of a circle. If a light and a heavy stone, attached to a string, take the same time to reach the lowest position (a quarter-period of the pendulum's oscillation), then these two stones must also take the same time to fall to the ground after being released from the same height. This conclusion stood in conflict with the generally accepted opinion of Aristotelian philosophy at that time, according to which heavy objects fall down faster than light ones. To prove his point Galileo dropped from the Leaning Tower of Pisa two spheres, one of wood and one of iron, and the unbelieving spectators below observed the two spheres hit the ground at the same moment. Historical research seems to indicate that this demonstration never actually took place and represents only a colorful legend. Neither is it certain that Galileo discovered the law of the pendulum while praying in the Cathedral of Pisa. But he certainly was dropping objects of different weight, maybe from the roof of his house, and was swinging stones attached to a string, maybe in his back yard.

### THE LAWS OF FALL

When one releases a stone it moves down faster and faster, and Galileo wanted to know what mathematical law governs that accelerated motion. But the free fall of objects proceeds too fast to study it in detail without the use of modern equipment such as, for example, fast photography. Thus, Galileo decided to "dilute the force of gravity" by making the ball roll down an inclined plane (Fig. II-4). The steeper the plane, the faster the ball rolls, and in the limiting case of a vertical plane the ball falls free alongside it. The main difficulty in performing that experiment was the measurement of time taken by the ball to cover different distances. Galileo solved it by using the water clock in which time was measured by the amount of water pouring out through a little opening near the bottom of a large container. Marking the positions of the ball at equal intervals of time, beginning from the start, he found that distances covered during these time intervals were standing in the ratios 1 : 3 : 5 : 7, etc. When the plane was steeper, the corresponding distances became longer but their ratios always remained the same. Thus, concluded Galileo, this law must also hold for the limiting case of free fall. The above-obtained result can be expressed in different mathematical form by saying that the total dis-

FIG. II–4.
Galileo investigating the accelerated motion of a ball rolling down the inclined plane.

tance covered during a certain period of time is proportional to the square of that time or, as one used to put it in Galileo's day, "double proportional" to time. Indeed, if we take as the unit length the distance covered by the ball during the first interval of time, the total distance covered at the end of consecutive intervals, according to the square law, will be $1^2$, $2^2$, $3^2$, $4^2$, etc., or 1, 4, 9, 16, etc. Thus the distances covered during each of the consecutive time intervals will be: $1; 4 - 1 = 3; 9 - 4 = 5; 16 - 9 = 7$ etc.*

* Algebraically, if the total distance covered by the end of the *n*th time interval is $n^2$, the distance covered during the last time interval is $n^2-(n-1)^2=n^2-n^2+2n-1=2n-1$.

From the observed dependence of distance traveled on time, Galileo concluded that the velocity of that motion must increase in simple proportion to the time. Let us give the proof of that statement in Galileo's own words:*

In the accelerated motion, the augmentation [of velocity] being continual, you cannot divide the degrees of velocity ["Values of velocity," in modern language], which continually increase into any determinate number, because, changing every moment, they are evermore infinite. Therefore, we shall be better able to exemplify our intention by describing

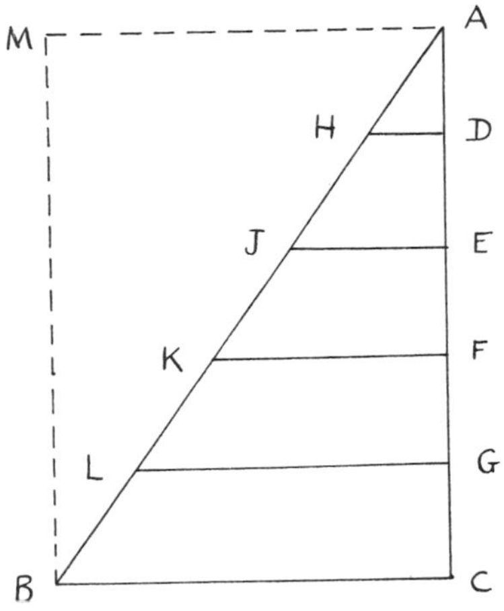


FIG. II–5.
Galileo's proof that in (uniformly) accelerated motion starting from rest, the distance covered by a moving body is one half of the distance the body would have covered had it been moving all the time with the same velocity.

a triangle ABC [Fig. II-5]. We take in the side AC as many equal parts as we please, AD, DE, EF, FG, GC and draw by the points D, E, F, G straight lines parallel to base BC. Now let us imagine the parts marked in the line AC to be equal times, and let the parallels drawn by the points D, E, F and G represent to us the degrees of velocity accelerated and increasing equally in equal times, and let the point A be the state of rest, departing from which the body has, for example, in the time AD acquired the degree of velocity DH; in the second time we will suppose

* Galileo Galilei. *Dialogue on the Great World Systems.* Chicago: Univ. of Chicago Press, 1953, pp. 244-45. Reprinted by permission.

that it has increased the velocity from DH to EI, and so on, in the succeeding times, according to the increase of the lines FK, GL, etc. But because acceleration is made continually from moment to moment, and not disjunctly from one certain part of the time to another, the point A being put for the lowest moment of velocity, that is, for the state of rest, and AD for the first instant of the time following, it is manifest that before acquiring the degree of velocity DH made at time AD the body must have passed through infinite other lesser and lesser degrees gained in the infinite instants that are in the time DA, answering the infinite points that are in the line DA. Therefore, to represent to us the infinite degrees of velocity that precede the degree DH, it is necessary to imagine lines successively lesser and lesser, which are supposed to be drawn by the infinite points of the line DA and parallel to DH. These infinite lines represent to us the surface of the triangle AHD. Thus we may imagine any distance passed by the body, with the motion which begins at rest, and accelerated uniformly, to have spent and made use of infinite degrees of velocity, increasing according to infinite lines that, beginning from point A, are supposed to be drawn parallel to the line HD, and to the rest IE, KF and LG, the motion continuing as far as one will.

Now let us complete the whole parallelogram AMBC, and let us prolong as far as to the side BM, not only the parallels marked in the triangle, but those infinite others imagined to be drawn from all the points of the side AC; and like as BC with the greatest of those infinite parallels of the triangle, representing to us the greatest degree of velocity acquired by the movable in the accelerate motion, and the whole surface of the said triangle was the mass and sum of the whole velocity, with which at the time AC it passed such a certain space: so the parallelogram is now a mass and aggregate of a like number of degrees of velocity, but each equal to the greatest BC. This mass of velocities will be double the mass of the increasing velocities in the triangle, even as the said parallelogram is double to the triangle, and therefore if the body that, falling, did make use of the accelerated degrees of velocity answering to the triangle ABC, has passed in such a time such a distance, it is very reasonable and probable that, making use of the uniform velocities answering to the parallelogram, it shall pass with an even motion in the same time a distance double that passed by the accelerated motion.

Although long-winded and cumbersome language, we must remember that it was written in 1632, and translated into English (by Thomas Salisbury) in 1661! Apart from being the first formulation of the law of free fall, the above-quoted passage from *Discorso* also contains the first step in the development of the so-called "integral calculus" in which the results are obtained by adding infinitely large numbers of infinitely small quantities. In the modern mathematical notations, Galileo's law of the uniformly accelerated motion can be written as:

$$\text{velocity} = \text{acceleration} \times \text{time}$$

and

$$\text{distance} = \tfrac{1}{2}\,\text{acceleration} \times \text{time}^2$$

For the free fall the acceleration, usually denoted by the letter $g$ (for gravity), is equal to 981 cm per sec per sec $\left(\frac{\text{cm/sec}}{\text{sec}} = \frac{\text{cm}}{\text{sec}^2}\right)$, meaning that every second after a body begins to fall, its velocity increases by 981 cm per sec. In Anglo-American units $g$ is equal to 32.2 ft per sec per sec. Just to give an example, a bomb dropped from an airplane will acquire in 10 sec the velocity of

$$981 \times 10 = 9{,}810\,\frac{\text{cm}}{\text{sec}} = 98.1\,\frac{\text{meter}}{\text{sec}} \quad \text{or } 32.2 \times 10 = 322\,\frac{\text{ft}}{\text{sec}},$$

and will fall by the distance of $\tfrac{1}{2} \times 981 \times 10^2 = 49{,}050$ cm $= 0.49$ km (or $\tfrac{1}{2} \times 32.2 \times 10^2 = 1{,}610$ ft).

Another important contribution of Galileo to the problems of dynamics was the idea of composite motion which can be demonstrated on the following simple example:

Suppose we hold a stone 5 ft above the ground and let it drop. According to the above-given formula the stone hits the ground 0.96 sec after being released since, indeed, $\tfrac{1}{2} \times 32.2 \times (0.96)^2 = 5$ ft. What happens if, releasing the stone, we communicate to it a horizontal velocity of, say, 10 ft per sec? Everybody knows from personal experience that in this case the stone will describe a curved trajectory and fall to the ground some distance away from one's feet. To draw the trajectory of the stone in this case, we must consider the stone as having two independent motions: 1) horizontal motion with the constant velocity which was communicated to it at the moment of release; 2) vertical motion of free fall with the velocity which increases proportionally to time.

The result of the addition of these two motions is shown in Figure II-6. On the horizontal axis we plot equal stretches corresponding to the distances traveled by the ball during the first second, the second second, etc. On the vertical axis we plot the distances which increase as the squares of the integer numbers in accordance with the law of free fall. Actual positions of the ball are shown by the small circles which lie on a curve known as a parabola. If we throw the ball with twice the velocity, it will cover in its horizontal motion distances which are twice as large, while its vertical motion remains the

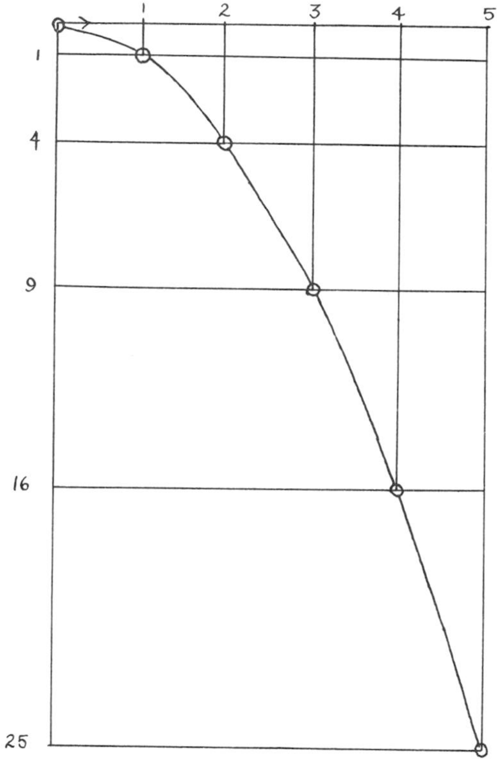


FIG. II–6.
The addition of a uniform motion in a horizontal direction and accelerated motion in a vertical direction. The resulting curve is known as a parabola.

same. As a result it will fall twice as far from our feet, but its time of flight through the air will be the same. (In all these considerations we neglect the friction of the air, which will deform slightly the trajectory of the thrown stone.)

An interesting application of the same principle is the problem of two boys playing jungle warfare (Fig. II-7). One boy stands on a branch of a tree while another shoots at him with a popgun. Suppose that the gunman points his gun directly at the boy on the tree, and that at the moment he pulls the trigger, the boy steps off the branch and starts falling to the ground. Did falling to the ground help? The answer is no, and this is why. In the absence of gravity, the bul-

FIG. II–7.

Since all bodies fall with the same acceleration, if a "bullet" from a popgun is shot by one boy playing jungle war directly at the "enemy" on a tree branch, the "bullet" will hit the enemy squarely on the nose if he jumps down at the moment of shooting.

let would proceed along the straight line *ABC* to the point where the boy was originally standing. Because of gravity, however, the bullet begins to fall down the moment it leaves the muzzle of the gun, and we have a twofold motion: a uniform motion along the line *ABC,* and an accelerated motion in the vertical direction. Since all material objects fall with the same acceleration, the vertical motion of the bullet and the boy are identical. Thus, when the bullet would have come to the point $B$, halfway to the original target, it would have fallen by a distance $BB'$, which is equal to the distance $CC''$ covered by the boy in his fall. When the bullet would have reached the point $C$ in the absence of gravity, it would have fallen the distance $CB''$ (twice the distance $BB'$) which is equal to the distance $CC''$ covered by the falling boy. Thus the boy will be hit squarely on the nose.

Instead of throwing a stone or shooting a bullet one can just drop an object from a moving vehicle. Suppose we release a stone from the top of the mast of a fast-moving, mechanically propelled ship (oar-propelled galley at Galileo's time). At the moment of release the stone will have the same horizontal velocity as the ship had, and will thus continue to move with this horizontal velocity after being released, remaining all the time just above the base of the mast. The vertical component of the stone's motion will be an accelerated free fall and, thus, it will hit the deck just at the base of the mast. The same will happen, of course, if we drop an object in a car of a moving train or in a cabin of a moving airplane, no matter how fast these vehicles move.

All this seems quite simple and evident to us in the present time, but it was not so when Galileo lived. At this time it was believed, according to the teachings of Aristotle which dominated scientific thought of that era, that the object moves only as long as it is being pushed and will stop as soon as the force disappears. According to this point of view, a stone released from the top of the mast will fall vertically down while the ship continues forward. Thus, it was expected that the falling stone would hit the deck closer to the stern. It is characteristic of medieval scholasticism that problems of that kind were discussed pro and con for ages, while nobody cared to climb the mast of a moving ship and to drop a stone!

The situation is colorfully illustrated by the following passages from Galileo's book *Dialogue on the Great World Systems,* published in 1632 in Florence. Following the tradition of the old Greek writers, Galileo composed his book as a conversation among three characters from the wonder-city of Venice: *Salviatus,* who speaks for the author himself; *Sagredus,* an intelligent layman; and *Simplicius,* a not too bright representative of Aristotle's school of thought. Here is an excerpt from their arguments concerning a stone falling from the mast of a moving ship, and from a tower erected on a moving (according to Copernicus) earth:

*Salv.:* . . . Aristotle says that is a most convincing argument of the Earth's immobility to see that projectiles thrown or shot upright return perpendicularly by the same line unto the same place from whence they were shot or thrown. And this holds true, although the motion be of a very great height. So that hither may be referred the argument taken from a shot fired directly upwards from a cannon, as also that other used by Aristotle and Ptolemy, of the heavy bodies that, falling from on high, are observed to descend by a direct and perpendicular line to the surface of the Earth. Now, that I may begin to untie these knots, I demand this of Simplicius: in case one should deny to Ptolemy and Aristotle that weights in falling freely from on high descend by a right and perpendicular line, that is, directly to the centre, what means would he use to prove it?

*Simp.:* The means of the senses, which assure us that that tower or other altitude is upright and perpendicular, and shew us that that stone slides along the wall, without inclining a hair's breadth to one side or another, and lights on the ground just under the place from where it was let fall.

*Salv.:* But if it should happen that the terrestrial globe did move round, and consequently carry the tower also along with it, and that the stone did then also graze and slide along the side of the tower, what must its motion be then?

*Simp.:* In this case we may rather say its motions, for it would have one wherewith to descend from the top to the bottom and should then have another to follow the course of the said tower.

*Salv.:* So that its motion should be compounded of two; from this it would follow that the stone would no longer describe that simple straight and perpendicular line but one transverse and perhaps not straight.

*Simp.:* I can say nothing of its nonrectitude, but this I know very well: that it would of necessity be transverse.

*Salv.:* You see then that, merely observing the falling stone to glide along the tower, you cannot certainly affirm that it describes a line which is straight and perpendicular unless you first suppose that the Earth stands still.

*Simp.:* True; for, if the Earth should move, the stone's motion would be transverse and not perpendicular.

*Salv.:* Aristotle's defense then consists in the impossibility, or at least in his esteeming it an impossibility, that the stone should move with a motion mixed of right and circular. For, if he did not hold it impossible that the stone could move at once to the centre and about the centre, he would have understood that it might come to pass that the falling stone might in its descent graze the tower as well when it moved as when it stood still. Consequently, he ought to have perceived that from this grazing nothing could be inferred touching the mobility or immobility of the

Earth. But this does not any way excuse Aristotle; because he ought to have expressed it, if he had had such a notion, it being so material a part of his argument. Also because it cannot be said that such an effect is impossible or that Aristotle did esteem it so. The first cannot be affirmed, for by and by I shall shew that it is not only possible but necessary; nor can the second be averred, for Aristotle himself grants that fire moves naturally in a right line, and moves about with the diurnal motion, imparted by the heavens to the whole element of fire and the greater part of the upper air. If therefore he held it possible to mix the straight motion upwards with the circular communicated to the fire and air from the concave of the sphere of the Moon, much less ought he to account impossible the mixture of the straight motion of the stone downwards with the circular which we presuppose natural to the whole terrestrial globe, of which the stone is a part.

Later in the *Dialogues,* Salviatus proposed a very interesting experiment designed to prove his point of view expressed in the previous discussions:

*Salv.:* Your making a greater scruple of this than of the other instances depends, if I mistake not, upon the birds being animated, and thereby enabled to use their strength at pleasure against the primary motion inbred in terrestrial bodies. For example, we see them fly upwards, a thing which ought to be altogether impossible for heavy bodies; whereas, when dead, they can only fall downwards. And therefore you hold that the reasons that hold for all kinds of missiles above named cannot hold for birds. Now this is very true, and, because it is true, therefore we see live birds behaving different from falling bodies. If from the top of a tower you let fall a dead bird and a live one, the dead bird shall do the same that a stone does, that is, it shall first follow the general diurnal motion, and then the motion of descent, just like a stone. But if the bird let fall be alive, what shall hinder it (there ever remaining in it the diurnal motion) from soaring by help of its wings to what point of the horizon it shall please? And this new motion, as being peculiar to the bird, and not participated in by us, must of necessity be visible to us. In short, the effect of the flight of birds differs from the missiles shot or thrown to any part of the world in nothing, except that the missiles are moved by an external projector, and the birds by an internal principle.

For a final proof of the nullity of all the experiments before alleged, I conceive it now a convenient time and place to demonstrate a way how to make an exact trial of them all. Shut yourself up with some friend in the largest room below decks of some large ship and there procure gnats, flies, and such other small winged creatures. Also get a great tub full of water and within it put certain fishes; let also a certain bottle be hung up, which drop by drop lets forth its water into another narrow-necked bottle placed underneath. Then, the ship lying still, observe how those small winged animals fly with like velocity towards all parts of the room; how the fishes swim indifferently towards all sides; and how the distilling drops all fall into the bottle placed underneath. And casting anything towards your friend, you need not throw it with more force one way than another, provided the distances be equal; and jumping broad, you will reach as far one way as another. Having observed all these particulars, though no man doubts that, so long as the vessel stands still, they ought to take place in this manner, make the ship move with what velocity you please, so long as the motion is uniform and not fluctuating this way and that. You shall not be able to discern the least alteration in all the forenamed effects, nor can you gather by any of them whether the ship moves or stands still. Of this correspondence of effects the cause is that the ship's motion is common to all the things contained in it and to the air also; I mean if those things be shut up in the room; but in case those things were above deck in the open air, and not obliged to follow the course of the ship, differences more or less notable would be observed in some of the forenamed effects, and there is no doubt but that smoke would stay behind as much as the air itself; the flies also and the gnats, being hindered by the air, would not be able to follow the motion of the ship, if they were separated at any distance from it; but keeping near thereto, because the ship itself, as being an anfractuous structure, carries along with it part of its nearest air, they would follow the ship without any pains or difficulty. For the like reason we see sometimes, in riding post, that the troublesome horseflies do follow the horses flying sometimes to one, sometimes to another, part of the body. But in the falling drops the difference would be very small and in the jumps and projections of grave bodies altogether imperceptible.

*Sagr.:* Though it came not into my thoughts to make trial of these observations when I was at sea, yet I am confident that they will succeed in the manner that you have related. In confirmation of this, I remember that being in my cabin I have wondered a hundred times whether the ship moved or stood still; and sometimes I have imagined that it moved one way, when it moved the other way. I am therefore satisfied and convinced of the nullity of all those experiments that have been produced in proof of the negative part.

There now remains the objection founded upon that which experience shows us, namely, that a swift whirling about has a faculty to extrude and disperse the matters adherent to the machine that turns round. On this fact many based the opinion, and Ptolemy amongst the rest, that, if the

Earth should turn round with such great velocity, the stones and creatures upon it should be tossed into the sky and that there could not be a mortar strong enough to fasten buildings to their foundations so that they should not suffer a like extrusion. . . .

The statement that it is impossible to find out whether a ship is lying at anchor or moving across the sea by performing any mechanical experiment in a closed cabin in its interior is now known as "Galileo's principle of relativity." It took almost three centuries of development of physics before this principle was extended by Albert Einstein to the case of optical and electromagnetic phenomena as observed in a closed, uniformly moving cabin. So much for Galileo's contributions to the science of mechanics.

## GALILEO THE ASTRONOMER

Besides being one of the first experimental and theoretical physicists, Galileo also made tremendous contributions to astronomy, opening to humanity the unlimited vistas of the universe around us. His attention was first drawn to the sky in the year 1604 when a brilliant new star (we now call these the "novae") suddenly appeared one night among the unchangeable constellations known to stargazers for thousands of years. Galileo, at that time 40 years old, had demonstrated that the new star was really a star and not some kind of meteor in the terrestrial atmosphere, and predicted that it would gradually fade away. The appearance of a new star in heaven, which was supposed to be absolutely unchangeable according to Aristotle's philosophy and the teachings of the Church, made Galileo many enemies among his scientific colleagues and among the high clergy. Only a few years after this first step in studying the sky, Galileo revolutionized astronomy by constructing the first astronomical telescope, which he described in the following words:

About ten months ago a rumor came to our ears that an optical instrument has been elaborated by a Dutchman, by the aid of which visible objects, even though far distant from the eye of the observer, were distinctly seen as if near at hand, and some stories of this marvellous effect were bandied about, to which some gave credence and which others denied. The same was confirmed to me a few days after by a letter sent from Paris by the noble Frenchman Jacob Badovere, which at length was the reason that I applied myself entirely to seeking out the theory and discovering the means by which I might arrive at the invention of a similar instrument, an end which I attained a little later, from considerations of the theory of refraction; and I first prepared a tube of lead, in the ends of which I fitted two glass lenses, both plane on one side, one being spherically convex, the other concave on the other side.

Having built the instrument he pointed it at the sky, and the marvels of the universe were unfolded before his eyes. He looked at the moon and found that:

The surface of the Moon is not perfectly smooth, free from inequalities and exactly spherical, as a large school of philosophers considers with regards to the Moon and other heavenly bodies, but on the contrary, it is full of inequalities, uneven, full of hollows and protuberances, just like the surface of the Earth itself, which is varied everywhere by lofty mountains and deep valleys.

He looked at the planets and found that:

The planets present their discs perfectly round, just as if described with a pair of compasses, and appear as so many little moons, completely illuminated and of a globular shape; but the fixed stars do not look to the naked eye [This must be the first use of that expression!] as if they were bounded by a circular circumference, but rather like blazes of light shooting out beams on all sides and very sparkling, and with the telescope they appear of the same shape as when they were viewed by simply looking at them.

He looked at Jupiter on the 7th day of January, 1610, and:

There were three little stars, small but very bright, near the planet; and although I believed them to belong to the number of the fixed stars, yet they made me somewhat wonder, because they seemed to be arranged exactly in a straight line, parallel to the ecliptic, and to be brighter than the rest of the stars, equal to them in magnitude. . . . On the east side there were two stars, and a single one towards the west. . . . But when on January 8th, led by some fatality, I turned again to look at the same part of the heavens, I found a very different state of things, for there were three little stars all west of Jupiter, and nearer together than on the previous night.

Thus Galileo decided that:

There are three stars in the heavens moving about Jupiter, as Venus and Mercury round the Sun.

He looked at Venus and Mercury and discovered that they have sometimes a crescent shape just as the moon does, from which he concluded that:

Venus and Mercury revolve round the Sun, as do also all the rest of the planets. A truth believed indeed by the Pythagorean school, by Copernicus, and by Kepler, but never proved by the evidence of our senses, as it is now proved in the case of Venus and Mercury.

He looked at the Milky Way and found that it is:

. . . nothing else but a mass of innumerable stars planted together in clusters.

Galileo's discoveries made by the use of the telescope gave undeniable proof of the correctness of the Copernican system of the world, and he went around jubilantly talking about it. But it was certainly more than the Holy Inquisition could permit; he was arrested and subjected to a long period of solitary confinement, and questioning, which did not seem, however, to change his fighting spirit! On January 15th, 1633, a few months before the final trial, Galileo wrote to his friend Ella Diodati:

When I ask: whose work is the Sun, the Moon, the Earth, the Stars, their motions and dispositions, I shall probably be told that they are God's work. When I continue to ask whose work is Holy Scripture, I shall certainly be told that it is the work of the Holy Ghost, *i.e.*, God's work also. If now I ask if the Holy Ghost uses words which are manifest contradictions of the truth so as to satisfy the understanding of the—generally uneducated—masses, I am convinced that I shall be told, with many citations from all the sanctified writers, that this is indeed the custom of Holy Scripture, containing as it does hundreds of passages that taken literally would be nothing but heresy and blasphemy, for in them God appears as a Being full of hatred, guilt and forgetfulness. If now I ask whether God, so as to be understood by the masses, had ever altered His works, or else if Nature, unchangeable and inaccessible as it is to human desires, has always retained the same kinds of motion, forms and divisions of the Universe, I am certain to be told that the Moon has always been round, even though it was long considered to be flat. To condense all this into one phrase: Nobody will maintain that Nature has ever changed in order to make its works palatable to men. If this be the case, then I ask why it is that, in order to arrive at an understanding of the different parts of the world, we must begin with the investigation of the Words of God, rather than of His Works. Is then the Work less venerable than the Word? If someone had held it to be heresy to say that the Earth moves, and if later verification and experiments were to show us that it does indeed do so, what difficulties would the church not encounter! If, on the contrary, whenever the works and the Word cannot be made to agree, we consider Holy Scripture as secondary, no harm will befall it, for it has often been modified to suit the masses and has frequently attributed false qualities to God. Therefore I must ask why it is that we insist that whenever it speaks of the Sun or of the Earth, Holy Scripture be considered as quite infallible?

On June 22nd, 1633, at the age of 69, he was brought in front of the judges of the Holy Office of the Church, and standing down on his knees he "confessed":

I, Galileo Galilei, son of the late Vincenzio Galilei of Florence, aged seventy years,* being brought personally to judgment, and kneeling before you, Most Eminent and Most Reverend Lords Cardinals, General Inquisitors of the Universal Christian Commonwealth against heretical depravity, having before my eyes the Holy Gospels which I touch with my own hands, swear that I have always believed, and, with the help of God, will in future believe, every article which the Holy Catholic and Apostolic Church of Rome holds, teaches, and preaches. But because I have been enjoined, by this Holy Office, altogether to abandon the false opinion which maintains that the Sun is the centre and immovable, and forbidden to hold, defend, or teach, the said false doctrine in any manner . . . I am willing to remove from the minds of your Eminences, and of every Catholic Christian, this vehement suspicion rightly entertained towards me, therefore, with a sincere heart and unfeigned faith, I abjure, curse, and detest the said errors and heresies, and generally every other error and sect contrary to the said Holy Church; and I swear that I will never more in future say, or assert anything, verbally or in writing, which may give rise to a similar suspicion of me; but that if I shall know any heretic, or any one suspected of heresy, I will denounce him to this Holy Office, or to the Inquisitor and Ordinary of the place in which I may be. I swear, moreover, and promise that I will fulfill and observe fully all the penances which have been or shall be laid on me by this Holy Office. But if it shall happen that I violate any of my said promises, oaths, and protestations (which God avert!), I subject myself to all the pains and punishments which have been decreed and promulgated by the sacred canons and other general and particular constitutions against delinquents of this description. So, may God help me, and His Holy Gospels, which I touch with my own

* This is the original text of Galileo's confession. He was actually 69 years, 4 months, and 7 days old.

hands, I, the above named Galileo Galilei, have abjured, sworn, promised, and bound myself as above; and, in witness thereof, with my own hand have subscribed this present writing of my abjuration, which I have recited word for word.

There is a story that immediately after the "confession" Galileo exclaimed: "Eppur si muove!" ("Nevertheless, it moves!"), but this is not true and it only gave ground to an old anecdote according to which Galileo was watching the wagging tail of a friendly dog which entered by mistake into the Holy Office of the Church. Having been convicted of heresy, Galileo was confined to his villa in Arcetri near Florence under what we would now call "house arrest." On January 8th, 1642, completely blind and tired of life, Galileo died.

# CHAPTER III *God Said, "Let Newton Be!"**

THE YEAR Galileo died in his Florentine seclusion, a premature baby christened Isaac was born into the family of a Lincolnshire farmer by the name of Newton. During the early years at school Isaac did not show any indication of future greatness. He was a sickly, shy boy, rather backward in his studies. What brought him out of this stage was a fist fight with a schoolmate who, being one of the best students in the class, was also very aggressive toward other boys. Being kicked in the belly by that bully (whose name was lost to history), Newton challenged him to fight, and beat him up because of his (Newton's) "superior spirit and resolution." Having won on the physical front, he decided to complete his victory in the battle of intelligence, and by hard work succeeded in becoming the first student in the class. Winning another battle against his mother, who wanted him to pursue a farming career, he entered Trinity College at the age of 18, and devoted himself to the study of mathematics. In the year 1665 Newton took his B.A. degree without any particular distinction.

### PROGRESS DURING THE PESTILENCE

In the midsummer of 1665 the Great Plague descended on London, and within a few months one of every ten Londoners had died of it. In the fall the University of Cambridge was closed because of its proximity to the plague center and all the students were sent

* From a verse by Alexander Pope (1688-1744):

"Nature and Nature's laws lay hid in night;
God said, Let Newton be! and all was light."

home. Thus, Newton went back to his parental home in Lincolnshire, and remained there for eighteen months until the university was reopened.

These eighteen months in rural seclusion were the most productive in his life and it can be said that during this period he conceived practically all the ideas for which the world is grateful to him.

To quote his own words:

> In the beginning of 1665 I found the . . . rule for reducing any dignity [power] of binomial to a series.* The same year in May I found the method of tangents . . . and in November the direct method of Fluxions [i.e., the elements of what is now known as differential calculus], and the next year in January had the Theory of Colours, and in May following I had entrance into the inverse method of Fluxions [i.e., integral calculus], and in the same year I began to think of gravity extending to the orb of the Moon . . . and . . . compared the force requisite to keep the Moon in her orb with the force of gravity at the surface of the Earth.

The rest of his scientific career was devoted to the development of the ideas conceived in Lincolnshire.

At the age of 26 he was appointed as a professor at the University of Cambridge and at 30 was elected as a Fellow of the Royal Society, the highest scientific honor in England. According to his biographers, Newton was a perfect example of the absent-minded professor. He would "never take any recreation or pastime either in riding out to take the air, walking, bowling, or any other exercise whatever, thinking all hours lost that were not spent in his studies." He often worked until the small hours in the morning, kept forgetting to eat his meals, and when he appeared once in a while in the dining hall of the college "his shoes were down at the heels, stockings untied, surplice on, and his head scarcely combed." Being always absorbed in his thoughts he was very naïve and impractical concerning everyday problems. There is a story that he once made a hole in the door of his house for his cat to come in and out. When the cat had kittens, he added to the big hole a number of smaller ones for each of the kittens.

As a person, Newton was not too pleasant and was often involved in controversy with his colleagues, which may have been the reflection of his fight with a schoolmate years before. He was involved in

* The so-called Newton binomial theorem which is now taught in high-school algebra.

bitter quarreling with another Cambridge physicist, Robert Hooke (the founder of the theory of elasticity), concerning his theory of color as well as the priority in the discovery of the law of universal gravity. There was another priority squabble with the German mathematician Gottfried Leibnitz concerning the invention of calculus, and with the Dutchman, Christian Huygens, about the theory of light. Astronomer John Flamsteed, who was hardly on speaking terms with Newton, described him as "insidious, ambitious, excessively covetous of praise, and impatient of contradiction . . . a good man at the bottom but, through his nature, suspicious."

Through his years at Cambridge, Newton worked on the development of his brilliant ideas conceived between the ages of 23 and 25, but kept most of his discoveries secret. This accounts for the fact that the full account was published much later in his life: the work on mechanics and gravity at the age of 44 and the work on optics at the age of 65.

## NEWTON'S *Principia*

In the preface (dated May 8, 1686) to his book *Philosophiae Naturalis Principia Mathematica* (*Mathematical Principles of Natural Philosophy**) Newton wrote:

> Since the ancients esteemed the science of mechanics of greatest importance in the investigation of natural things, and the moderns, rejecting substantial forms and occult qualities, have endeavored to subject the phenomena of nature to the laws of mathematics, I have in this treatise cultivated mathematics as far as it relates to [natural] philosophy. The ancients considered mechanics in a twofold respect; as rational, which proceeds accurately by demonstration, and practical. To practical mechanics all the manual arts [engineering] belong, from which mechanics took its name. But as artificers do not work with perfect accuracy, it comes to pass that mechanics is so distinguished from geometry that what is perfectly accurate is called geometrical; what is less so, is called mechanical. However, the errors are not in the art, but in the artificers. He that works with less accuracy is an imperfect mechanic; and if any could work with perfect accuracy he would be the most perfect mechanic of all. . . .
>
> I consider [natural] philosophy rather than arts and write not concerning manual but natural powers, and consider chiefly those things which relate to gravity, levity [buoyancy], elastic force, the resistance of fluids, and the like forces, whether attractive or impulsive; and therefore I

* At that time, "natural philosophy" meant the study of the laws of nature.

offer this work as the mathematical principles of [natural] philosophy, for the whole burden of philosophy seems to consist in this—from the phenomena of motions to investigate the forces of nature, and then from these forces to demonstrate the other phenomena. . . .

I wish we could derive . . . the phenomena of Nature . . . from mechanical principles, for I am induced by many reasons to suspect that they all may depend upon certain forces by which the particles of bodies, by some causes hitherto unknown, are either mutually impelled towards one another, and cohere in regular figures, or are repelled and recede from one another. These forces being unknown, philosophers have hitherto attempted the search of Nature in vain; but I hope the principles here laid down will afford some light either to this or some truer method of [natural] philosophy.

In the above-quoted words, Newton laid down the program of the so-called *mechanistic interpretation* of all physical phenomena, a point of view which dominated physics until the beginning of the present century, and succumbed only under the impact of the theory of relativity and the theory of quanta. Having formulated his aim, he proceeded to develop mathematical treatment of mechanical phenomena in terms so clear and precise that they can be used unchanged in any modern book of classical mechanics. We reproduce here the opening passages of Newton's *Principia* with only a few comments (in brackets) to clarify the modern meaning of the 17th-century scientific terminology.

## *DEFINITIONS*

*Definition I. The quantity of matter* [mass] *is the measure of the same, arising from its density and bulk* [volume] *conjointly.*

Thus air of a double density in a double space [volume] is quadruple in quantity; in a triple space [volume], sextuple in quantity. The same thing is to be understood of snow, and fine dust or powders, that are condensed by compression or liquefaction, and of all bodies that are by any causes whatever differently condensed. . . . [In modern language, we say that the mass of any given object is a product of its density times its volume.]

*Definition II. The quantity of motion is the measure of the same, arising from the velocity and quantity of matter conjointly.* [In modern language, the amount of motion, now usually called "mechanical momentum" or simply "momentum," is the product of the velocity by the mass of the moving object.]

The motion of the whole is the sum of the motion of all the parts; and therefore in a body double in quantity [twice as massive], with equal velocity, the motion [mechanical momentum] is double; with twice the velocity, it is quadruple.

*Definition III. The* vis insita, *or innate force of matter, is a power of resisting, by which every body, as much as in it lies, continues in its present state, whether it be of rest, or of moving uniformly forwards in a right* [straight] *line.*

This force is always proportional to the [mass of the] body whose force it is and differs nothing from the inactivity of the mass, but in our manner of conceiving it. A body, from the inert nature of matter, is not without difficulty put out of its state of rest or motion. Upon which account, this *vis insita* may, by a most significant name, be called "inertia" (*vis inertiae*) or force of inactivity. . . .

*Definition IV. An impressed force is an action exerted upon a body, in order to change its state, either of rest, or of uniform motion in a right* [straight] *line.*

This force consists in the action only, and remains no longer in the body when the action is over. For a body maintains every new state [of motion] it acquires, by its inertia only. But impressed forces are of different origins, as from percussion, from pressure, from centripetal force.

Having defined the notions of *mass, momentum, inertia,* and *force,* Newton proceeded with the formulation of the basic laws of motion:

*Law I. Every body continues in its state of rest, or of uniform motion in a right* [straight] *line, unless it is compelled to change that state by forces impressed upon it.* [Fig. III-1a.]

Projectiles continue in their motion, so far as they are not retarded by the resistance of the air, or impelled downwards by the force of gravity. A top, whose parts by their cohesion are continually drawn aside from rectilinear motions, does not cease its rotation, otherwise than as it is retarded by the air. The greater bodies of the planets and comets, meeting with less resistance in freer spaces, preserve their motions both progressive and circular for a much longer time.

*Law II. The change of motion* [i.e., of mechanical momentum] *is proportional to the motive force impressed; and is made in the direction of the right line in which that force is impressed.* [Fig. III-1b.]

If any force generates a motion, a double force will generate double the motion, a triple force triple the motion, whether that force be impressed altogether and at once, or gradually and successively. And this motion (being always directed the same way with the generating force) if the body moved before, is added to or subtracted from the former motion, according as they directly conspire with or are directly contrary to each other; or obliquely joined, when they are oblique, so as to produce a new motion compounded from the determination of both.

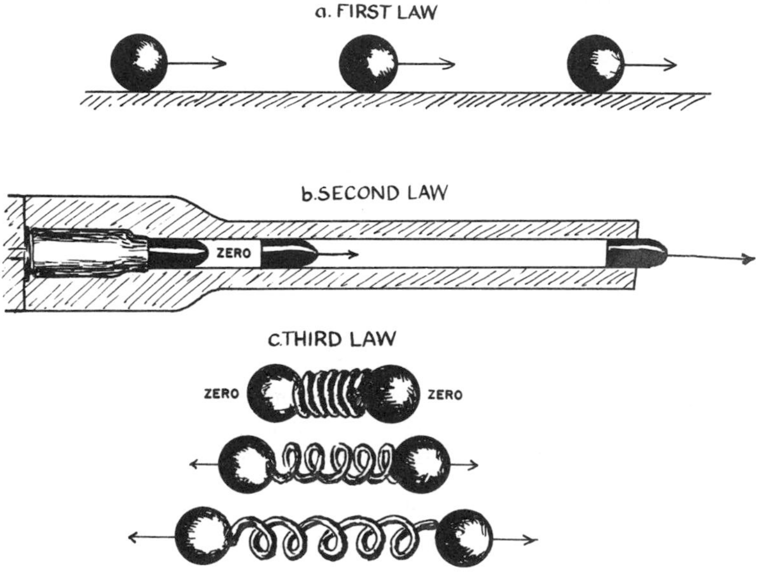


FIG. III–1.

The three laws of Newton: (a) A ball on the horizontal plane, with no force acting on it in the direction of its motion, moves along a straight line with constant speed. (b) A projectile in the gun barrel, being pushed by the powder gases, is moving with continuously increasing speed. (c) Two balls are pushed with equal force by a compressed spring placed between them. If it is assumed in the drawing that they have equal masses, they will move in opposite directions with equal speeds.

The second law of Newton can be formulated in a somewhat different way. Since the amount of motion is a product of the mass of the moving object times its velocity, the rate of change of motion is the product of the mass times the rate of change of velocity, i.e., the acceleration. Thus one concludes that the acceleration of an object acted upon by a certain force is directly proportional to that force, and inversely proportional to the mass of the object. On the basis of that law, we can introduce a unit of force, defining it as a force which, acting on an object having the mass of 1 gm, communicates to it the acceleration of 1 cm per second each second. That unit of force is called a *dyne,* and is rather small, about the force with which an ant can pull its load. In engineering one often uses a unit which is $10^5$ times larger and is called a *newton.*

When a given force acting on a certain object displaces it by a certain distance, the product of that force by the distance is known as *work* done by it. If the force is expressed in dynes and the distance in centimeters, the work will be measured in units known as *ergs.* For engineering purposes, a much larger unit of energy known as a *joule* is used; 1 joule is equal to $10^7$ ergs. One can also introduce the unit of power telling us how much work is done per unit time; it is usually measured in ergs per second and has no special name. In engineering one uses *watts,* which is 1 joule per second or $10^7$ ergs per second, and horsepower, which is 751 watts, or 0.751 kilowatts.

*Law III. To every action there is always opposed an equal reaction: or, the mutual actions of two bodies upon each other are always equal, and directed to contrary parts.* [Fig. III-1c.]

Whatever draws or presses another is as much drawn or pressed by that other. If you press a stone with your finger, the finger is also pressed by the stone. If a horse draws a stone tied to a rope, the horse (if I may so say) will be equally drawn back towards the stone; for the distended rope, by the same endeavor to relax or unbend itself, will draw the horse as much towards the stone as it does the stone towards the horse, and will obstruct the progress of the one as much as it advances that of the other. . . .

Why then, one can ask, is the horse pulling the stone, and not the stone pulling the horse? The answer is, of course, that the difference lies in the friction against the ground. The four horseshoes cling more strongly to the ground than the stone that the horse pulls, and if it were not so, the stone would remain in place, and the horse's hoofs would slide. Putting rollers under the stone will reduce the friction against the ground and will make the horse's job much easier. If the friction is absent, which is approximately true on the icy surface of a frozen pond, the motion of two objects pulling or pushing one another will also not be the same unless they have exactly equal masses, since for a given force the acceleration is inversely proportional to the mass. If a thin man and a big fat one standing face to face on the icy surface would give each other a push, the thin man would slide back with much higher speed than the fat one. Similarly, the recoil velocity of a rifle is much smaller than the velocity of the (much lighter) bullet expelled from its muzzle.

The principle of recoil is used in constructing all kinds of rockets. The gases resulting from the burning of the rocket fuel stream backward through the nozzle with a high velocity, as a result of which the body of the rocket is pushed forward. The final velocity obtained by a rocket at burnout depends on the weight ratio of the rocket to that of the fuel, and for better performance one should make that ratio as small as possible. In modern rockets the ratio of empty rocket weight to the weight of the fuel is about the same as the weight ratio of an empty eggshell to the body of the egg.

It is not the place here to discuss engineering problems of modern rocketry, and we will limit ourselves to mentioning an incident which took place at the large rocket proving ground at Cape Canaveral, Florida. At the beginning of the first lesson in the first grade of a local elementary school the teacher wanted to know what the boys and girls knew about the three R's. "I can count," volunteered little Johnny. "Go ahead," said the teacher, "count." "Ten, nine, eight," started Johnny, "seven, six, five, four, three, two, one, . . . nuts!"

But to return to Newton without abandoning abruptly the problems of space flight, one should mention that he was the first to have the idea of an earth satellite. In the third part of *Principia* we read:

That by means of centripetal forces the planets may be retained in certain orbits, we may easily understand, if we consider the motions of projectiles; for a stone that is projected is by the pressure of its own weight forced out of the rectilinear path, which by the initial projection alone it should have pursued, and made to describe a curved line in the air, and through that crooked way is at last brought down to the ground; and the greater the velocity is with which it is projected, the farther it goes before it falls to the earth. We may therefore suppose the velocity to be so increased, that it would describe an arc of 1, 2, 5, 10, 100, 1,000 miles before it arrived at the earth, till at last, exceeding the limits of the earth, it should pass into space without touching it. Let AFB [Fig. III-2] represent the surface of the earth, C its center, VD, VE, VF, the curved lines which a body would describe, if projected in an horizontal direction from the top of an high mountain [somewhere in the Scottish highlands, no doubt] successively with more and more velocity; and, because the celestial motions are scarcely retarded by the little or no resistance of the spaces in which they are performed, to keep up the parity of cases, let us suppose either that there is no air about the earth, or at least that it is endowed with little or no power of resisting; and for the same reason that the body projected with a lesser velocity describes the lesser arc VD, and with a greater velocity of the greater arc VE, and, augmenting the velocity, it goes farther and farther to F and G, if the velocity was still more and more augmented, it would reach at last quite beyond the circumference of the earth, and return to the mountain from which it was projected. . . .

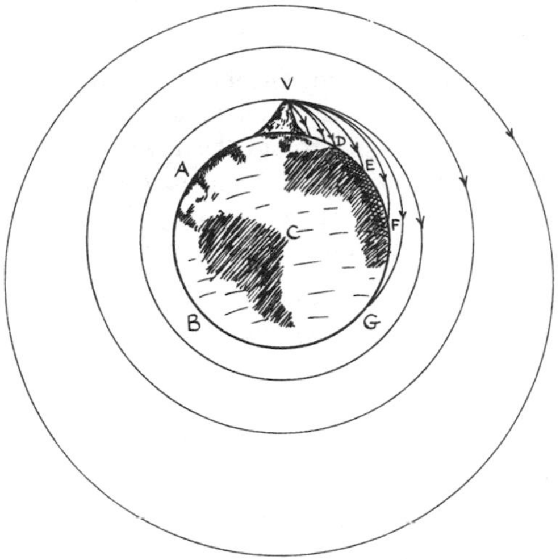


FIG. III–2.
The trajectory of the earth's satellite as a limiting case of the trajectories of projectiles falling farther and farther away from the base of the mountain from which they were shot. (Adaptation from the original drawing in Newton's *Principia.*)

But if we now imagine bodies to be projected in the directions of lines parallel to the horizon from greater heights, as of 5, 10, 100, 1,000 or more miles, or rather as many semidiameters of the earth, these bodies, according to their different velocity, and the different force of gravity in different heights, will describe arcs either concentric with the earth, or variously eccentric, and go on revolving through the heavens in those orbits just as the planets do in their orbits.

This passage includes the idea that one and the same force, the force of gravity, is responsible both for the fall of a stone and for the motion of celestial bodies, the idea which Newton allegedly first got by observing an apple falling down from a tree. Whether the "apple theory" is authentic or not, it led to an interesting verse given below:

Sir Isaac, walking deep in thought,
Was by a farmer-neighbor caught
And, wrenched from gravitation's laws,
Persuaded by the man to pause
And chat a while. Along the breeze
Pale apple-blossoms from the trees
Of Newton's orchard-owning friend
Blew down the road from end to end.

To Newton said the neighbor: "Stay!
I'd like a word with you today.
The talk is going round the town
That you've acquired some renown
From watching apples as they fall.
Pray tell me more, sir—tell me all."

"Why yes," said Newton; "Yes, of course.
Well, don't you see, the selfsame force—
Which is decreasing as the square
Of *r*, the distance up to there—
That acts upon our faithful moon
Acts on the apple. Late or soon. . . ."

"Please!" said the neighbor, "let it go!
That isn't what I want to know.
The only thing that interests me
About the blooming apple tree
And all its apples, one by one,
That ripen in the gentle sun
Along this quiet country road
Is *How much do you charge per load?"*

*Rendered into English verse by B.P.G., from the unpublished Russian verse by an anonymous author.*

In order to establish the dependence of the force of gravity on the distance from the center of the earth, Newton decided to compare the fall of a stone (or an apple) on the surface of the earth with the motion of the moon, which can also be considered as an endless fall according to the above-given argument. In this way Newton was able to compare the "astronomical" force acting on the moon with the "terrestrial" force acting on the objects which we handle in everyday life.

His argument, in a somewhat modified form, is represented in Figure III-3 which shows the moon *M* revolving around the earth *E* along a (nearly) circular orbit. At position *M* the moon possesses a certain velocity which is perpendicular to the radius of the circle.

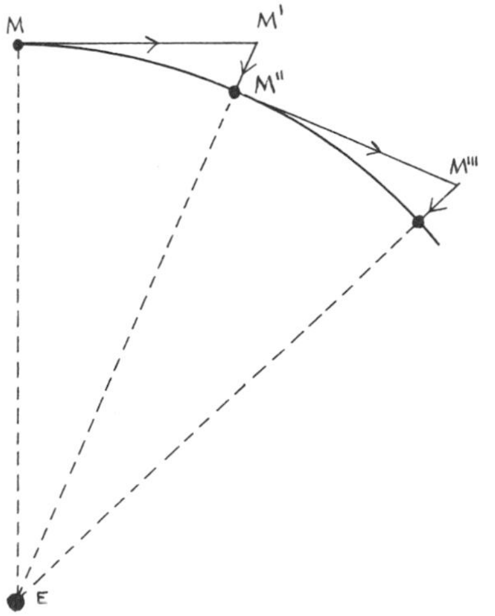


FIG. III–3.

Considering the circular motion of the moon around the earth as a continuous fall (cf. Fig. III–2), Newton could calculate the acceleration caused by the force of gravity acting on the moon. The above diagram shows how this is done.

If there were no forces, the moon would proceed along a straight line and, a unit of time later, would move to position $M'$. Since, however, it arrives at position $M''$, the stretch $M'M''$ should be considered as the distance covered by the moon during a unit time, in its *free fall* toward the earth. According to the Pythagorean theorem: $M'M'' = \sqrt{(EM^2 + (MM')^2} - EM$ (since $EM'' = EM$) which can be algebraically shown to be closely equal (for $MM' \ll EM$) to: $\frac{(MM')^2}{2EM}$ or $\frac{1}{2}\left(\frac{MM}{EM}\right)^2 \times EM$ where $MM'/EM$ is apparently the *angular velocity* of the moon in its motion around the earth, i.e., the change of angular position of the moon on its orbit in the course of 1 sec. Since the moon describes a complete circle in a month, the angular velocity is equal to $2\pi$ divided by the length of a month expressed in seconds $= 2.66 \times 10^{-6}$. But, in the discussion of accelerated motion, we have already seen that the distance covered during the first second is equal to one half of the quantity known as "acceleration," so that we conclude that the acceleration due to the force holding the moon on its circular orbit is $(MM'/EM)^2 \times EM$. Using the above given value for angular velocity and substituting for the distance to the moon the values of 384,400 km or $3.844 \times 10^{10}$ cm, Newton obtained, for the acceleration due to gravity at the distance of the moon, the value: 0.27 cm/sec$^2$, which is much smaller than the acceleration of gravity on the surface of the earth (981 cm/sec$^2$). There exists, however, a very simple correlation between these two quantities on one hand and the distances of the moon and of a falling apple from the center of the earth on the other. Indeed, the ratio of 981 to 0.27 is 3,640, which is exactly equal to the square of the number representing the ratio of the radius of the moon's orbit to the radius of the earth. Thus, Newton arrived at the result that *the forces of terrestrial gravity decrease as the inverse square of the distance from the center of the earth.*

Generalizing this finding to all material bodies in the universe he formulated the universal law of gravity according to which: *all material bodies attract each other with a force directly proportional to their masses and inversely proportional to the square of the distance between them.* Applying this law to the motion of the planets around the sun, he derived mathematically the three laws of Kepler, described in the previous chapter.

The development of Newton's work carried out by great mathematicians of the 18th and 19th centuries led to the birth of a large branch of astronomy known as "celestial mechanics," which permits us to calculate with great precision the motion of the planets of the solar system under the action of mutual gravitational attraction. One of the biggest triumphs of celestial mechanics came in 1846 with the discovery of a new planet Neptune, the existence and the orbit of which were predicted independently by the French astronomer U. J. J. Leverrier and the British astronomer J. C. Adams on the basis of perturbations of the motion of Uranus caused by the gravitational attraction of the, at that time, unknown planet. A similar event occurred in 1930 when a trans-Neptunian planet, later named Pluto, was discovered as the result of theoretical calculations.

Applying his law of gravity to the motion of the terrestrial globe, Newton gave the first explanation of the phenomenon of the "precession of equinoxes" known since the days of Plutarch. He showed

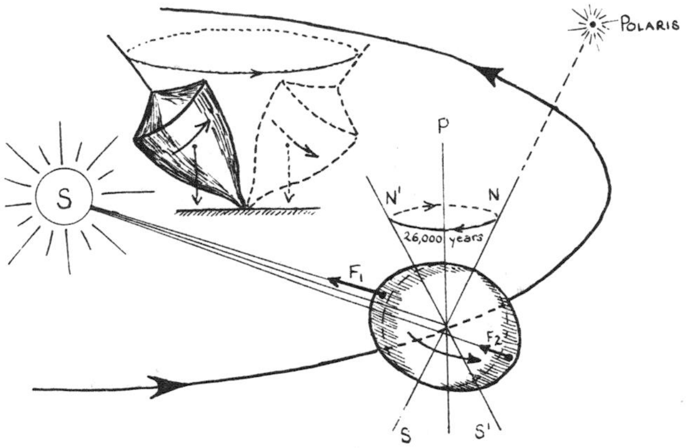


FIG. III–4.

Newton's explanation of the precession of the earth's axis of rotation. Since the force of gravity decreases with the distance, the force $F_1$ acting on the equatorial bulge turned toward the sun is larger than the force $F_2$ acting on the bulge turned away from it. Thus, the combined action of the two tends to exert a "righting force" on the earth's rotational axis, i.e., to make it perpendicular to the plane of the orbit. The situation is similar to that encountered in the case of a spinning top with an inclined axis, where the force of gravity, i.e., its weight, tends to bring its axis to a horizontal position. And, just as the spinning top does not drop down on its side as long as it spins, but remains upright with its axis describing a conical surface around the vertical, the earth's axis does not become perpendicular to the orbit but describes a conical surface around that direction.

that since the rotation axis of the earth is inclined to the plane of its orbit (ecliptics), the gravity forces of the sun, acting on the equatorial bulges of the globe, must cause the slow rotation of the earth's axis around a line vertical to the ecliptics within a period of about 26,000 years (Fig. III-4). This explanation met a strong opposition among contemporary astronomers, since at that time it was believed, on the basis of erroneous measurements, that our earth does not have the shape of a pumpkin, being broader at the equator, but rather that of a watermelon, with the distance between the poles larger than the equatorial diameter.

To settle the argument, the French mathematician P. L. M. de Maupertuis organized an expedition to Lapland to measure the length of one degree of meridian in northern latitudes, and had a lot

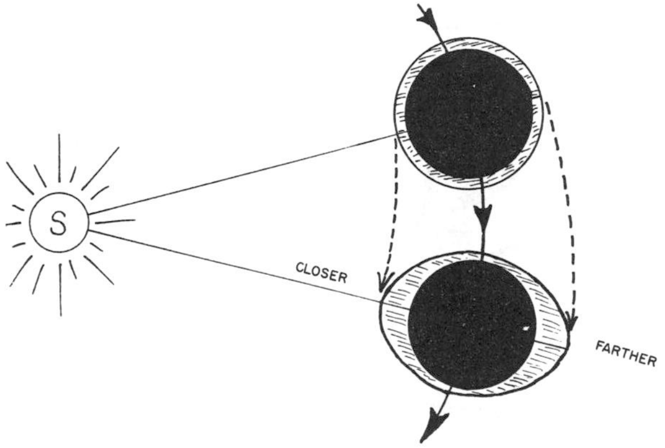


FIG. III–5.

Newton's explanation of ocean tides. Since the force of gravity decreases with the distance from the sun, the force acting on the ocean water on the day side of the earth is somewhat *larger* than the force acting on the solid body of the earth. Similarly, the force of gravity acting on the ocean water on the night side is somewhat *smaller* than that acting on the solid body. As a result of these differences, the water surface on the day side has the tendency to rise higher over the ocean bottom, while on the night side the bottom of the ocean is, so to speak, "pulled from under" the surface of the ocean. Both effects result in the formation of two water bulges which, in conjunction with the earth rotation around its axis, are observed as two tidal waves running around the earth with the period of 24 hr.

of adventures with a pack of wolves. His measurement proved that Newton's views were correct and Voltaire wrote to him, jokingly:

> Vous avez confirmé dans les lieux pleins d'ennui
> Ce que Newton connût sans sortir de chez lui.*

Along the same lines, Newton explained the phenomenon of ocean

* You have confirmed, in the lands full of boredom
what Newton knew without leaving home.

tides as being due to the unequal gravitational force exerted by the sun on the terrestrial hemispheres turned toward and away from it (Fig. III-5).

The 626 pages of Newton's *Principia* are packed with information on all branches of solid and fluid dynamics, but we give here only one more problem because it is simple and amusing. It pertains to the motion of projectiles thrown or shot with a certain initial velocity through a resistant medium like air or water. How far will they move before coming to rest?

The situation is shown graphically in Figure III-6, in which a projectile shot from a gun moves through the air or water as the case may be. While moving through the medium, the projectile

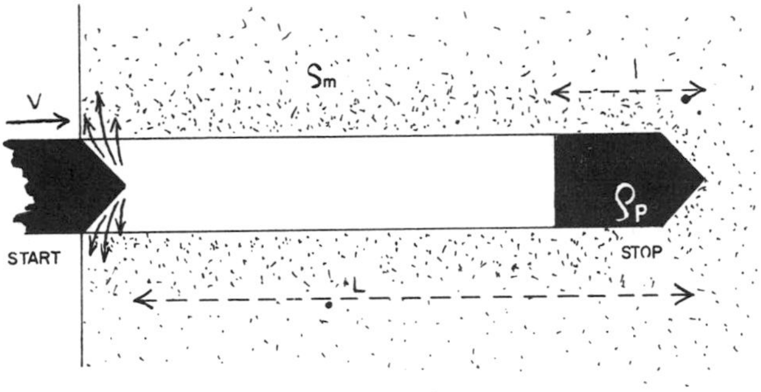


FIG. III–6.
Newton's theory of penetration of projectiles into a medium.

must evidently push aside the medium in order to bore a tunnel to move forward. At high velocities the friction forces are of comparatively small importance, and the main energy losses by the projectile are due to the necessity of communicating high velocities to the medium which is being pushed aside. It is easy to see that this sidewise velocity of the medium is about the same as the velocity of the advancing projectile. Thus, the projectile will stop when the mass of the medium moved aside is of the same order of magnitude as its own mass. We conclude, therefore, that the length of the tunnel must stand in the same ratio to the length of a projectile as the density of the projectile's material to the density of the medium,

$$\frac{L}{l} = \frac{\rho_p}{\rho_m} \text{ (approximately)}$$

which, of course, is true only very approximately. But even so, we can get from it several interesting results. If we shoot a steel projectile (density about 10 times more than that of water) through the air (density about 1,000 times less than that of water), the projectile is expected to stop after covering about 10,000 times its length (if it does not drop to the ground before doing so). Thus, large naval artillery shells which may be 5 ft or more in length, are expected to cover about 50,000 ft or over 10 miles. On the other side, a ½-inch-long bullet from a lady's revolver would hardly fly more than 400 ft. In water, which is only about ten times less dense than metal, a bullet will lose most of its energy after traveling only ten times its length; that is why skin divers use long metal arrows to hunt their underwater prey. It is interesting that the length of penetration does not depend on the initial velocity of the projectile (provided that this velocity is sufficiently high). This is the fact that puzzled the U.S. military experts who were dropping from different heights the explosive missiles which were supposed to burrow deep into the ground before busting up. The penetration did not seem to change with the height from which the missiles were dropped (thus hitting the ground with different velocities) and the experts were scratching their heads until somebody pointed out to them a theory on that subject in Newton's *Principia.*

### FLUID STATICS AND DYNAMICS

The studies of Sir Isaac on the equilibrium and motion of fluids were complemented and expanded by the French mathematician Blaise Pascal, who was 19 years old when Newton was born, and the Swiss physicist Daniel Bernoulli, who was 27 years old at the time Newton died. Pascal's law, which along with Archimedes' law forms the basis of hydrostatics, states that a fluid (be it a liquid or a gas) squeezed inside of a closed container presses with equal force per unit area at any part of the enclosure. Pascal's principle finds wide application in constructing various hydraulic gadgets. In fact, if we have two cylinders A and B (Fig. III-7), with different diameters connected by a thin tube and supplied with movable pistons, the total force acting on the piston in the wider tube will be larger than that acting on the piston in the narrower one, in proportion to their areas. Thus, a comparatively weak force applied by hand to the piston in the narrow cylinder results in a much larger force acting on the piston in the wider cylinder and can raise a heavy carriage. The pay-off is, however, that the displacement of the piston in the wider tube will be correspondingly smaller than that of the piston in the narrower one.

Bernoulli's law, or principle, as it is often called, pertains to the motion of fluids through pipes of varying diameter, and seems at

FIG. III–7.
According to Pascal's principle, a force exerted by a hand can raise a heavy carriage.

first sight to contradict common sense. Imagine a wide horizontal pipe which narrows down at some place, and then widens up again (Fig. III-8a). Water is flowing through the pipe, and its pressure in different sections can be measured by the heights of water columns in the vertical pipes attached in various places to the main horizontal one. It would seem at first sight that pressure will be higher in the narrow section of the pipe, since water has to "squeeze" through it. The direct experiment indicates, however, that the situation is exactly opposite, and that the water pressure in a narrow section is lower than in a broader one. The explanation can be obtained by considering the change in the velocity of flow in different sections

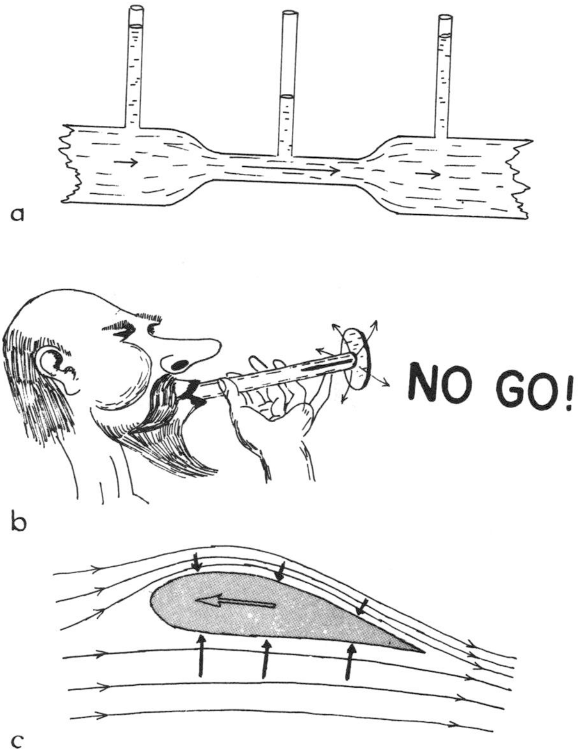


FIG. III–8.
Bernoulli principle. (a) A simple demonstration. (b) A tricky tube. (c) How an airplane wing works.

of the pipe. In the broad section the water moves comparatively slowly and is speeded up while entering the narrower section. To accelerate the motion of water, there must be a *force* acting in that direction, and the only force one can think about here is *the difference of pressure* between the broad and narrow pipes. Since the speed of the water increases after the water enters the narrow pipe, the force must act in the direction of the flow, the pressure in the wider pipe must be higher than that in the narrower pipe.

One can demonstrate that fact without calling in a plumber, simply by procuring a piece of glass pipe (a cigarette holder would probably do, too), a cardboard disc, and a pin (Fig. III-8b). Drive the pin through the center of the disc, and put it into the pipe as shown in the figure, so that the weight of the disc presses it to the pipe's rim. If now one blows into the other end of the pipe, one would expect that the disc would be easily blown off. Try it, and you will see that it is not at all true, and that the more strongly one blows, the more tightly the disc will be pressed to the pipe's end. The explanation is based on Bernoulli's principle. The air blown into the tube has to escape through the narrow circular slit between the end of the tube and the cardboard pressing at it. This passage is much narrower than the tube itself, so that the air pressure here is much lower than that of the atmospheric air. Thus the outer air pressure pushes the cardboard to the tube's end.

Bernoulli's effect also explains the forces which support the wings of a flying airplane. As shown in Figure III-8c, the profile of the wing is such that the distance from its leading edge to the rear edge is longer when the air moves over the top of the wing rather than under it. Consequently the air masses moving over the wing have higher velocity and, according to Bernoulli, lower pressure than the air masses moving below the wing. The difference between these two pressures accounts for the lift of the airplane.

## OPTICS

But we must definitely stop here the discussion of Newton's mechanics, in order to have some place left for the discussion of his optics. Here, Newton's main contributions lie in the studies of colors, and the basic proof that the white light actually is a mixture of the rays of different colors from red to violet. Newton's studies in optics actually preceded his basic work in mechanics described in *Principia.* At the age of 23 he bought a glass prism "to try therewith the phenomena of colours," and probably all his fundamental discoveries in that field date back to that period of his life. However, sometime in February, 1692, he left a light burning in his room when he went to chapel, and the fire that started by accident destroyed his papers, including a large work on optics, which contained the experiments and research of twenty years. Thus, the first edition of Newton's *Optics* appeared only in 1704, and one can only wonder whether that delay was actually due to fire and not to Newton's reluctance to publish his ideas in the face of the opposition of his pertinacious antagonist Robert Hooke, who died just a year before Newton sent his *Optics; or a Treatise of the*

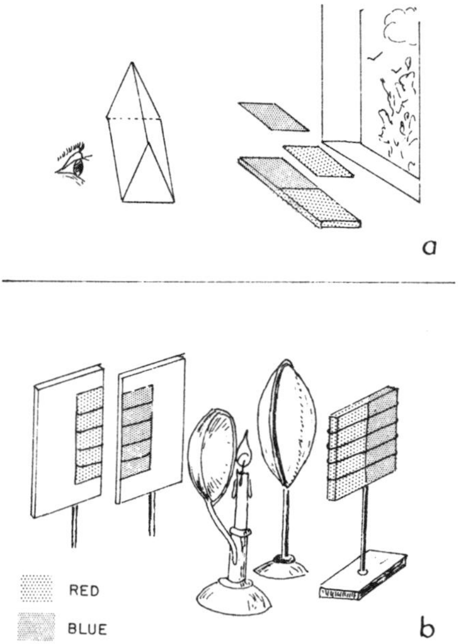


FIG. III–9.
Newton's experiments on refraction of light.

*Reflections, Refractions, Inflections, and Colours of Light* to press. Early in that book he describes a simple experiment which proves that light of different colors has different *refractivity*.

To prove this, he took a long piece of cardboard, one half of it painted bright red and another blue, and, placing it near the window, looked at it through a glass prism (Fig. III-9a). In Newton's own words he "found that if the refracting angle of the prism is turned upwards so that the paper may seem to be lifted upwards by the refraction, its blue half will be lifted higher by the refraction than its red half. But, if the refracting angle of the prism be turned downwards, so that the paper may seem to be carried lower by the refraction, its blue half will be carried something lower than its red half." He decided on the basis of this experiment that blue light is refracted more strongly than red light, and concluded that a lens must focus blue and red rays at different distances from it. To prove this conclusion, he took a piece of paper, one side of which was painted blue and another red, illuminated by a candle ("because the experiment was tried in the night"), and using a lens, tried to get a sharp image on a piece of paper (Fig. III-9b). To judge the sharpness of the image he had several black threads stretched across it. True to his expectations, he could not focus both sides of the colored paper simultaneously. "Noting as diligently as I could the places where the images of the red and blue half of the coloured paper appeared most distinct, I found that where the red half of the paper appeared distinct, the blue half appeared confused, so that the black threads upon it could scarcely be seen; and on the contrary, where the blue half appeared most distinct, the red half appeared confused, so that the black lines upon it were scarcely visible." And, as expected, the image of the blue part of the paper looked sharp at the smaller distance than that at which the red part looked sharp.

The next experiment was to see what happens when the white sunlight passes through a prism. Making a small hole in the window shade, Newton placed a prism in the way of the narrow beam of sunlight passing through it, and a white screen somewhat behind. Instead of a round (pinhole camera) image of the sun on the screen, as would be the case without the prism, he observed an elongated image which was showing a slight bluish tint on its top and a slight reddish tint on its bottom. This brought him to an idea that the white sunlight may consist of the rays of different colors: from the most refrangible blue rays to the last refrangible red rays. If so, the elongated image on the screen would be formed by many overlapping images of the sun in different colors and only two extreme positions would be pure blue and pure red. In order to eliminate the overlapping of the sun images on the screen, he introduced into the light beam a lens which would focus the image of the little hole in the window shade on the screen (Fig. III-10), and was satisfied to observe a vertical band of brilliant colors: red, orange, yellow, green, blue, and violet, with all intermediate shades. That was the first "spectroscope" and the first proof of the fact that the white light is composed of rays of different colors possessing different refractivity.

For the modern reader, Newton's experiments with the prism may seem very childish since, indeed, every child can easily perform them today. But it was quite different in his time, when it was generally believed that the coloring of white sunlight which passes through the beautiful stained glass windows of old cathedrals is

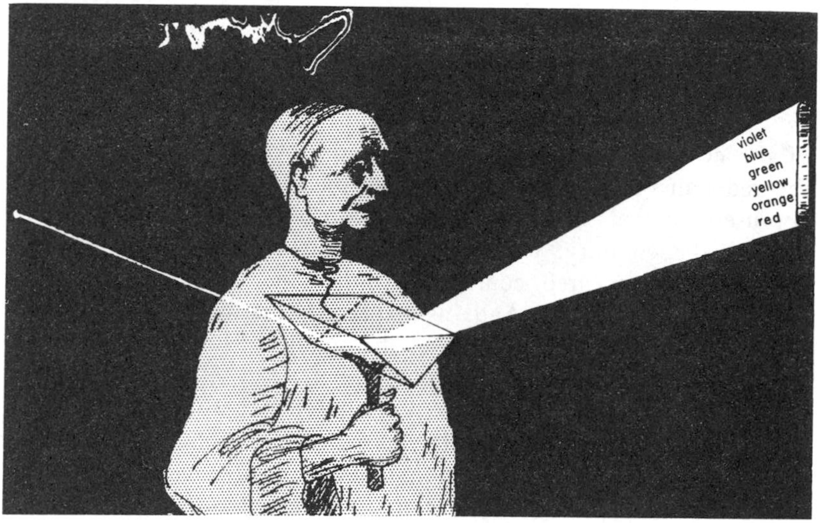


FIG. III–10.
Sir Isaac Newton demonstrating the decomposition of white light in many spectral colors.

something similar to the coloring of white cloth by dipping it into a solution of different dyes. We know now that the retina of the human eye contains three kinds of color-sensitive nerve cells: those responding to red light, to green light, and to blue light. When all the spectral colors are present in the same proportion as they are in the solar light under which the organ of vision was developed through hundreds of millions of years of organic evolution, we get the sensation of "ordinary" or as we call it "white" light. When only a part of the spectrum is present we get the sensation of different colors.

One of the important applications of Newton's discovery that rays of different color have different refractivity was his theory of the rainbow. This beautiful display of colors appears in the sky when the sun shines on one side of it, while the opposite side is covered with heavy rain clouds. According to Newton's explanation, what we see in this case is actually sun rays reflected by the tiny rain droplets in the cloud or down below. The drawing in Figure III-11, adapted from Newton's original drawing in his *Optics,* shows what actually happens. The rays of white light from the sun (shown by black lines in the figure)* fall on the water

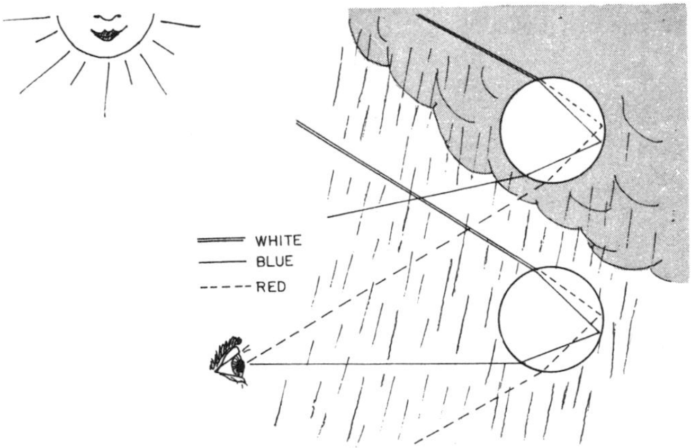


FIG. III–11.
Newton's explanation of rainbow colors.

droplets and are refracted while passing into them. This is followed by the internal reflection, and by the second refraction on the way out of the droplet. As a result, the rays of different color fan out at their exit from the droplet, and the eye of the observer standing on the ground with his back toward the sun observes different colors from different directions in the sky. The existence of several concentric rainbows is explained by the assumption that, instead of being reflected once within the rain droplet, light rays from the sun are reflected several times. We should also mention

* The reason for representing white rays by black lines is that white lines would not show on white paper. Besides, as we shall see later in the book, white light is often called by physicists "black body radiation" because it is emitted at its best from white-hot black bodies (such as carbon).

here the so-called "hallows," colorless arches which are sometimes observed around the sun, and especially around the moon. In contrast to the rainbow, they are due to reflection (*not* refraction) of the light rays from the tiny ice crystals which form the high altitude clouds known in meteorology as *cirrus* clouds.

Having shown that light of different colors has different refractibility, Newton concluded erroneously that lenses have an intrinsic defect in forming sharp images of objects, since the rays of different colors cannot be focused at the same distance from it. This brought him to the decision that the telescopes using glass lenses, like the one built by Galileo, cannot be perfected any more and should be replaced by the telescope based on the reflection of light which

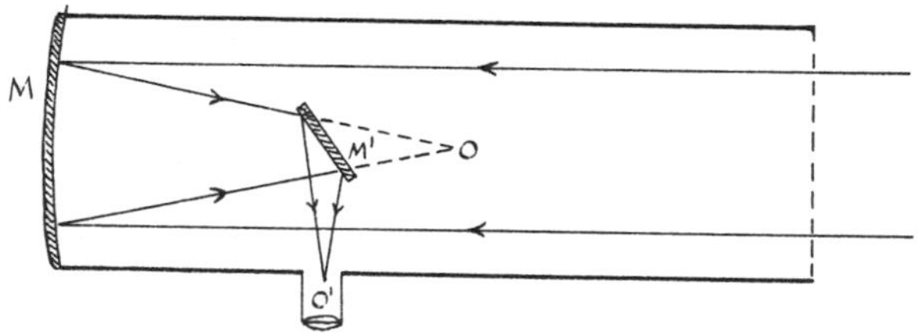


FIG. III–12.
Newton's reflecting telescope.

is independent of color. Thus, in the year 1672 he constructed a reflecting telescope (or simply a "reflector") shown in Figure III-12. It consisted of a parabolic mirror *M*, which was forming the image of a celestial object in some point *O* within the tube. Before the light rays came to focus at *O* they were reflected by a small mirror *M′* placed at the axis of the tube, and were deflected to the point *O′* outside of the tube where the image could be observed. The error of Newton in this case resulted from his belief that different transparent materials refract different colors in a similar way. Only after his death was it found that this assumption was not correct, and that it is actually possible to focus red and blue light in the same point by using compound lenses made of different kinds of glass (crown glass, flint glass, etc.). Nevertheless, reflecting telescopes, which use large parabolic mirrors instead of lenses, have many other practical advantages, and, as a matter of fact, the two most powerful astronomical telescopes today (100-

inch at Mt. Wilson, and 200-inch at Palomar Mt.), are the reflectors.

Another colorful discovery of Newton was the so-called "Newton's rings" which appear around the contact point when a convex lens is placed on a plain piece of glass. He describes this work in the following words:

It has been observed by others, that transparent Substances, as Glass, Water, Air, &etc., when made very thin by being blown into Bubbles, or otherwise formed into Plates, do exhibit various Colours according to their various thinness, altho' at a greater thickness they appear very clear and colourless. (Earlier in this Book) I forbore to treat of these Colours, because they seemed of a more difficult Consideration, and were not necessary for establishing the Properties of Light there discoursed of. But because they may conduce to farther Discoveries for compleating the Theory of Light, especially as to the constitution of the parts of natural Bodies, on which their Colours or Transparency depend; I have here set down an account of them. . . .

I took two Object-glasses, the one a Plano-convex for a fourteen Foot Telescope, and the other a large double Convex for one of about fifty Foot; and upon this, laying the other with its plane side downwards, I pressed them slowly together, to make the colours successively emerge in the middle of the Circles, and then slowly lifted the upper Glass from the lower to make them successively vanish again in the same place. The Colour, which by pressing the Glasses together, emerged last in the middle of the other Colours, would upon its first appearance look like a Circle of a Colour almost uniform from the circumference to the center and by compressing the Glasses still more, grow continually broader until a new Colour emerged in its center, and thereby it became a Ring encompassing that new Colour. And by compressing the Glasses still more, the diameter of this Ring would increase, and the breadth of its Orbit or Perimeter decrease until another new Colour emerged in the center of the last: And so on until a third, a fourth, a fifth, and other following new Colours successively emerged there, and became Rings encompassing the innermost Colour, the last of which was the black Spot. And, on the contrary, by lifting up the upper Glass from the lower, the diameter of the Rings would decrease, and the breadth of their Orbit increase, until their Colours reached successively to the center, and then they being of a considerable breadth, I could more easily discern and distinguish their Species than before. And by this means I observ'd their Succession and Quantity to be as followeth.

Next to the pellucid central Spot made by the contact of the Glasses succeeded blue, white, yellow, and red. The blue was so little in quantity, that I could not discern it in the Circles made by the Prisms, nor could I well distinguish any violet in it, but the yellow and red were pretty copious, and seemed about as much in extent as the white, and four or five times more than the blue. The next Circuit in order of Colours immediately encompassing these were violet, blue, green, yellow, and red: and these were all of them copious and vivid, excepting the green, which was very little in quantity, and seemed much more faint and dilute than the other Colours. Of the other four, the violet was the least in extent, and the blue less than the yellow or red. The third Circuit or Order was purple, blue, green, yellow, and red; in which the purple seemed more reddish than the violet in the former Circuit, and the green was much more conspicuous, being as brisk and copious as any of the Colours, except the yellow, but the red began to be a little faded, inclining very

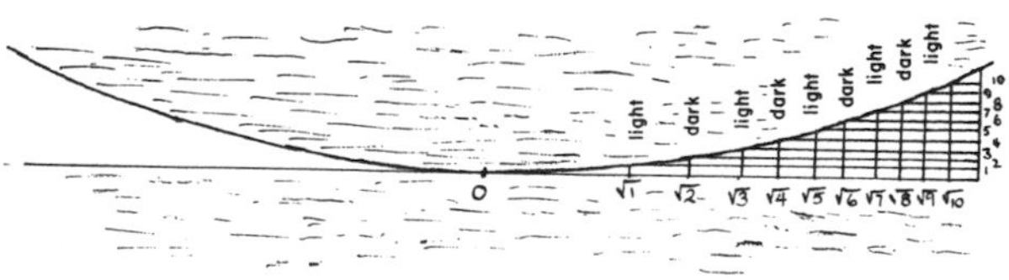


FIG. III–13.
Formation of Newton's rings.

much to purple. After this succeeded the fourth Circuit of green and red. The green was very copious and lively, inclining on the one side to blue, and on the other side to yellow. But in this fourth Circuit there was neither violet, blue, nor yellow, and the red was very imperfect and dirty. Also the succeeding Colours became more and more imperfect and dilute, till after three or four revolutions they ended in perfect whiteness.

[Plate I, *upper,* shows a photograph of Newton's rings obtained with monochromatic light, a single wave length.]

By measuring the radii of the first six rings (in their brightest parts), Newton found that their squares form an arithmetical progression of odd numbers: 1, 3, 5, 7, 9, 11. On the other hand, the squares of the radii of dark rings formed a progression of even numbers: 2, 4, 6, 8, 10, 12. The situation is shown in Figure III-13, representing the cross section of convex and flat glass surfaces near the point of contact. On the horizontal axis are plotted the distances to the square roots of integer numbers: $\sqrt{1} = 1$; $\sqrt{2} = 1.41$;

$\sqrt{3} = 1.73$; $\sqrt{4} = 2$; $\sqrt{5} = 2.24$, etc., at which Newton observed alternative maximum and minimum of light. We notice from that figure, and it can also be proved mathematically, that the vertical distances between the two glass surfaces increase as a simple arithmetical progression: 1, 2, 3, 4, 5, 6, etc. Knowing the radius of the convex lens, Newton could easily calculate the thickness of the air layer at the places where bright and dark rings appear. Writes he:

... $\left(\frac{1}{89000}\right)$th Part of an Inch is the Thickness of the Air at the darkest Part of the first dark Ring made by perpendicular Rays; and half this Thickness multiplied by the Progression 1,3,5,7,9,11 etc. gives the Thickness of the Air at the most luminous Parts of all the brightest Rings, *viz.*: $\frac{1}{178000}, \frac{3}{178000}, \frac{5}{178000}, \frac{7}{178000}$ &etc., their arithmetical Means: $\frac{2}{178000}, \frac{4}{178000}, \frac{6}{178000}$, &etc. being its Thickness at the darkest Parts of all the dark ones.

Contrary to Newton's above quoted assertion that the rainbow colors of thin layers "are not necessary for establishing the Properties of Light," Newton's rings represent one of the best proofs of the wave nature of light, the truth of which Newton did not want to recognize until his death. The rings are the result of so-called "interference" between two beams of light reflected by the two glass surfaces separated by varying distance. When a narrow beam of light falls from above on the boundary between the glass of the upper lens and the layer of air between the two lenses, a part of it is reflected while the rest enters into the air. A second partial reflection takes place when the beam enters the glass of the lower lens and the two reflected beams travel together upward into the eye of the observer. What happens in this case is illustrated graphically in Figure III-14. For convenience in drawing, the waves are represented by shaded and white stretches corresponding to their crests and hollows. Also, the light beams are shown not exactly perpendicular to the interface, to avoid the overlap; this is anyway the case in actual observation since the light source and the head of the observer cannot be on the same line. In Figure III-14a we see what happens when the thickness of the air layer is equal to one-half of the wave length of the incident light (in the figure wave length corresponds to the combined length of one white and one shaded stretch). In this case the wave reflected from the surface of the lower lens joins the wave reflected from the upper one in such a way that the crust of the first wave coincides with the hollows of the second and vice versa. If the waves are of the same intensity, they will completely cancel one another; otherwise the intensity will be considerably reduced. In Figure III-14b we have a case where the air thickness is equal to one half (or $\frac{2}{4}$) a wave length. The two reflected beams now propagate crust to crust, and hollow to hollow, and we have increase of intensity. In Figure

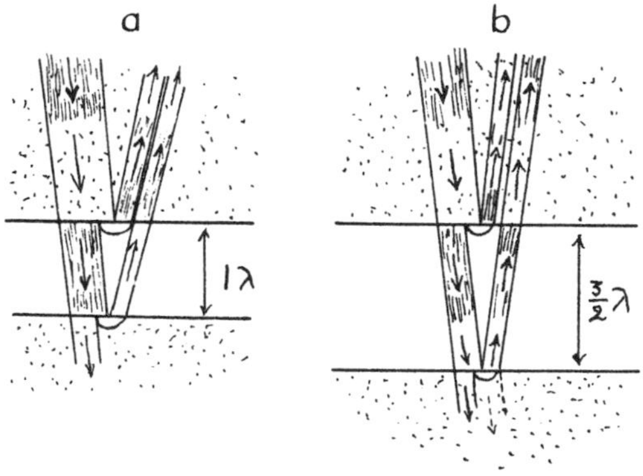


FIG. III–14.
Young's explanation of Newton's rings.

III-14b the air layer is $\frac{3}{4}$ of the wave length thick and the situation is similar to that shown in Figure III-14a. For larger thicknesses of air layer, we will have alternatively light and shaded for each change increase of the thickness by $\frac{1}{4}$ of the wave length. In the case of Newton's arrangement the thickness increases continuously outwards from the contact point so that alternative dark and light rings will be observed. Since the light of different colors corresponds to different wave lengths, the radii of different color rings will be somewhat different and we will observe the rainbowlike rings as Newton did. Using the above-given figures of Newton for the thicknesses of air we find that the wave length of light which would produce the rings of that radii must be $\frac{4}{178000}$ in. or $0.58 \times 10^{4}$ cm.

As we know now, this is just the wave length of the yellow light, the brightest part of the visible spectrum.

But Newton was violently opposed to the wave theory of light, mostly because he could not see how it could possibly explain rectilinear propagation of light rays. He insisted that light must be a stream of particles rushing at high speed through space. Thus, in order to explain the appearance of interference rings, he invented a complicated theory of "fits of easy reflection and transmission" according to which:

> . . . every Ray of Light in its passage through any refractive Surface is put into a certain transient Constitution or State, which in the progress of the Ray returns at equal Intervals, and disposes the Ray at every Return to be easily transmitted through the next refracting Surface, and between the returns to be easily reflected by it.

Newton's "length of fit" evidently corresponds to what we now call wave length, and he concluded that this "length of fit" is longer for red light and shorter for blue. But, he writes:

> . . . What kind of action or disposition this is, whether it consists in a circular or a vibrational motion of the Ray, or of the Medium or something else, I do not here inquire.

Newton's opponent in the discussions concerning the nature of light, and the man whose theory subsequently won out, was the Dutch physicist Christian Huygens, 13 years his senior. The reasons why Huygens preferred to consider light as waves propagating through some universal medium filling the entire space rather than a beam of fast-moving particles is best summarized by a passage from his book: *Traite de la lumière,* published in 1690:

> *On the Rectilinear Propagation of Light Rays*
>
> The procedures of proof in optics, just as in all other sciences in which geometry is applied to matter, are based on truths derived from experience; *e.g.,* the fact that light rays are propagated in straight lines, that the angle of reflection is equal to the angle of incidence, and that refraction obeys the sine rule, so well-known today and no less certain than the others.
>
> The majority of those who have written on the different parts of optics, have been content to take these truths for granted. Some of the more enquiring strove to discover their origins and causes, seeing that they considered them as inherently wonderful effects of nature. Since, however,

the opinions offered, although ingenious, are not such that more intelligent people would need no further expanations of a more satisfying nature, I wish here to present my thoughts on the subject so that, to the best of my ability, I might contribute to a solution of that part of science which, not without reason, is considered to be one of the most difficult. I acknowledge my great indebtedness to those who were the first to start dispelling the strange gloom surrounding these things, and who aroused the hope that they might yet be explained rationally. But, on the other hand, I am not a little surprised to find that very often they considered as most certain and proven, conclusions that were only too flimsy; for to my certain knowledge no one has as yet offered a satisfactory explanation of even the first and most important phenomenon of light, *viz.*, why it is propagated precisely in straight lines, and how light rays arriving from infinitely varied directions cross without impeding one another.

In this book I shall therefore attempt, according to the principles held in contemporary philosophy, to give clearer and more probable reasons for the properties, first of the rectilinear propagation of light, and second of the reflection of light when it encounters other bodies. Then I shall explain those phenomena of the rays, which in traversing different kinds of transparent bodies undergo so-called refraction; and in this I shall also treat of the effects of refraction in the air arising out of differences in the density of the atmosphere.

I shall continue by investigating the strange refraction of light of a particular crystal brought from Iceland. Finally I shall treat of the different forms of transparent and reflecting bodies, by means of which the rays are either made to converge on one point, or else are deviated in most different ways. In this it will be seen with what ease our new theory will lead to the discovery not only of ellipses, hyperbolae and other curves, which Descartes had ingeniously suggested for this effect, but also of those figures which form one surface of a glass, when the other surface is known to be spherical, plane, or of any other shape. . . .

Since, now, according to this philosophy, it is held as certain that the sense of vision is only stimulated by the impression of a certain motion of a material acting on the nerves at the back of our eyes, this is a further reason for believing that light consists of a motion of the matter between us and the luminous body. If, furthermore, we pay attention to, and weigh up, the extraordinary speed with which light spreads in all directions, and also the fact that coming, as it does, from quite different, indeed from opposite, directions, the rays interpenetrate without impeding one another, then we may well understand that whenever we see a luminous object, this cannot be due to the transmission of matter which reaches us from the object, as for instance a projectile or an arrow flies through the air, for this is too great a contradiction of the two properties of light, and the second in particular. Thus it must spread in a different way, and precisely our knowledge of the propagation of sound in air can lead us to an understanding of this way.

We know, that by means of air, which is an invisible and impalpable body, sound spreads through the whole of space surrounding its source by a motion which advances gradually from one air particle to the next, and since the propagation of this motion takes place with equal speed in all directions, spherical surfaces must be formed that spread out further and further, finally to reach our ears. Now it is beyond doubt that light also reaches us from luminous bodies by means of some motion which is imparted to the intermediate matter, for we have already seen that this could not have happened by means of the translation of a body that might have reached us from there. If now, as we shall soon investigate, light needs time for its path, it follows that this motion imparted to matter must be gradual, and that, like sound, it must spread in spherical surfaces or waves; I call them waves because of their similarity to those which we see being formed in the water when a stone is thrown into it, and because they enable us to observe a like gradual spreading-out in circles, although they are due to a different cause and only form in a plane surface. . . .

Considering the propagation of waves, be it on the surface of water, in the air, or in the mysterious "world ether," the carrier of light waves, Huygens was basing his arguments on a simple principle which now carries his name. Suppose, to use the most familiar and obvious case, we drop a stone into the quiet surface of a pond. We see a circular wave, or rather a train of waves, spreading around the point where the stone broke the surface. Given the position of the wave at a certain moment, how do we find its position a short time thereafter? According to the Huygens' principle, *each point at the front of a propagating wave can be considered as a source of a new wave, or wavelet, and the new position of the wave front is a common convolute of these little wavelets emitted from all points of the wave front in its previous position.* This idea is illustrated in Figure III-15 for the simplest cases of a circular and a plain wave.

The most brilliant application of Huygens' principle was his explanation of the refraction of light, shown in Figure III-16. Suppose a plain wave front falls from the upper left on the interface between air and glass (or any other two media). When this wave

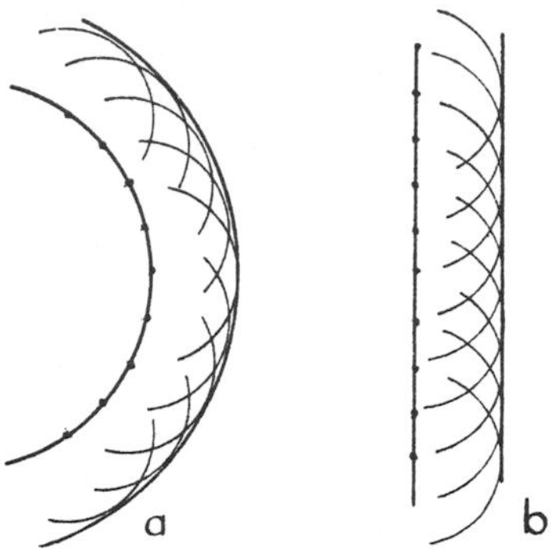


FIG. III–15.
Huygens' principle of wave propagation.

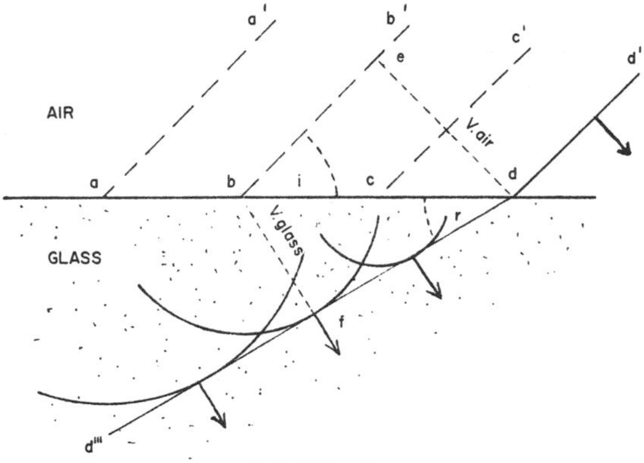


FIG. III–16.
Huygens' explanation of the refraction of light.

front is in the position *aa'* and touches the interface in point *a*, a spherical wavelet starts propagating into the glass from that point. As the wave front progresses in the air other wavelets are emitted consecutively from points *b, c,* etc. The drawing corresponds to the moment when the advancing wave front is in the position *dd'* and the wavelet in the glass is just starting from the point *d*. In order to find the position of the wave front in glass, we have to drive a convolute to all the wavelets which will be in this case a straight line. If, as it is assumed in the drawing, the velocity of light in the glass is smaller than in the air (i.e., if the radii of spherical wavelets in glass are smaller than the distances between the successive positions of the wave front in the air), the wave front in the glass will be tilted downward and the refracted rays will be closer to the vertical than the incident ones; this is what actually happens when light passes from air into glass. If the velocity of light in glass would be larger than in air, the opposite situation would take place. In order to find the relation between the angle of incidence *i*, and the angle of refraction *r*,* we consider two rectangular triangles *bde* and *bdf* which have a common hypotenuse. According to the definition of sine:

$$\sin i = \frac{ed}{bd}; \qquad \sin r = \frac{bf}{bd}$$

Dividing the first equation by the second we get:

$$\frac{\sin i}{\sin r} = \frac{ed}{bf} = \frac{V\text{-air}}{V\text{-glass}}$$

where *V*-air and *V*-glass are the velocity of light in these two media. This is exactly Snell's law, with the amendment that the ratio of the two sines, known as *refractive index,* is equal to the ratio of light velocities in the two media. It follows that the velocity of light in denser media (like glass) is smaller than in rarer media (as air).

It is interesting to notice that Newton's corpuscular theory of light would bring us to exactly opposite conclusions. In fact, in order to explain the bending of rays entering from air into water on the basis of the corpuscular theory, it would be necessary to assume that there is some force perpendicular to the interface which pulls the light particles in when they cross it. In this case, of course, the velocity in glass would be larger than in air.

## THE VICTORY OF THE WAVE THEORY OF LIGHT

In spite of apparent advantages of Huygens' wave theory of light over Newton's corpuscular theory, it did not gain recognition for a very long period of time. This was due partially to the great authority of Newton among his contemporaries, and particularly to the inability of Huygens to develop his views with sufficient mathe-

* Both angles can be defined either as the angles between the direction of the rays and the perpendicular to the boundary between two media, *or* as the angles between wave fronts and this boundary.

matical precision to make them invulnerable against any opposition. Thus, the question about the nature of light was hanging in the air for a century until the appearance in 1800 of a paper by an English physicist, Thomas Young, entitled: "Outlines of Experiments and Inquiries Respecting Sound and Light." In this paper Young explains the phenomenon of Newton's rings on the basis of the wave nature of light, and describes his own experiment in which the interference of two light beams can be shown in a more elementary way. In this experiment (Fig. III-17), he had two holes made close together in the screen covering the window of a darkened

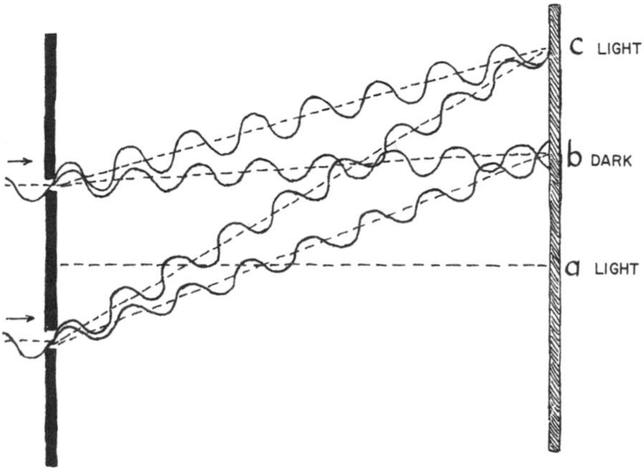


FIG. III–17.
Young's interference experiment.

room. When the holes are comparatively large, the sunlight passing through them forms two patches of light on the screen placed some distance away. But, when the holes are very small, the light beams passing through them spread out in accordance with Huygens' principle, and the two patches spread out, and partially overlap one another. In the region where the screen receives the light from both holes, Young observed a series of rainbow-colored fine bands separated by dark intervals, quite similar to Newton's rings. When the holes in the screen were 1 mm apart, and the screen 1 meter away, the bands were about 0.6 mm wide. The explanation of this phenomenon is based on the interference of light waves, just as it is in the case of Newton's rings. The point *a* on the screen which is located exactly halfway between the centers of the two images is equidistant from two holes *O* and *O′* and the light waves arrive here "in phase," i.e., crest to crest and hollow to hollow. The two wave motions add together and we have the increase of illumination. The same is true for the point *c*, the distances of which from *O* and *O′* differ by one wave length. On the other hand, in the points *b* and *d* for which *bO-b′O* and *dO-d′O* differ by ½ wave length and by 1½ wave lengths rsp., the oncoming light waves are "out of phase" and the crests overlap the hollows. Here one observes dark bands.

The works of Thomas Young and his great contemporary, a Frenchman, Augustin Jean Fresnel, firmly established the validity of the wave theory of light, and so Huygens won his lifelong argument with Newton post mortem.

## A CRYSTAL FROM ICELAND

Another problem tackled but unsolved both by Newton and by Huygens was that of the polarization of light. It was discovered in 1669 by the Danish philosopher Erasmus Bartholin that the crystals of a transparent mineral called Iceland spar have the peculiar property of splitting rays of light passing through it in a certain direction into two separate rays (Plate I, *lower*). If the crystal is rotated around the direction of incident light ray, one of the two emerging rays, called the *ordinary ray,* remains stationary while the other, the *extraordinary ray,* moves around as the crystal is rotated. Huygens interpreted this phenomenon by assuming that a light wave entering into a crystal of Iceland spar (and some other crystals) is split into two waves: one which propagates with the same velocity in all directions through the crystal, and another the velocity of which depends on its direction in respect to the crystalline axis. Huygens' idea of how this difference in the propagation velocities leads to the formation of two rays is shown in Figure III-18 and is based, of course, on Huygens' principle. When the beam of light falls vertically on the surface of the Iceland spar crystals, two sets of wavelets are formed, the spherical and the ellipsoidal ones. The spherical wavelets lead to a wave front which is continuous in the same direction as the incident one, while the ellipsoidal wavelets cause the resulting wave front to shift continuously sidewise thus forming the extraordinary ray. After both rays emerge from the crystal, only spherical waves are formed in the air, and the two beams become parallel. Although this explanation given by Huygens was completely correct, he could not explain why light waves in crystal propagate in two different manners. That is because he believed that the oscillations in light waves take place in the direction of their propagation (longitudinal vibration) as in the case of sound, in which case there should not be any difference if one rotates the crystal around the direction of the incident beam. Newton, on the other hand, not believing in Huygens' waves and wavelets, sought to explain this phenomenon (known as "double refraction") by assuming that the particles forming the ordinary

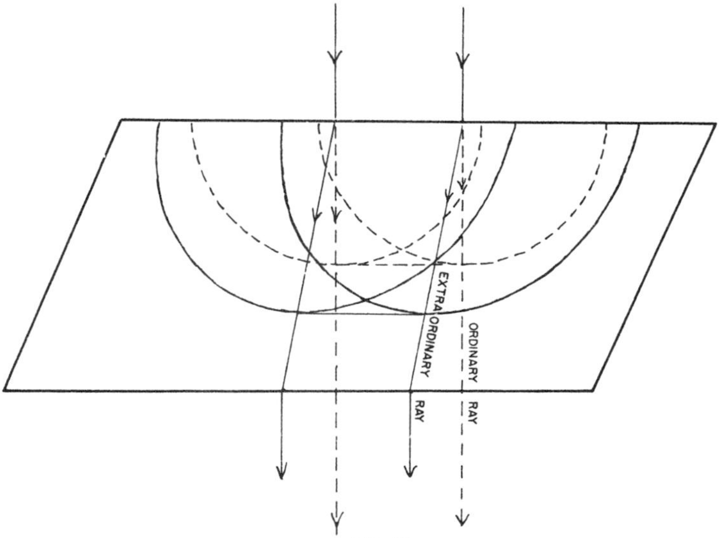


FIG. III–18.
Huygens' explanation of double refraction.

and extraordinary rays are differently oriented in the direction perpendicular to the ray. In the second edition of his *Optics* he compares the difference between two rays with the difference between two long rods, one with the circular and another with the rectangular cross section. If one rotates the first rod around its axis, no difference will be noticed, which is certainly not the case for the second rod. "Every Ray of Light," writes Newton, "has therefore two opposite Sides, originally endowed with Property on which the unusual Refraction depends, and the other two opposite sides not endowed with that property."

Clearly realizing that light rays must have certain transversal properties (i.e., perpendicular to the direction of propagation), Newton was unable to visualize what they could be.

It was only much later, owing to the work of the French physicist Étienne Malus (1775-1812) and others, that Huygens' and Newton's ideas concerning this subject were put together into one unified point of view. There is no doubt that light is nothing but wave propagation through space, but the vibrations of the medium are taking place not in the direction of the propagation, as Huygens thought, but perpendicular to it. The difference between the ordinary and the extraordinary ray in the Iceland spar is that in the first case the vibrations take place in the plane passing through the ray and the axis of the crystal, while in the other case they are perpendicular to it.

The discovery of the transversal nature of light vibrations caused no end of headaches to the physicists of the following generation. In fact, transversal vibrations can exist only in solid bodies which resist shearing and bending. That meant that world ether, the hypothetical carrier of light, was not a very rarefied gas as Huygens had imagined it, but a solid body! If all penetrating ether is solid, how could the planets and other celestial bodies move through it without practically any resistance? And, even if one would assume that world ether is a very light, easily crushable solid material, like the Styrofoam used today in many connections, the motion of celestial bodies would bore so many channels in it that it should soon lose its property of carrying light waves over long distances! This headache was pestering physicists for many generations until it was finally removed by Albert Einstein, who threw ether out of the window of the physics classrooms.

## NEWTON'S ECLIPSE

At the age of 50, Newton decided to abandon academic life, and started looking for a position which would bring him a better income. He was offered the position of headmaster of the Charterhouse* in London, but he did not think highly of it. In his letter rejecting the job he wrote:

* A fashionable school for British aristocracy.

I thank you for putting me in mind of Charterhouse, but I see nothing in it worth making a bustle for: besides a coach [which he was evidently offered], which I consider not, it is but £200 per annum, with a confinement to the London air, and to such a way of living as I am not in love with; neither do I think it advisable to enter into such a competition as that would be for a better place.

In 1696, at the age of 54, he was appointed first a Warden and later Master of London's Mint and began making money, both literally and actually. In 1705 he was knighted and became Sir Isaac, and received many other honors. But the last quarter century of his life (he died in 1727 at the age of 85) was devoid of any important discoveries, which were pouring in as if from the horn of plenty when he was under 25. Some of his biographers say that this was due to senility, some say that it was because he had exhausted all the possible ideas which could have been come upon in his era. Anyway, he did enough!

# CHAPTER IV *Heat as Energy*

THE FIRST studies of the phenomenon of heat were made by the prehistoric cave man who learned how to build a fire to keep warm during periods when the sun did not supply him enough heat. His close collaborator, the prehistoric cave woman, made an additional important discovery—that different food items held for a period of time over the flame or in boiling water taste much better and are more digestible. The notions of "hot" and "cold" are inbred into man, as well as into all other living beings, and the temperature of the surrounding medium is recorded and signaled into the brain by billions of nerves ending at the surface of the skin. But phsysiological response to the temperature is often misleading, and a blindfolded man cannot tell whether his hand was burned by a red-hot iron or frozen with a piece of dry ice. In both cases the sensations are the same because they are both just the physiological response to the injury of the tissue.

## THERMOMETERS

The first really scientific instrument for temperature measurement was invented in 1592 by Galileo, who used for that purpose a glass flask with a very narrow neck. The flask was half-filled with colored water and placed upside down into a bowl of the colored water. With the change of temperature the air contained in the bulb would expand or contract, and the column of water in the neck would move down or up. Galileo did not care to introduce any temperature scale, so that his gadget should be called a "thermoscope" rather than a "thermometer." The modification of Galileo's thermoscope was proposed by Ray in 1631; it was simply the inverted Galileo flask in which heating and cooling were registered by the expansion of water.

In the year 1635, Duke Ferdinand of Tuscany, who was interested in science, built a thermometer, using alcohol (which freezes at a lower temperature than water) with the top of the tube sealed off so that the alcohol could not evaporate. Finally, in 1640, the scientists in the Academia Lincei in Italy built a prototype of the modern thermometer, using mercury and, at least partially, removing air from the upper part of the sealed tube. It is interesting to notice that the entire development took about half a century as compared with only a few years which passed between the discovery of electromagnetic waves and the construction of the first radiotelegraph, or the discovery of uranium fission and the first atomic bomb.

## GAS LAWS

While Newton was pondering in Cambridge on light and gravity, another Englishman, Robert Boyle, was working in Oxford on mechanical properties and the compressibility of air and other gases. Having heard about Otto von Guericke's invention of the air plumb, Boyle considerably improved its design and started on a series of experiments measuring the volume of air at various low and high pressures. This work resulted in what is now known as "Boyle's law," which states that *the volume of a given amount of any gas at a given temperature is inversely proportional to the pressure to which it is subjected* (Fig. IV-1).

Almost a century later a Frenchman, Joseph Gay-Lussac, studying the expansion of gases when they were heated, found another important law which states that *the pressure of any gas contained in a given volume increases by* $\frac{1}{273}$*d of its initial value for each degree (centigrade) of the temperature*. The same law was discovered by another Frenchman, Jacques Charles, two years earlier and is therefore often called "Charles' law."

## GAS THERMOMETER AND ABSOLUTE TEMPERATURE

These two laws underline the simplicity of the internal structure of gases, since the compressibility and the thermal expansion of liquids and solids are subject to more complicated laws and depend essentially on the nature of material. The simplicity of laws governing the behavior of gases independently of their chemical nature makes the gas "thermoscope" built by Galileo a much more rational instrument for measuring the temperature than any other gadget built afterward. Different liquids such as water, alcohol, mercury, etc. (as well as solids, which can also be used in construction of thermometers) expand in somewhat different fashion with the rise in temperature; water even contracts instead of expanding when

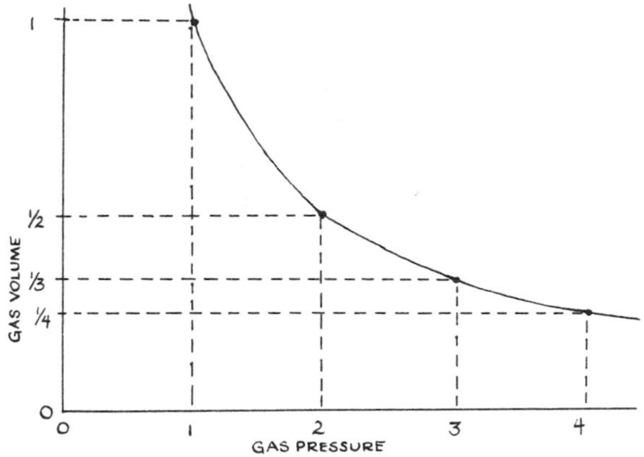


FIG. IV–1.

Graphic presentation of Boyle's law of inverse proportionality between gas volume and gas pressure.

the temperature increases from the freezing point to a few degrees above it. Thus, if one builds two thermometers using different liquids, marks on them the positions of the column at two different temperatures (say the freezing and boiling point of water), and divides the distance between these two marks into an equal number of intervals (100° in the case of the centigrade scale), these two thermometers will show somewhat different values in-between the two end points. On the other hand, since all gases expand in exactly the same way when heated, they represent much more standard material for the purpose of temperature measurements. Using the gas thermometer, as Galileo did, one does not have to specify whether the gas is ordinary air, hydrogen, helium, or anything else. The modern version of the gas thermometer is shown in Figure IV-2, and is based on measuring the pressure rather than the volume of the heated gas. As the temperature increases, the gas expands and pushes down the mercury in the left glass. Raising the right glass tube, one brings the gas to the original volume and measures the temperature by the difference *h* between the two

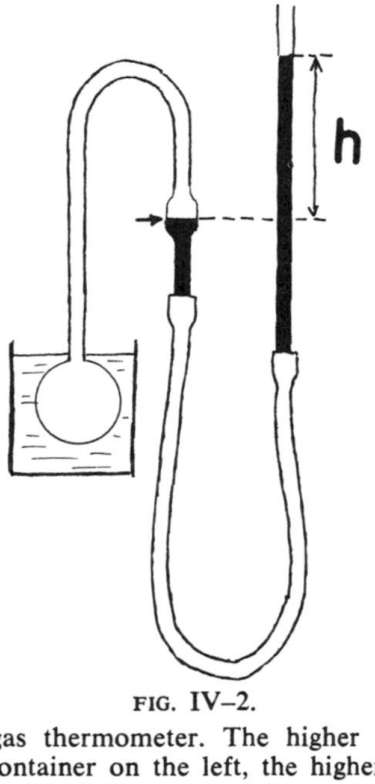


FIG. IV–2.

Principle of the gas thermometer. The higher the temperature of the liquid in the container on the left, the higher the level *h* of the mercury in the movable tube on the right must be in order that the level of the mercury in the middle tube remain in the same place indicated by the arrow.

mercury levels. Having established the temperature scale on the basis of the gas thermometer, one can then graduate to all other thermometers, using the gas thermometer as a standard. Using the gas thermometer and starting with the atmospheric pressure (when two mercury columns in glass tube are at the same level), one finds, as mentioned above, that the pressure of gas increases or decreases by $\frac{1}{273}$d of its original value when its temperature rises or falls by 1° C. Thus, if we start with 0° C (the freezing point of water) and cool the gas to 273° below it, the gas pressure is expected to drop to zero, and the gas should be squeezed to the zero volume (Fig. IV-3). The point at which this is supposed to happen is known as *absolute zero of temperature,* and temperatures counted from that point are *absolute temperatures* (T° abs = 273 + T° C). Of course the cooled gases never collapse into a mathematical point having no volume, and shortly before absolute zero is reached condense into liquids which cannot be squeezed much more. Nevertheless the absolute zero point of temperature plays a very important role in thermophysics, as the temperature at which the collapse of the gas into a mathematical point would happen if

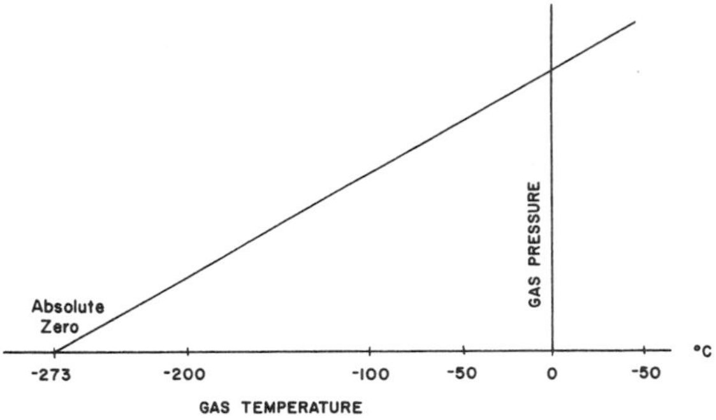


FIG. IV–3.

A graph showing the dependence of the pressure of a gas (contained in a given volume) on the temperature. At the temperature 273° C below the freezing point of water the gas pressure becomes zero.

gas molecules were infinitely small in size, and there were no intermolecular attractive forces. (Both conditions are very closely satisfied for "rare gases" such as helium, neon, argon, etc.)

## HEAT FLUID

Although people have spoken about heat from time immemorial —about too much of it in southern lands and not enough in northern lands—the first man who spoke about heat as a definite physical entity, the amount of which can be measured just as we measure the amount of water or kerosene, was probably a Scotch physician with an interest in physics and chemistry, named James Black (1728-1799). He visualized heat as a certain imponderable fluid which he called "calor," which can interpenetrate all material bodies increasing their temperature. Mixing a gallon of boiling water with a gallon of ice cold water, he noticed that one finds the temperature of the mixture just halfway between two initial temperatures, and he interpreted this fact by saying that, after the mixing, the excess of "calor" in hot water is equally distributed between the two portions. He defined the unit of heat as the amount necessary to raise the temperature of 1 lb of water by 1° F (in the modern metric system we speak of *calorie*, which is the amount of heat it takes to raise the temperature of 1 gm of water by 1° C). He concluded that equal weights of different materials heated to the same temperature contain different amounts of "calor" since, indeed, by mixing equal weights of hot water and cold mercury, one gets a temperature which is much closer to the original temperature of water than that of mercury. Thus, he argued, cooling by 1° of a certain amount of water lets free more heat than is necessary to heat by 1° an equal weight of mercury. This led him to the notion of the *heat capacity* of different materials characterized by the amount of heat needed to raise their temperature by 1°. Another important notion introduced by Black was that of *latent heat,* which is the amount of heat needed to turn ice into ice water (both at 0° C), or boiling water into water vapor (both at 100° C). He thought that adding a given amount of the imponderable heat fluid to a piece of ice loosens up its structure, making it liquid, and that, in a similar way, adding more heat to the hot water loosens its structure still more, turning it into vapor.

The analogy between heat and a fluid was carried further by a young Frenchman, Sadi Carnot (Fig. IV-4), who died in 1832 at the age of 36. Carnot compared the steam engine, in which the mechanical work is produced by the heat flowing out from a hot boiler, with a water wheel, in which the work is done by water falling from a high level. This analogy brought him to the conclusion that, just as in the case of the water wheel, where the amount of work provided by a given amount of water increases in proportion to the difference between the water levels above and below the wheel, the amount of mechanical energy which can be produced by a steam engine must be proportional to the temperature difference between the boiler where the steam is produced and the cooler where it is condensed. He believed, however, that, just as in the case of the water wheel, the amount of heat coming into the cooler is equal to that taken from the boiler, and that mechanical work is done because a certain amount of heat "falls" from the high temperature into the low temperature region. We now know that this assumption was wrong, that steam engines transform a part of heat flowing through them into mechanical energy, and that the amount of heat coming into the condenser is less by the amount of heat so transformed.

FIG. IV–4.
Ludwig Boltzmann (left), Sadi Carnot (center), and Josiah Gibbs (right), the founders of the modern theory of heat.

### HEAT IS MOTION

The idea that heat is some kind of internal motion of a material body, and not a special substance as Black and others thought, occurred first to a professional soldier and was substantiated by experiments carried on in a gun factory. Benjamin Thompson was born in Massachusetts, and during his youthful days participated in the Revolutionary War. Later on he changed his affiliation to England and soon became undersecretary of state in the ministry of the colonies. Still later he went to Bavaria as the minister of war, and was given the title of Count Rumford for reorganization of the German army. Between all these military activities, he was deeply interested in the problems of science and in particular in the nature of heat. He was not satisfied with the contemporary view that heat is a certain substance, not unlike all other chemical substances, which unites with ice to produce water (ice + heat = water), or is liberated in various combustion processes. The reason for his doubts was the fact that heat is produced "from nothing" in friction processes which have apparently nothing to do with chemical transformation. Watching the boring of cannons at the munitions factory in Munich he was wondering why the casting became so hot, especially when the borer was blunt. He considered the possibility that material bodies may have a larger capacity for caloric fluid when they are in one solid block than when they are broken into small fragments; this would explain the liberation of heat during cannon boring when a large number of gun-metal turnings are produced. He carefully measured the heat capacity of a solid metal block and an equal weight of metal turnings and found them exactly the same. He tried to compare the weight of hot bodies with their weight when cold in an attempt to discover the weight of the escaping heat fluid but came to a negative result.

According to the figures given in his article in the *London Philosophical Transactions* (1799), a calorie cannot weigh more than 0.000013 mg. We know now that any form of energy possesses a ponderable mass which is obtained, according to the famous Einstein relation, by dividing it by the square of the velocity of light. The weight of a calorie of heat is actually: 0.00000000004 mg, which is well below the precision of any measurements. All this brought him to a conclusion that heat cannot be an ordinary substance and must be some kind of motion. He writes: "What is heat? It cannot be material substance. It appears difficult, if not quite impossible, for me to imagine heat to be anything else than that which in this experiment [cannon boring] was supplied continuously to the piece of metal, as the heat appeared, namely motion."

## MECHANICAL EQUIVALENT OF HEAT

The ideas of Count Rumford were further developed several decades later by a German physician, Julius Robert Mayer, in his article "Remarks on the Forces of Inanimate Nature," published in 1842. Mayer arranged an experiment at a paper factory where the pulp contained in a large cauldron was stirred by a mechanism powered by a horse going around the circle. Measuring the rise of pulp temperature, he obtained a figure for the amount of heat produced by a given amount of mechanical work done by a horse. However, being too busy with his medical practice, he never pursued this line any further by carrying out more precise experiments, and the honor of measuring exactly the mechanical equivalent of heat went to an Englishman, James Prescott Joule. In his experi-

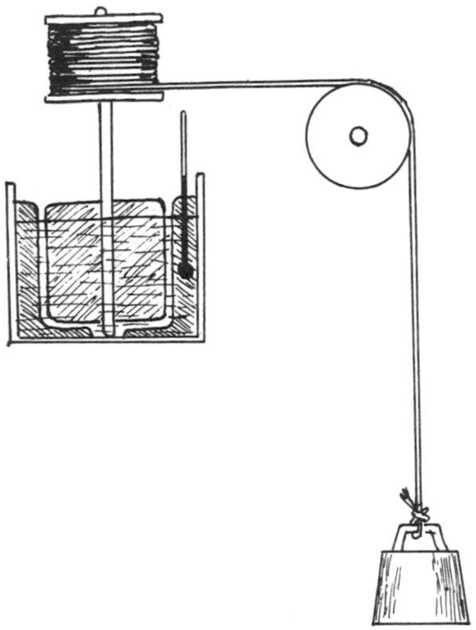

FIG. IV–5.

Joule's experiment on transformation of mechanical energy into heat. The descending weight rotates the paddles in a water-filled container and the temperature of water is increased by internal friction. Comparing the work done by descending weight with the increase of the heat content of water, Joule obtained the value of the mechanical equivalent of heat.

ments Joule used an apparatus schematically shown in Figure IV-5, which consisted of a water-filled vessel containing a rotating axis with several stirring paddles attached to it. The water in the vessel was prevented from rotating freely along with the paddles by special vanes attached to the walls of the vessel which increase the internal friction. The axis with the paddles was driven by a weight suspended across a pulley, and the work done by the descending weight was transformed into friction-heat communicated to the water. Knowing the amount of water in the vessel, and measuring the rise of temperature, Joule could calculate the total amount of heat produced. On the other hand, the product of the driving weight and the distance of its descent provided the value of mechanical work. Repeating this experiment many times, and under different conditions, Joule established that there is a direct proportionality between the work done and the heat produced. Announcing in 1843 the result of his studies, he wrote: "The work done by the weight of one pound through 772 feet in Manchester will, if spent in producing heat by friction of water, raise the temperature of one pound of water by one degree Fahrenheit." This is the figure which, being expressed in these or other units, is now used universally whenever one has to translate heat energy into mechanical energy or vice versa.

## THERMODYNAMICS

When the equivalence of heat and mechanical energy, now known as the first law of thermodynamics, was firmly established, the time came to extend the work of Sadi Carnot concerning the laws of transformation of one form of energy into another. The pioneering work in this direction was done during the second half of the last century by the German physicist Rudolph Clausius and the British physicist Lord Kelvin. From everyday experience we know that heat always flows from hotter bodies into colder bodies, and never in the opposite direction. We also know that mechanical energy can be completely transformed into heat, for example by friction, while a complete transformation of heat into mechanical energy represents a physical impossibility. Indeed, as it was already realized by Sadi Carnot, production of mechanical work is associated with a "descent" of a certain amount of heat from a high temperature level to a low temperature level. While Carnot believed (erroneously) that heat remains intact passing from the boiler to the cooler, the first law of thermodynamics states that some of it is lost, and its equivalent amount appears as mechanical work done by the engine. The situation is analogous to that of a house on a hill which draws its supply of water from a creek running below. Instead of operating the pump by an electric motor, the inhabitants of the house decide to run the pump by a water wheel powered by the same creek as shown in Figure IV-6. Thus, while part of the creek's water supply falls down and turns the wheel, the other part is pumped uphill into the house.

FIG. IV–6.
A hydrodynamic analogy of the thermal engine that transforms into mechanical energy a part of the heat flowing from a high temperature region into a low temperature region.

It is clear that one cannot pump up the entire water supply of the creek since there will be no water left to operate the pump. The work delivered by falling water or required to raise it up is equal to the amount of water multiplied by the height, so that the best we can do is to arrange things in such a way that the amount of water left in the creek is just enough to pump the rest of it into the house.

If for example the height of the dam is 3 ft and the house is 12 ft above the pump, we write, calling $x$ the fraction of water pumped into the house:

$$12x = 3(1 - x)$$

from which follows:

$$x = \frac{3}{12 + 3} = \frac{1}{5}$$

Thus such an arrangement cannot possibly pump more than one fifth of the water into the house. As will be shown later, in the case of heat flowing from a hot region into a cold region, being partially transformed into mechanical energy, the fraction of the heat which can be transformed into work is given by the expression:

$$\frac{T_1 - T_2}{T_1}$$

where $T_1$ and $T_2$ are absolute temperatures of boiler and condenser. The temperature of boiling water is 100° C or 373° abs, and if the condenser is cooled by ice its temperature is 0° C or 273° abs. Thus the maximum efficiency of the steam engine is $100/373 = 26\%$. Actually, because of heat losses and for other practical reasons, the efficiency of the steam engine is even lower.

The statement that *it is impossible to turn heat into mechanical energy without having more heat "descending" from a hotter place to a cooler place* is known as the "second law of thermodynamics." It is equivalent to the statement that *heat would not flow by itself from a cooler place to a hotter one*. Indeed, if we could persuade the heat to flow by itself from the cooler to the boiler, we would have a vicious heat cycle and steam engines would operate without any fuel. A similar mechanical device would be water flowing all by itself uphill, and then pouring down on the water-mill wheel!

In the mathematical treatment of thermodynamics the notion of "entropy," is introduced, usually denoted by $S$ and defined as the amount of heat received or lost by the body, divided by the body's (absolute) temperature. Using the notion of entropy one can reformulate the above-stated second law of thermodynamics, by saying that *the entropy of any "isolated system"* (i.e., a system which is not in any thermal or mechanical interaction with its surroundings) *can only increase or remain constant*. If we put an ice cube into a glass of warm water, heat could flow from ice to water, cooling the ice cube well below zero, and heating water to the boiling point. According to the second law of thermodynamics this does not happen because it would correspond to the decrease of the entropy of the ice cube–water system.

Indeed, let $T_1$ be the temperature of warm water and $T_2$ the temperature of an ice cube, so that $T_1 > T_2$. Suppose a certain amount of heat, call it $Q$, would flow spontaneously from the ice cube into the warm water surrounding it. The amount of heat received by the water would be $+Q$, and the change of its entropy $\Delta S_1 = +\frac{Q}{T_1}$. The amount of heat received by the ice cube would be $-Q$, since ice loses heat, and the change of entropy of the ice cube would be $\Delta S_2 = -\frac{Q}{T_2}$. Thus, the total change of entropy in the water–ice system would be $\Delta S_1 + \Delta S_2 = \frac{Q}{T_1} - \frac{Q}{T_2} = Q\left(\frac{1}{T_1} - \frac{1}{T_2}\right)$. Since $T_1 > T_2$, it follows that $\frac{1}{T_1} < \frac{1}{T_2}$, and consequently the value in the bracket is negative. Thus, the flow of heat from the ice cube into the water would correspond to the decrease of entropy, which contradicts the second law of thermodynamics. If, however, heat flows from warm water into ice, the signs are reversed, the change of entropy becomes positive, and the process is in accord with the rules of thermodynamics. This argument applies, of course, only to "isolated" systems, i.e., the systems into which no energy is supplied from the outside. In the case of a kitchen refrigerator or a window air conditioner, the heat is pumped from the ice compartment or from the room into the warmer air outside, but in this case the decrease of entropy is compensated for by the work done by the electric current driving the motor.

The law of increasing entropy also permits us to derive in a rather simple way the expression for the efficiency of the thermal engine which was mentioned on p. 98. Let the temperatures of the boiler and the cooler be $T_1$ and $T_2$, and suppose that a certain amount of heat, $Q_1$, is taken from this boiler. The cooler will receive a smaller amount of heat, $Q_2$, and the difference, $Q_1 - Q_2$, will be transferred into mechanical energy. Thus the entropy of the boiler will decrease by the amount $\frac{Q_1}{T_1}$, while that of the cooler will increase by $\frac{Q_2}{T_2}$. Since the increase of the entropy in the cooler must be larger than, or at least equal to, its decrease in the boiler, we may write:

$$\frac{Q_1}{T_1} \leq \frac{Q_2}{T_2}$$

from which follows that:

$$\frac{Q_1}{Q_2} \leq \frac{T_1}{T_2}, \text{ or } \frac{Q_2}{Q_1} \geq \frac{T_2}{T_1}$$

Using simple algebra, the above can be rewritten as:

$$\frac{Q_1 - Q_2}{Q_1} \leq \frac{T_1 - T_2}{T_1}$$

which is the formula mentioned earlier.

### DUNKING BIRDS

An ingenious device based on the principle of the thermal engine is the Japanese dunking bird (Fig. IV-7). It consists of an evacuated glass container made from two spheres connected by a long tube. Inside the container there is a certain amount of ether which quickly evaporates at room temperature. The vapor of ether originally filling the body sphere rises into the head sphere, which is cooler because the hydroscopic layer covering it is kept continuously wet. Condensing ether collects itself in the lower part of the head sphere, being unable to flow down because the tube extends to the center of the sphere. When sufficient ether is accumulated, the head becomes heavier than the body and the bird swings on the pivot to an almost horizontal position, thus permitting the ether to run back to the body sphere, which makes the bird straighten up again. Each time the bird bends down, its nose dunks into water, which keeps its head continuously cool.

If, instead of water, we fill the glass with vodka or, even better, pure alcohol, the cooling of the head will be more intensive and the bird will operate faster. If, on the other hand, we cover the bird with a glass dome, the air inside will be quickly saturated with vapor and the motion will stop. The birds operate less effectively when the humidity of the atmospheric air is high; in fact, the author was unable to make them go at all during a typical summer day in Washington, D.C.

In connection with this toy, working on the principle of the evaporation of water, we can raise an interesting physical question. If we attach some kind of cogwheel mechanism to the axis around which the bird swings, we can get a certain amount of mechanical energy and operate a pump, bringing water into the glass from the sea below. How high above sea level can we place the bird so that it

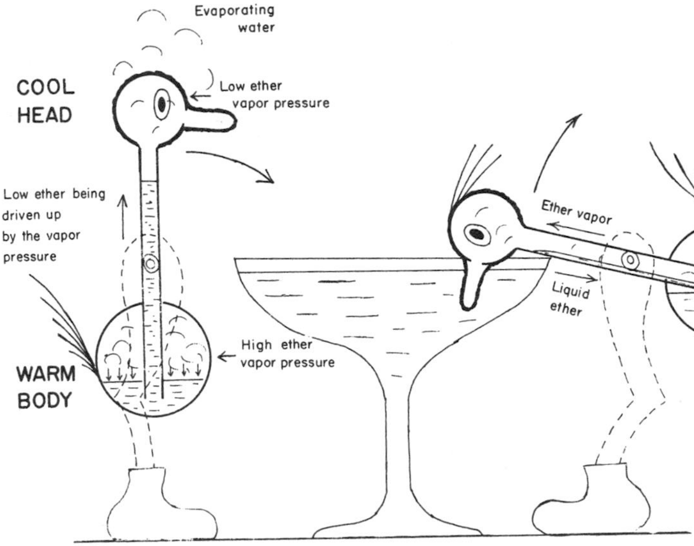


FIG. IV–7.
Japanese dunking birds.

would still be operating? We can consider it as a thermal machine, with the heat flowing from the bird's warmer body to the cooler head, and turning partially into mechanical energy. The latent heat of evaporation of water (from the cool head of the bird) is 539 cal per gm, which is equivalent to $2 \times 27 \times 10^{10}$ erg of mechanical energy. This figure must also represent the amount of heat flowing from the warmer air into the body of the bird, while 1 gm of water is evaporated from its head (because there is no accumulation or loss of heat in the bird's body). The efficiency of a thermal machine in transforming heat into mechanical energy is $\frac{T_1 - T_2}{T_1}$. In our case both $T_1$ and $T_2$ are about 300° K (room temperature), while the difference $T_1 - T_2$ is only just a few degrees. Taking this difference to be, say, 3° C we find that the efficiency is about 1%, so that the evaporation of 1 gm of water from the bird's head will produce about $2 \times 10^8$ erg. To raise 1 gm (of water) to the height of 1 cm, one has to do work equal to the acceleration of gravity which has the value of about 1,000 $\left(981 \frac{\text{cm}}{\text{sec}^2}\right)$ so that 1 gm of water evaporating from the bird's head results in mechanical work which can bring another gram of water to replace it from sea level to the height of $2 \times 10^5$ cm or 2 km above it. Of course, the above calculations are very crude and various energy losses will considerably reduce this figure, but the fact is that dunking birds can drink the water from the sea while sitting at rather high altitudes!

### PERPETUAL MOTION MACHINES OF THE FIRST AND SECOND KIND

In olden times people were dreaming about machines which would work endlessly without any fuel or any other energy supply from the outside. The endless chain of Stevinus discussed in Chapter II was often used as a possible design for such a machine before the said Stevinus had shown that it would not operate if one used the correct laws of mechanical equilibrium on the inclined plane.

While a perpetual motion machine of the first kind contradicts the first law of thermodynamics, i.e., the law of conservation of energy, one can think about a perpetual motion machine of the second kind, which would contradict the second law of thermodynamics. Indeed, if we could turn heat 100% into mechanical energy, mechanical engineering would have the upper hand over all highly advertised atomic energy projects. One would be able to build ocean liners that pump in sea water, extract the heat from it to run the engines, and throw overboard the resulting ice blocks. One would be able to build automobile and airplane engines which suck in atmospheric air, use its heat content for propelling power, and blow out ice cold jets from the exhaust pipes. One would be able to . . .

But all these marvelous possibilities are prohibited by the second law of thermodynamics, the law of ever-increasing entropy!

THERMODYNAMIC ARGUMENTATION

Once we accept the laws of thermodynamics, we can use them for the discussion of various physical phenomena, and for proving many important statements about them. Let us consider, for example, a dish of water with a capillary tube protruding vertically over the water surface (Fig. IV-8). In order to isolate this system from the surroundings, we cover it with a glass case and pump the air out.

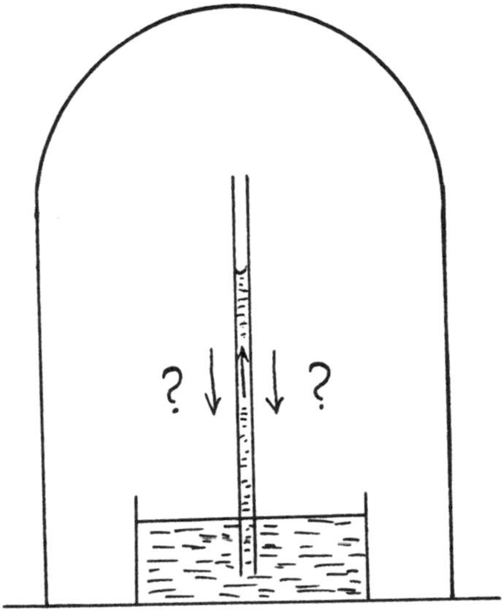


FIG. IV–8.

An example of thermodynamic argument. If the vapor pressure above the concave meniscus in the capillary tube were the same as above the plane water surface in the dish, the water would be in a state of perpetual motion in the directions indicated by the arrows.

It is known that the water rises in the capillary tube, forming a concave meniscus. Now let us ask ourselves what will happen. First of all, part of the water in the dish will turn into water vapor which will fill the interior of the glass case. Owing to gravity, the density and pressure of vapor will be higher at the bottom and lower at the top, just as they are in the case of terrestrial atmosphere. Now we know that, for any given temperature, there is a certain pressure of the vapor at which it is "in equilibrium" with the liquid. If vapor pressure is too high, part of it will condense into liquid; if it is too low, some liquid will evaporate, producing more vapor. We are going to prove now by thermodynamic argumentation that the vapor pressure over the concave capillary liquid surface is smaller than over the flat surface. Suppose, indeed, that the above statement is not true and the vapor pressure is independent of the curvature of the liquid surface. What would happen in that case? Since, because of gravity, vapor pressure at the surface of the meniscus is lower than that at the water surface in the dish, water would evaporate inside of the capillary tube and condense in the dish. This would cause an upward flow of water in the capillary tube and this motion would continue without end. We would be able to place in the capillary tube some kind of water mill and operate the device endlessly, in contradiction to the second law of thermodynamics. Since this law should not be contradicted, we conclude that *vapor pressure over the concave surface of a liquid is lower than over the flat surface.* Similarly (taking a waxed capillary tube, in which case the meniscus is below the water surface in the dish and is convex), we conclude that *vapor pressure over the convex surface of a liquid is higher than over the flat surface.* The narrower the capillary tube, the larger the difference of heights and, consequently, the larger the change of the vapor pressure. Using the numerical values of the surface tension constant (which determines the height of the water column in the capillary) and the density of water vapor (which determines the difference of pressure between the dish level and the capillary tube level) we can obtain a formula for the dependence of vapor pressure on the curvature of the water surface. If this formula were not true, water would flow continuously through the capillary and we would have a perpetual motion machine of the second kind.

The conclusion reached above has an important bearing on the understanding of the phenomenon of rain. The clouds floating high in the sky are formed by innumerable tiny water droplets (fog) which are so small and light that they have practically no tendency to come down. Some of these droplets are larger; others are smaller. What is the effect of the difference in the size of the droplets? As we have just seen, vapor pressure over convex surfaces is higher than over plane surfaces, and the pressure difference increases with the decreasing radius of the curvature. Thus, vapor pressure will be higher over the surface of smaller droplets and lower over the surface of larger ones. As a result of this pressure difference, the vapor will flow from the smaller droplets toward the larger ones, condensing on their surface and further increasing their size. On the other hand, smaller droplets will gradually evaporate and finally vanish. The growing droplets will soon become too large to float in the air, and will rain down on our heads and umbrellas.

## KINETIC THEORY OF HEAT

Further development of the theory of heat and correlation of the basic law of thermodynamics with the idea that heat is the energy of motion of tiny particles, the molecules, from which all material bodies are formed, were carried out during the last quarter of the last century mostly by Ludwig Boltzmann in Germany, James Clerk Maxwell in England, and Josiah Gibbs in the United States (Figs. IV-4 and V-16). Considering the motion of innumerable tiny molecules forming material bodies, it is, of course, impossible (and also useless) to follow exactly the trajectory of each individual particle. What we want to know is the *average* behavior of molecules under different physical conditions, which brings us to the use of the laws of statistics. Statistical methods are always used in human connection when large numbers of individuals are involved. Insurance companies, government agencies dealing with food production by farmers, etc., base their policies on statistical data, and are not interested in the details of the death of Mr. John Doe, or particulars concerning the farm run by Jeremiah Smith. Considering that the population of the United States is approximately 170,000,000, whereas the number of molecules in 1 cc of air is 20,000,000,000,000,000,000, we see that the laws of statistics must hold much more closely in the case of molecules than in the case of people.

It is easiest to apply statistical considerations to the case of gases in which, in contrast to liquids and solids, the molecules rush freely through the space, colliding with one another and with the walls of the container. The walls of a vessel containing a gas are subject to a continuous bombardment by molecules bouncing back from them, which averages up into a steady force, the pressure of the gas. Suppose the same amount of gas is contained in a vessel with only half the volume. Since in this case the number of molecules per unit volume will be twice as large, twice as many molecules will bounce every second from a given area of the wall, and thus the pressure of gas will be doubled. This explains the law of the inverse proportionality of gas pressure and gas volume, discovered by Robert Boyle.

Now consider what happens if the molecules move faster. It will have two effects: 1) More molecules will reach any given area of the wall per second; 2) The strength of each impact determined by the mechanical momentum ("amount of motion," in Newton's terminology) of the molecules will increase. Since both effects are proportional to molecular velocity, the pressure will increase as the square of that velocity, or, what is the same, as the kinetic energy of molecules. We have seen that, according to the Charles–Gay-Lussac law, the pressure of gas kept at a constant volume is proportional to its absolute temperature, from which it follows that *absolute temperature is simply the measure of the energy of thermal motion of the molecules*. It does not matter which kind of molecules we are speaking about, since one of the fundamental laws of statistical mechanics, known as the "law of equipartition of energy," states that *in case of a mixture of a large number of particles of two or more different masses, the average kinetic energy per particles remains the same.* Thus, for example, in a mixture of hydrogen molecules and the molecules of oxygen, which are sixteen times heavier, the velocity of oxygen molecules is four times smaller than that of hydrogen molecules, so that the product of their mass times the square of their velocity is the same. At room temperature, i.e., at about 300° abs, the energy of thermal motion is about 0.0000000000000002 erg, which, in case of air molecules, corresponds to the velocity of 50,000 cm/sec (about 1,000 miles/hr).

The energy of thermal motion determined by the absolute temperature is, of course, only an average for a large number of particles, and as is always the case for statistical phenomena, the energies of individual particles may show large deviation from the mean value. Due to the randomness of their mutual collisions, some of the molecules may attain for a short while much larger velocities, while the others may be temporarily slowed down. Using the laws of statistical mechanics, one can calculate the percentages of molecules in a gas which deviate by different degrees from having exactly an average velocity. This velocity distribution curve, which was first calculated by Maxwell and carries his name, is shown in Figure IV-9.

Another important notion in the statistical theory of gases is the notion of the "mean free path," i.e., the average distance traveled by molecules between two collisions. In the atmospheric air the mean free path of the molecules is very short, only about 0.00001 cm, while in the case of very rarefied gas filling interstellar space a molecule may travel many miles before encountering another one. The shortness of free path accounts for the fact that molecules, moving as fast as they do, take quite a long time to travel from one end of a room to another; in fact, they are in the position of a football player with the ball in his hands rushing toward the goal line and being tackled at almost every step by his opponents. Of course, while a football player has the goal line as his aim and tries to run toward

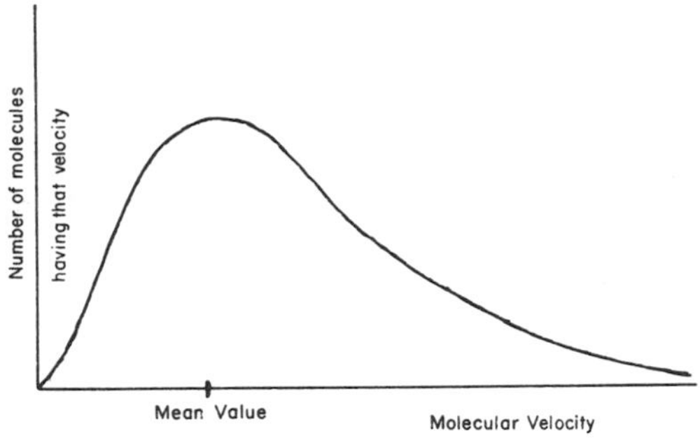


FIG. IV–9.
Maxwell's velocity distribution curve.

it, molecules are blind in their motion and after each new collision bound off in any odd direction. It can be shown mathematically that, in case of such a motion, known as "random walk," the average distance covered after many steps is equal to the length of each step multiplied by the square root of the total number of steps, not by the number of steps as it would be if they all were in the same direction. Thus, we have the formula:

$$\text{distance traveled} = \text{length of step} \times \sqrt{\text{number of steps}}$$

In the case of air molecules the length of each step is 0.00001 cm, and, if they must travel a distance of, say, 10 meters (1,000 cm), the above formula tells us that the total number of (random) steps must be equal to

$$\left(\frac{1000}{0.00001}\right)^2 = 10{,}000{,}000{,}000{,}000{,}000.$$

At the speed of 50,000 cm/sec, each step will be made in $\frac{0.00001}{50,000}$ = 0.0000000002 sec, so that the total travel time will be 10,000,000,-000,000,000 × 0.0000000002 = 2,000,000 sec or 20 days!

How does the kinetic theory of heat account for the basic law of thermodynamics, which states that in all thermal processes the entropy must always increase? What is the meaning of entropy anyway, from the point of view of the statistical theory of molecular motion? Why does heat always flow from hotter bodies into cooler ones, and why can we not transform a certain amount of heat completely into mechanical energy, while there is no problem of transforming mechanical energy into heat? The answer to all these questions comes quite naturally if we visualize what happens to the molecules in these cases. Consider a container divided into two halves by a thermoinsulating partition: Fill one half with a hot gas and another with a cool gas, and remove the partition. What will happen? Apparently the fast-moving molecules of the hot gas will lose their energy in collisions with the slower molecules of the cooler gas, and the process will go on until the equipartition of energy between all molecules, i.e., the equality of temperatures in both halves of the container, will be reached. The situation is similar to the case of a bucket, the lower half of which is filled with black beads, and the upper part with white ones. If we shake the bucket, the beads will get mixed together, so that both the white and the black ones will be evenly distributed through the bucket from the bottom to the top. Can we separate them again by more shaking? Theoretically yes. There is, in fact, no reason why such a separation should not take place; but it is very improbable that it would! We may have to shake the bucket for centuries, or, maybe, for millions of years until, by sheer chance, all the black beads collect again at the bottom and all the white ones at the top. The same is true in the case of gas molecules. It is possible in principle that one half of the molecules will be slowed down by random collisions well below the average speed, while the other half will be correspondingly speeded up. But it is very improbable!

A similar situation exists in the case of the transformation of mechanical energy into heat and vice versa. Imagine a bullet which hits a steel wall. While the bullet is flying toward the target, all its molecules are moving together in the same direction and with the same speed. (This common motion of molecules overlaps, of course, their irregular movements, due to the initial temperature of the bullet.) When the bullet is stopped by the wall, this organized motion is turned into irregular motion of individual particles, increasing the original thermal agitation of the molecules forming the bullet and the wall. Here again we can imagine the reverse process in which the molecules forming one end of a metal bar heated in the flame will, by sheer chance, have their thermal velocities oriented in the same direction, so that this metal slug will fly away as if shot out of a gun. But, again, this would be an extremely improbable event. Thus we see that the law of increasing entropy is simply the statement that *in all natural processes the organized motion of molecules has a tendency to become disorganized or random.* All processes go in the direction from the less probable pattern of molecular motion to the more probable one, and the increase of entropy corresponds to the increase of the probability of the pattern of molecular motion.

One can derive the relation between the probability of a given pattern of molecular motion and entropy in the following simple way, first proposed by Ludwig Boltzmann. Consider two thermodynamic systems, *A* and *B,* which may be two containers filled with two different gases under two different pressures, or any other more complicated systems containing liquids, their vapors, solid crystals, their solutions in the liquids, etc. If the two systems have the same temperature *T,* and we bring them into thermal contact with each other, no heat flow will take place in either direction, and the two systems will remain in the same states in which they were when they were separate. Suppose a certain amount of heat flows into the systems from the outside, system *A* gaining $Q_A$ calories and system *B* gaining $Q_B$ calories. If we consider the two systems separately, their entropy increases will be given by $\frac{Q_A}{T}$ and $\frac{Q_B}{T}$. If we consider them as a single composite system, the total increase of entropy will be

$$\frac{Q_A + Q_B}{T}.$$

Since

$$\frac{Q_A + Q_B}{T} = \frac{Q_A + Q_B}{T},$$

we conclude that *the entropy of a composite system is equal to the sum of entropies of its parts.*

How does this situation look from the point of view of the probabilities of various patterns of molecular motion? How does one express that probability for the composite system $A$ and $B$ in terms of the probabilities for $A$ and $B$ alone? According to the mathematical theory of probabilities, *the probability of a composite event* (i.e., an event which must satisfy several independent conditions) *is given by the product of the probabilities of the individual events from which it is composed.* Thus if a coed going out on a blind date hopes that he will be "tall, dark, and handsome," the probability that her hope will be fulfilled is the product of the probabilities that he is tall, that he is dark, and that he is handsome. If the chance for a man to be tall is $1/4$ (i.e., one out of four), the chance to be dark $1/3$, and the chance to be handsome $1/50$, the probability that all three conditions will be fulfilled is:

$$1/4 \times 1/3 \times 1/50 = 1/600$$

i.e., one out of 600.

Thus we see that, while in a composite thermodynamic system the entropies must be added, the probabilities must be multiplied. Which kind of mathematical dependence between two quantities satisfies that condition? Well, of course, the logarithmic dependence, because in order to multiply two numbers we must add their logarithms. Thus, the entropy must vary as the logarithm of probability and we must write:

$$S = k \log P$$

where $k$ is a numerical coefficient named after Boltzmann.

The above formula forms a bridge between classical thermodynamics and the kinetic theory of heat, and permits us to calculate all thermodynamic quantities on the basis of statistical considerations.

## MAXWELL'S DEMON

A very important personage in statistical physics is Maxwell's "demon," a product of the imagination of James Clerk Maxwell (Fig. V-16), who contributed so much to that field of science. Imagine a tiny and very active demon (Fig. IV-10) who can see individual molecules and is fast enough to handle them as a champion handles tennis balls. Such a demon could help us beat the law of increasing entropy, by manipulating a small window in a wall separating two gas chambers, *A* and *B*. The window's shutter is supposed to slide without any friction, and the demon opens it when he sees a particularly fast molecule heading that way, and closes it when the approaching molecule is slow. Thus all the fast molecules from Maxwellian distribution get into Chamber *B*, while only the slow ones will be left in Chamber *A*. *B* will become hotter and *A* cooler, with heat flowing in the wrong direction, violating the second law of thermodynamics.

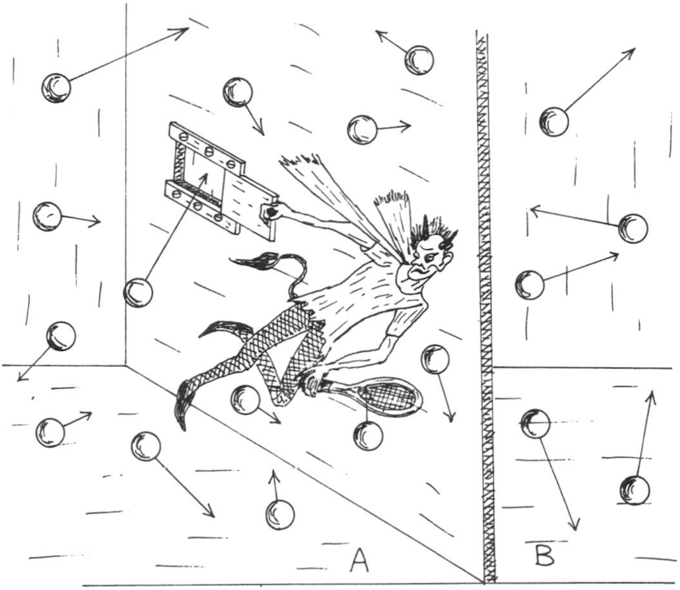


FIG. IV–10.
Maxwell's demon, who allegedly can separate the fast and slow molecules.

Why cannot this be done, not with the help of a real demon, of course, but by the use of some tiny, ingeniously constructed physical gadget, which would act in the same way? To understand the situation let us remember a cryptic question asked by the famous Austrian physicist Erwin Schrödinger (Fig. VII-19) in his very interesting booklet *What Is Life?*:* "Why are atoms so small?" At first sight, this question looks quite nonsensical, but it makes sense and becomes

* Cambridge University Press, 1944.

answerable if one reverses it and asks: "Why are we so large (as compared with atoms)?" The answer is simply that such a complex organism as a human being, with its brain, muscles, etc., cannot be constructed just from a few dozen atoms, in the same way as one cannot build a Gothic cathedral from just a few stones.

Maxwell's demon, and any mechanical device which would substitute for him, would have to be built from a small number of atoms and could not possibly carry the complicated tasks assigned to them. The smaller the number of particles, the larger the statistical fluctuations in their behavior, and an automobile in which one of the four wheels may spontaneously jump up and become the steering wheel, while the radiator becomes the gasoline tank and vice versa, is not a reliable vehicle to drive! Similarly, a Maxwell demon, real or mechanical, will make so many statistical mistakes in handling the molecules that the entire project will fail completely.

## MICROSCOPIC THERMAL MOTION

The very large and the very small figures quoted above for the molecular world are the result of calculation, since molecules and their motions are too small to observe even through the best microscopes. It happens, however, that we can bridge the gap between the invisible molecules and large bodies which we encounter in our everyday experience by observing the behavior of tiny particles, 1 micron ($\mu$) or so in diameter, which are on the one hand sufficiently small to show noticeable thermal motion, and on the other hand sufficiently large to be seen through a good microscope. It was first noticed by the British botanist Robert Brown that plant spores floating in water are never at rest, but are involved in a kind of "tarantella," jumping irregularly to and fro as if being constantly kicked by some invisible agent (Fig. IV-11). Brown himself and his contemporary scientists were unable to explain this jittery behavior of the tiny particles, and it was almost a century later that it was interpreted by a French physicist, Jean Perrin, as the result of numerous impacts received by them from water molecules involved in thermal motion. Perrin's studies of Brownian motion gave an indisputable proof of the correctness of the kinetic theory of heat, and permitted physicists to observe directly the statistical laws of motion which were before this just theoretical conjectures. The exact mathematical theory of Brownian motion was developed by the young Albert Einstein in one of the three articles which he published in 1905. The other two were on the theory of light quanta and on the theory of relativity. Today, the statistical theory of heat, more generally called "statistical physics," compares only with Newtonian mechanics in its completeness and clarity.

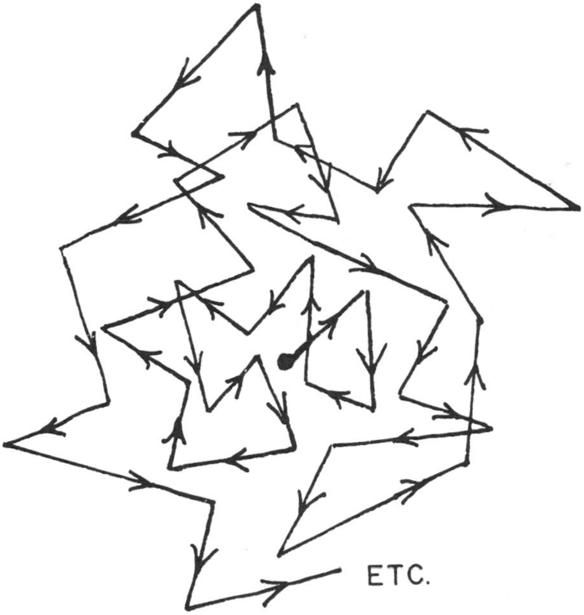


FIG. IV–11.
"Random walk" is the motion of a body in which the direction of motion changes often and irregularly as the result of collision with other bodies, be it a molecule colliding with other molecules or a drunkard colliding with the lampposts. It is clear that in such walking the body makes less progress than in walking a straight line, and it can be shown that the average distance from the starting point is in this case equal to the length of each straight step multiplied by the square root of the number of steps.

## THERMAL MOTION AND THE PROPAGATION OF SOUND

It is well known that sound is nothing more than the compression wave propagating through the air and other materials. Experimental studies revealed an amusing fact—that the velocity of sound is independent of the density of the air, and that at sea level it is the same as in the rarefied upper atmosphere. On the other hand, this velocity depends on the temperature of the air, being directly proportional to the square root of the absolute temperature. How can one explain these facts from the point of view of molecular structure and thermal motion?

To do this we must remember that air consists of a multitude of molecules rushing at random through space with the velocities increasing with temperature. When a compression sound wave is emitted, let us say by a vibrating tuning fork, air molecules nearest to the prongs are pushed in the direction of motion, and in the collision with other molecules located further away (in the next thin layer of air) communicate the push to them. Those, in their turn, push the molecules in the next layer, and thus the compression propagates through the air, forming a sound wave. Since air molecules must fly a comparatively long distance (the so-called *free path*) before they hit molecules in the next layer, the velocity of propagation is essentially determined by the thermal velocity of the molecules. This dynamic picture explains the two above-quoted facts concerning the velocity of sound. Indeed, thermal velocity of molecules remains the same for a given temperature, no matter how much the gas is compressed or rarefied. On the other hand, since the kinetic energy of molecules is proportional to absolute temperature, their velocity increases as the square root of the temperature. And what is true for the velocity of molecules must also be true for the velocity of sound.

An entirely different situation occurs when the velocity of the object producing the compression in the gas exceeds the velocity of molecular thermal motion under given conditions. This happens, for example, when hot gases formed in an explosion push on the surrounding air, or when the air is pushed apart by the wings and body of a supersonic airplane or missile. In this case, the thermal velocity of molecules is not high enough to escape from the advancing "pusher" and they begin to be piled one upon the other with the resulting increase of density. The difference between this and the previously discussed case is illustrated schematically in Figure IV-12. The advancing front of highly compressed gas forms what is called a *shock wave*. Because of the highly increased density, the shock waves possess correspondingly high overpressure, which accounts for their destructive effects. In the case of explosions, the expansion of hot gases slows down, the compression of air separates from the "pusher," and continues to travel as a shock wave. In the case of supersonic airplanes and missiles which travel at a constant speed, being pushed forward by their motors, the shock wave remains stationary in respect to the moving body and is thus known as a "standing shock."

### LIGHT EMISSION BY HOT BODIES

It is well known that all material bodies become luminous when heated to a sufficiently high temperature. This is how light was pro-

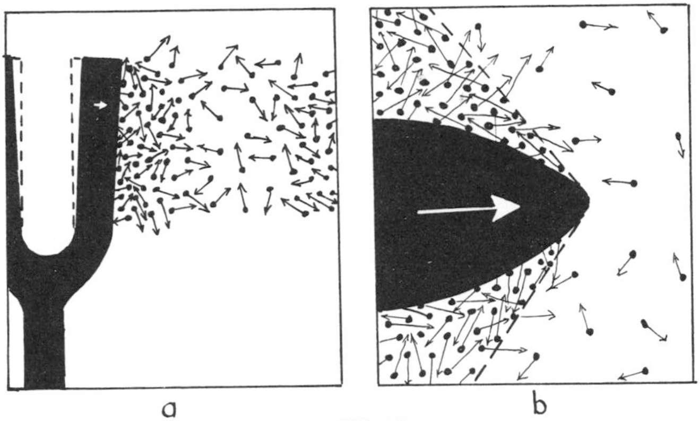


FIG. IV–12.
Formation of a sound wave in a case when the "pusher" moves more slowly than the molecules (a), and the formation of shock waves (standing shocks) when the "pushers" move faster (b).

duced by the flame of the old-fashioned gas burners; this is how light is produced by the hot filaments of modern electric bulbs. On the cosmic scale, sun and stars emit light because their surfaces are very hot. It is a common experience that at comparatively low temperatures, as in the case of room heating units, one gets radiant heat but no visible light. The kitchen range heating unit at the temperature 600° to 700° C is "red hot," glowing with a faint reddish light. The filament of an electric bulb, heated to a temperature of over 2,000°, emits a bright light which, however, looks yellowish as compared with the light of the brilliant electric arc which operates at a temperature between 3,000° and 4,000°. The surface of the sun at a temperature of about 6,000° emits light which is richer in blue rays than the light emitted from all previously mentioned sources. Thus, *as the temperature goes up, the emitted radiation becomes rapidly more intensive, and richer in the short wave lengths.* Figure IV-13 shows the observed distribution of intensity between different wave lengths in the radiation emitted by material bodies at different temperatures. At

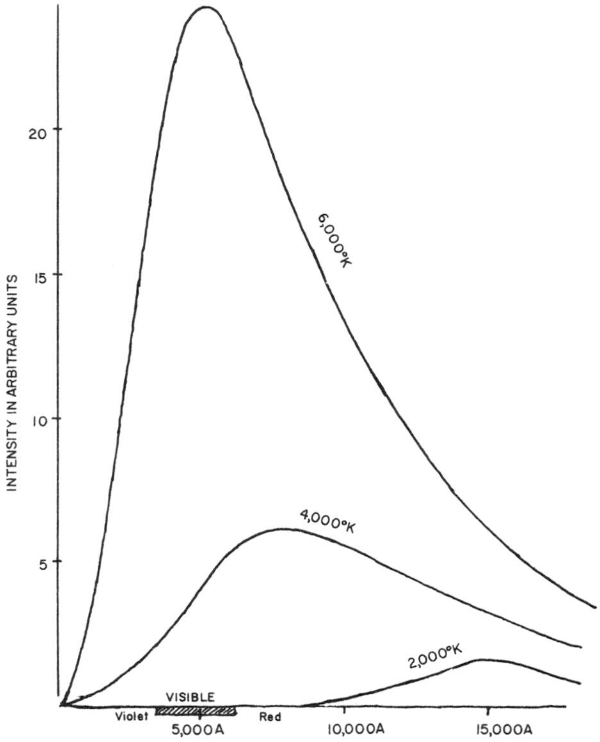


FIG. IV–13.

Distribution of energy in a continuous spectrum emitted by bodies of three different temperatures.

2,000° K all energy is in the region of long wave heat rays with zero intensity in the visible light region shown by dashed stretch. At 4,000° K, there is some visible light emitted, but the intensity of red rays exceeds considerably the intensity of yellow, green, and blue rays. At 6,000° K (the temperature of the sun's surface) the maximum of intensity falls into the yellow region of the spectrum, and we perceive the color mixture as the "white light." At still higher temperatures the intensity maximum moves into the region of invisible ultraviolet rays; there are stars which are so hot (several hundred thousand degrees) that most of the light emitted by them falls into this invisible ultraviolet region.

Light emission by hot bodies is subject to two important laws discovered during the second half of the last century:

*Wien's law* was established by the German physicist Wilhelm Wien (1864–1928). It states that *the wave length corresponding to the maximum intensity in the spectrum is inversely proportional to the (absolute) temperature of the emitting hot body*. We notice from Figure IV-13 that, while at 6,000° K the maximum of intensity is at 5,000 Å, it shifts to 15,000 Å at the temperature of 2,000° K.

The *Stefan-Boltzmann law* was discovered by the German physicist Josef Stefan (1835–1893) and then derived theoretically by using thermodynamic arguments by Ludwig Boltzmann, who was mentioned earlier in the text. The law states that *the total amount of energy emitted by a hot body is proportional to the fourth power of its (absolute) temperature*. Indeed, the area under the curve marked 6,000° K in Figure IV-13 is $3^4 = 81$ times larger than the area under the curve marked 2,000° K.

## LIGHT EMISSION BY HOT GASES

The discussion of light emission by hot bodies given in the previous section pertains to the case of solid or liquid materials such as the tungsten filament of an electric bulb or molten iron in a foundry.* We encounter an entirely different situation in the case of light emitted by hot gases. If we look through a prism at the light emitted by a gas (or a kerosene) lamp, we will see a continuous spectrum running all the way from red to violet. But it can be shown that this continuous spectrum is not actually due to hot gases in the flame, but rather to tiny solid particles of soot in it. If one achieves a complete burning of the gas, as in a Bunsen burner, developed by a German physicist, Robert Wilhelm Bunsen (1811–1899), one obtains a flame which,

* The material of the sun is at a gaseous state because of extremely high temperatures which run from 6,000° K on the surface to 20,000,000° K at the center. However, except for a thin outer layer known as the chromosphere, the gases forming the body of the sun are compressed to a density comparable with that of ordinary solid and liquid materials and are therefore emitting a continuous spectrum.

being very hot, emits very little light. Bunsen used his burner to study the emission of various material in a gaseous form. If one introduces into the flame of a Bunsen burner a small amount of sodium (which can be used in the form of sodium chloride, i.e., ordinary

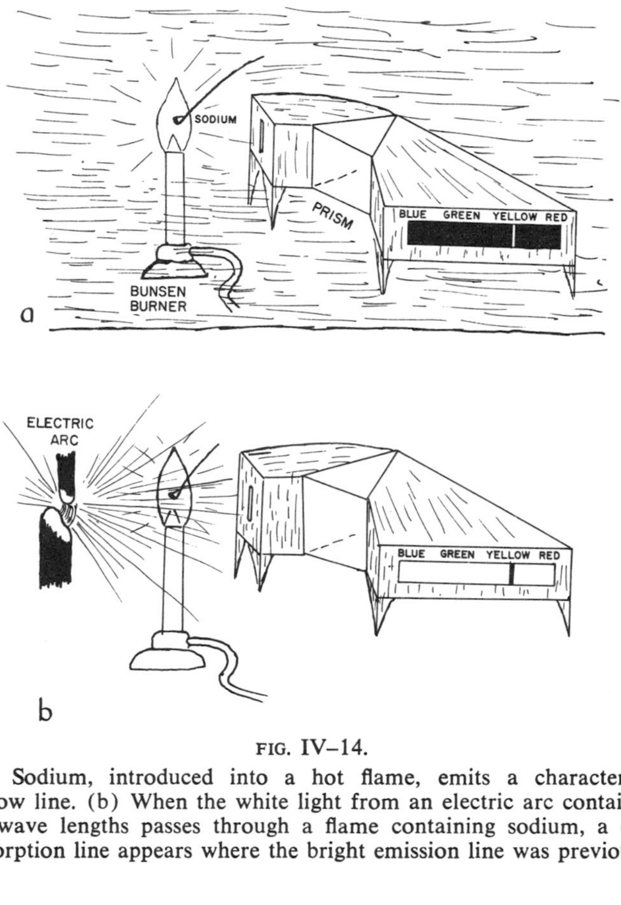


FIG. IV–14.

(a) Sodium, introduced into a hot flame, emits a characteristic yellow line. (b) When the white light from an electric arc containing all wave lengths passes through a flame containing sodium, a dark absorption line appears where the bright emission line was previously.

table salt), the flame becomes brilliantly yellow. Analyzing that light by means of a prism in the old Isaac Newton fashion, one finds that the spectrum consists of one single yellow line with all other wave lengths completely absent (Fig. IV-14a). A similar experiment with potassium, which gives to the flame a bright red color, shows a red line farther to the right in the spectrum. Other materials turned into vapor in the hot flame of the Bunsen burner give rise to other lines, sometimes single, sometimes in larger numbers.

Why is it that hot gases emit the light of strictly selected wave lengths (or, as is the same, of strictly selected frequencies) while hot solid or liquid materials emit the entire gamut of wave lengths forming a continuous spectrum? As we shall see later in the book, an atom or a molecule can be compared with a musical instrument, the only difference being that they emit light waves instead of sound waves. A musical instrument, be it a modest tuning fork or a grand piano, is built in such a way that it produces only an assortment of selected sound frequencies (one for the tuning fork, many for a grand piano) which, sounded one after another, form a melody. The atoms and molecules are also emitting a selection of light wave lengths, typical for each of them. Gases, atoms, or molecules fly freely through space, colliding with each other now and then. At each collision they are "excited" (if the temperature is high enough) and fly on, vibrating and emitting light waves characteristic for them. Thus, the vapor of sodium, copper, iron or whatnot, emit characteristic line spectra by which they can be recognized. In the solids, on the other hand, the atoms are tightly packed together, and the situation resembles more that of a large bag into which all the instruments of a symphonic orchestra are piled higgledy-piggledy, one upon the other. If we shake the bag, we will hear the noise which contains all audible frequencies and has nothing to do with the particular properties of the instruments which it contains. Similarly, the atoms piled in a piece of metal or any other solid (or liquid) substance completely lose their pure tone properties, and light emitted by red hot iron does not differ much from that of the red hot copper or red hot anything else.

The specificity of light emission by different substances represents the basis for an important method of spectral analysis which permits us to find the chemical composition of any given material simply by watching the light emitted by its vapor.

## ABSORPTION OF LIGHT

Let us now return to our experiment (Fig. IV-14a) with a Bunsen burner containing some sodium in its flame. Suppose we place behind the flame a very strong source of light with continuous spectrum such as an electric arc (Fig. IV-14b). The light from the white hot electrode of the arc will pass through the flame and, falling on the slit, produce a rainbow band in the spectroscope. But we will notice that the continuity of color is interrupted by a narrow dark line just exactly on the same place where the yellow line of sodium was. This effect is due to an important phenomenon called *resonance* which occurs in all cases where we deal with some kind of oscillation. Consider a child in a swing at the playground being given a ride by his father. If the father pushes the swing rhythmically at time intervals equal to the swing's own oscillation period, the amplitude of the motion will become larger and larger and the child will either be pleased or scared. But if the father is distracted by a pretty nurse nearby, and does not deliver his pushes within the proper periods, no good will come out of his efforts. Sometimes he will push the swing as it moves away from him, and this will help; sometimes he will push it as it comes toward him, and this will hurt. To increase the amplitude of any oscillation, the force should be applied with the period equal to the proper period of the oscillating object. If we place two identical tuning forks close to each other, and start one of them vibrating by hitting it with a hammer, sound waves coming out of it will soon bring the other fork into a state of motion. But if the two tuning forks have different periods of vibration, nothing will happen. Similarly, tuning a radio receiver or a TV set to the desired station, one turns the knob which makes the vibration frequency of the receiving set equal to that of the broadcasting station.

Our experiment with the sodium-containing flame falls into the same category. The atoms of sodium resonate with that particular wave length in the continuous spectrum of the arc which they can emit themselves, and scatter these waves in all directions, thus weakening the original beam. The black absorption line in this case is, of course, not quite black. In fact it may even be brighter than the original emission line, but it looks very dark in contrast to other parts of the continuous spectrum of the arc. The law that *all substances absorb the same light frequencies which they can emit* was discovered by a German physicist, Gustav Kirchhoff (1824–1887), and carries his name. This law is of great importance in many walks of physics, chemistry, and astronomy. One of its most important applications is the study of chemical composition of the sun and all other stars.

In the beginning of the 19th century a German physicist, Joseph von Fraunhofer (1787–1826), repeating Newton's experiments on the solar spectrum but using prisms of much better quality, was surprised to see that the color rainbow was intersected by a large number of very thin black lines. The origin of these "Fraunhofer lines" can easily be understood on the basis of what was said earlier in this section. We have stated earlier that, although the body of the sun is made entirely of gaseous material, it emits a continuous spectrum simply because the atoms are squeezed so tightly together that they have "no elbow room to draw their bows without interfering with neighboring players." But the very outermost layer of the sun's body, known as the chromosphere, is formed of very rarefied hot gases and *does* produce pure optical tones. When the continuous spectrum from the *photosphere* (i.e., dense body of the sun) passes through the chromosphere the wave lengths which correspond to chemical elements present in it are absorbed and scattered, and Fraunhofer's dark lines appear in the originally spotless rainbow. The use of spectral analysis in astronomy led to tremendous advances in our knowledge of the sun and stars, and opened to the human eye the limitless vistas of the universe in which we live. In Plate II we reproduce Fraunhofer's spectrum of the sun, the visible part (a), and the far ultraviolet part (e), being obtained by modern instrumentation.

# CHAPTER V *The Age of Electricity*

## EARLY DISCOVERIES

AS WE STATED in the first chapter, the phenomena of electricity and magnetism were known to the ancient Greeks and, probably, to the rest of the ancient world. However, the first systematic studies of these phenomena were undertaken only in the beginning of the renaissance of arts and sciences. Sir William Gilbert, the personal physician of Queen Elizabeth I and the contemporary of Galileo, carried out careful studies of magnetic interactions and published his results in a book, *De Magnete,* which contains a description of all essential qualitative properties of magnets. Gilbert was an enthusiastic adherent of the Copernican system of the world, and hoped that the forces which hold the planets in their orbital motion around the sun could be explained as being the result of magnetic attraction. To study these problems more closely he made spheres of magnetite (magnetic iron ore) and studied the field around them by tiny compass needles placed in different positions and at different distances around the spheres. He found that at one point on the sphere there is a maximum attraction of one end of the needle, and at the opposite point a maximum attraction of the other end. At various points on the surface of the sphere the needle always oriented itself in a definite position along a large circle connecting the points of maximum attractions or the magnetic poles of the sphere. This was strikingly similar to the behavior of compass needles in various points on the surface of the earth, and Gilbert concluded that our globe can be considered as a giant magnet with the poles located near the geographical north and south poles. This concept survived through the centuries and, being expanded mathematically by the great German mathematician Karl Friedrich Gauss, is today a standard notion in the theory of terrestrial magnetism. On the other hand, Gilbert's attempts to make magnetic forces responsible for the motion of planets around the sun have failed completely, and it was more than half a century later that Newton explained this motion by the forces of universal gravity, which have nothing to do with magnetism.

When Newton had already conceived his ideas on universal gravity but was still keeping them secret, German physicist Otto von Guericke, best known for his experiments with so-called Magdeburg hemispheres (two metal hemispheres, evacuated of air and put together, which could not be pulled apart by two teams of horses), tried to explain the attraction between the planets and the sun by the electric interactions. Although he failed in this task, just as Gilbert did in his, he made many important discoveries concerning the properties of electric charges. He found that, whereas rubbed amber would attract and pick up light objects like small pieces of paper, two light bodies which have been touched by rubbed amber would repel each other. He also found that an electric charge can be transferred from one body to another, not necessarily by direct contact but also by a wet string or better a metallic wire stretched between them. Further studies of electric phenomena carried out by Du Fay early in the 18th century led to the discovery that there are two kinds of electricity: that produced by rubbing amber, sealing wax, hard rubber, and other resin-like substances, and that produced by rubbing vitreous substances such as glass or mica. These two kinds of electric fluids were called "resinous" and "vitreous" and it was established that like kinds of electric charges repel each other while unlike kinds attract each other. Electrically neutral bodies were supposed to contain balanced amounts of both electric fluids, while electrically charged bodies had an excess of either resinous or vitreous electricity. The phenomena first observed by Otto von Guericke were explained as due to the interaction between these two kinds of electric fluid. Suppose we rub a hard rubber sphere so that it becomes charged by resinous electricity. If we bring to its vicinity a small uncharged body in which both kinds of electricity are balanced up, resinous electricity will be pushed away to the far end of the body while vitreous electricity will be pulled to the near end. Since electric interactions decrease with distance, the attractive force acting on vitreous charges will be larger than the repulsive force action of resinous charges, and the net effect will be attraction between the two bodies. If, instead of a hard rubber sphere we take a glass sphere, the result will be the same except that V's and R's will be interchanged. Thus, neutral bodies will be always attracted by charged bodies. The phenomenon of separating the charges in an originally uncharged body is known as electric "polarization" or "induction." If now we bring two small bodies in contact with a large

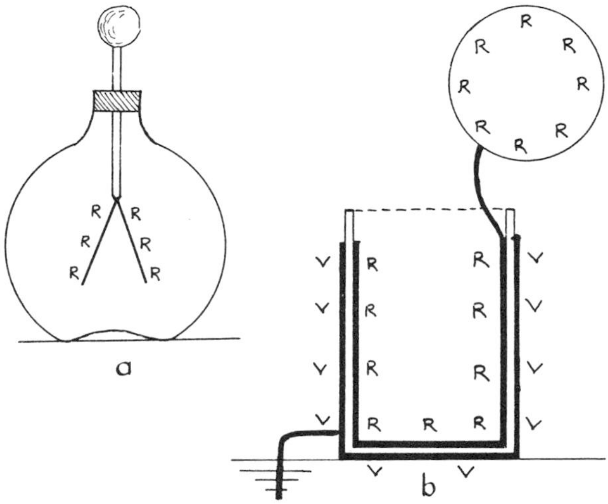


FIG. V–1.
(a) Leaf electroscope. (b) Leyden jar.

charged body, they will be charged by like electricities and will repel each other when removed.

During these early studies of electric phenomena there were developed two very important electric gadgets: a *leaf electroscope,* and a *Leyden jar.* The electroscope (Fig. V-1a), i.e., an instrument which shows the presence of electric charge, was first built in 1705 by Haukesbee and consisted of two straws suspended side by side to the lower end of a metal bar. When the bar was charged with either resinous or vitreous electricity, both straws became charged with like electricity and came apart from one another. We are still using this gadget, except that for straws are substituted much lighter gold leaf. The Leyden jar (Fig. V-1b), built in 1745 by a group of scientists at the University of Leyden (Holland), was designed to collect large amounts of electricity. It was made of an ordinary cylindrical glass jar, the inside and outside of which were coated by a thin silver foil. If the outer foil is grounded (i.e., connected with the ground) and the inner one is connected with a charged body or vice versa, the electricity (either resinous or vitreous) tries to escape into the ground but is stopped by the layer of glass. Thus large amounts of electricity are collected in the jar, and one can extract impressive sparks by connecting the inside and outside foils with a wire. The old-fashioned Leyden jar has developed today into various types of condensers, consisting of a large number of metal plates separated by thin layers of air, glass, or mica. Such condensers, which can store very large amounts of electricity, are used in all walks of physics and electrotechnics. In particular, the first atom smasher, constructed in 1930 at Cambridge University by John Cockroft and E. T. S. Walton, consisted of a battery of such condensers which could be charged up to 1 million volts. When the condensers were discharged through a glass tube containing hydrogen, they produced "atomic projectiles" of such high energy hitting the atoms of lithium target placed at one end of the tube that they broke them in two.

To the same period belongs the work of the great American statesman and writer Benjamin Franklin, who became interested in physics at the ripe age of 40. He was not satisfied by tiny sparks that could be obtained by rubbing a galosh against a fur coat, and wanted to play with much larger sparks, which Zeus throws down from the clouds during thunderstorms. Thus, he was sending up kites into the thunderclouds to collect electricity from them. The wet string holding the kite served as a perfect conductor of electricity, and he was able to charge the Leyden jars with it, and extract sparks from them afterward. His studies, collected in the book *Experiments and Observations on Electricity Made at Philadelphia in America* (1753), earned him a fellowship in the Royal Society of London and an associate membership in the Royal Academy of Science in Paris. Challenging Zeus in his experiments, he did not do so well in the theoretical interpretation of electric phenomena by introducing a one electric fluid hypothesis. He postulated that "vitreous" electricity was the only single kind of electric fluid and that two different kinds of electrification correspond to the *excess* or the *lack* of that ponderless fluid. Accordingly, he called the body with an *excess* of vitreous electricity (as a rubbed glass stick) a *positively* charged body, while a body with a *lack* of it (as a rubbed rubber stick) a negatively charged body. When two bodies, one containing an excess and another a deficiency of electric fluid (vitreous university) come together, the electric current must be flowing from the first body, where it is in excess, to the second, where it is lacking. These ideas of Benjamin Franklin led to the modern terminology in which the electric current flows from the positive electrode (anode) to the negative one (cathode). We know now that Du Fay's idea of two electric fluids is closer to reality than that of Franklin, although the situation is much more complicated than visualized by either of them. There are both positively and negatively charged particles, and for each particle carrying normally either positive or negative charge there exists a corresponding "anti-particle" carrying an opposite charge. Franklin was closer to truth in the case of the electric current in metallic wires, where the transport of electricity is due exclusively to the motion of electrons, except that electrons carry resinous and not vitreous electricity. One sometimes hears the proposal today that the names of positive and negative electricity should be interchanged so that conventional direction of current from + to − would coincide with the direction in which the electrons move. If it were done, however, there would be trouble with the atom smashers which shoot high energy protons at atomic targets; instead of going out of the atom smasher's muzzle the electric current would flow in from the target. And in the case of liquids where electricity is carried equally by positive and negative ions moving in opposite direction, such change of terminology would not help at all.

### THE LAW OF ELECTRIC AND MAGNETIC FORCES

During the second half of the 18th century, physicists in many countries were employed in quantitative studies of electric magnetic forces. One of the first important discoveries in this line was made by a Frenchman, Charles Augustin de Coulomb, who developed the so-called "torsion balance" for measuring very weak forces. As can be seen from Figure V-2, which represents the sketch of that instrument made by Coulomb, it consists of a light bar, which is suspended on a long thin thread, with two balanced spheres at each end. When no forces act on the spheres, the bar assumes a certain equilibrium position. If one of the spheres is electrically charged, and another charged sphere is placed in its vicinity, the electric force acting on the movable sphere will cause the bar to turn around the suspension point until the torque in the thread will balance the acting force. Since

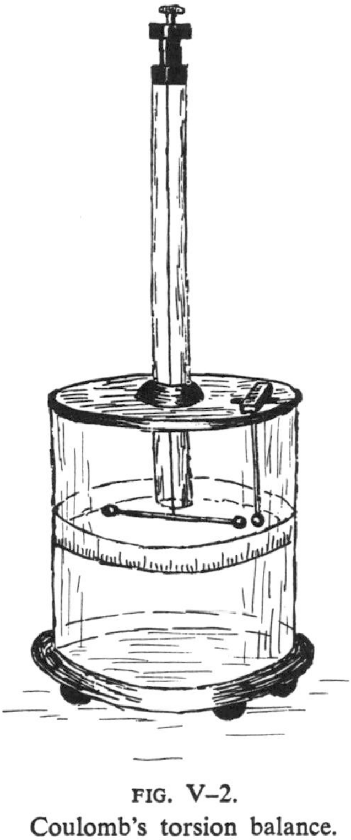

FIG. V–2.
Coulomb's torsion balance.

the thread is very thin, a small force acting on the sphere will produce a considerable deviation of the bar from its original position, the angle of rotation being proportional to the force. Charging the movable and immovable spheres with various amounts of electricity and varying the distance between them, Coulomb discovered the law carrying his name, according to which the forces of electric attraction and repulsion are directly proportional to the product of two charges and inversely proportional to the square of the distance between them (Fig. V-3). Using this law, one can define an *electrostatic unit of charge* as a charge which acts with the force of 1 dyne on the equal charge placed 1 cm away. In practice one uses a much larger unit of electric charge called one *coulomb,* which is equal to 3 billion of the smaller electrostatic units defined above. Using the same torsion balance, and suspending a magnet on the thread with another magnet placed vertically through the top of the case, Coulomb proved that

$$\left(\frac{1}{1}\right)^2 = 1$$

$$\left(\frac{1}{2}\right)^2 = \frac{1}{4}$$

$$\left(\frac{1}{3}\right)^2 = \frac{1}{9}$$

etc.

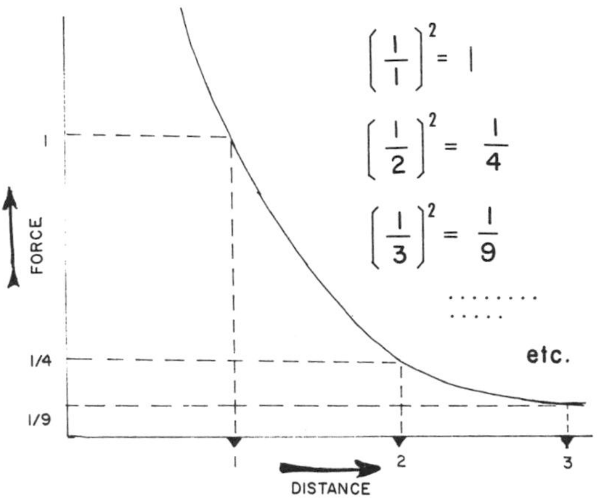

FIG. V-3.
A graph of Coulomb's law.

the same law holds for magnetic interaction. Accordingly, a unit of magnetization is defined as the strength of magnetic pole which attracts or repels, with a force of 1 dyne, a pole of equal strength placed 1 cm away.

At about the same time there lived in England a very reclusive character named Henry Cavendish, the son of a British peer. He had no close friends, was afraid of women, and the women servants in his large house in Clapham Common, a section of London, were ordered to keep out of his sight, getting the orders for his meals by notes left daily on the hall table. He was disinterested in music or art of any kind, and spent all of his time carrying on experiments in physics and chemistry in a private laboratory located in his large mansion. His work was interrupted only by traditional strolls for the purpose of health and by occasional attendance at the Royal Society Club dinners, for getting information on what other physicists and chemists were doing. During his long life (he died at the age of 79), he published only a handful of comparatively unimportant papers. But after his death about a million pounds sterling were found in his bank account and twenty bundles of notes in his laboratory. These notes remained in the hands of his relatives for a very long time, but when they were published, about a hundred years later, it became clear that Henry Cavendish was one of the greatest experimental scientists that ever lived. He discovered all the laws of electric and magnetic interactions at the same time as Coulomb, and his work in chemistry matches that of Lavoisier. Furthermore he applied a balance for the study of extremely weak gravitational forces between small objects, and, on the basis of these experiments, he arrived at the exact value of the mass of the earth (Fig. V-4). No physical unit is called by his name but the Cavendish Laboratory in Cambridge is one of the world's most renowned centers of scientific studies.

### A SHOCK FROM AN ELECTRIC EEL

The natives of Africa and South America were long familiar with a peculiar tropical river fish which delivers painful shock to anyone who attempts to handle it. In the middle of the 18th century a British ship brought several samples of the fish to London and the biologists began studying it. It was found that a shock was obtained only if the top of the fish's head and the underside of its body were touched with one hand. This fact, and the sensation of the shock, were reminiscent of the effect of the Leyden jar, newly invented at that time, and the fish was called *sirius electronicus* or electric eel. When it was proved that the fish could be used to charge the Leyden jar, no doubt was left but that electric discharge was being dealt with here. Electricity produced by fish attracted the attention of Italian physiologist Luigi Galvani, who was studying the phenomenon of the muscular contraction of frogs' legs, a favorite delicacy in Bologna restaurants. He happened to notice (so the story goes) that the severed frogs' legs, suspended from copper hooks on the iron balustrade of his balcony, jerked as if alive when they touched the iron bars of the balustrade. To check "under controlled conditions," Galvani conducted an experiment, dated September 20, 1786, in his laboratory notebook, in which he used a fork with one iron and one copper prong with which he touched the nerve and the muscle of

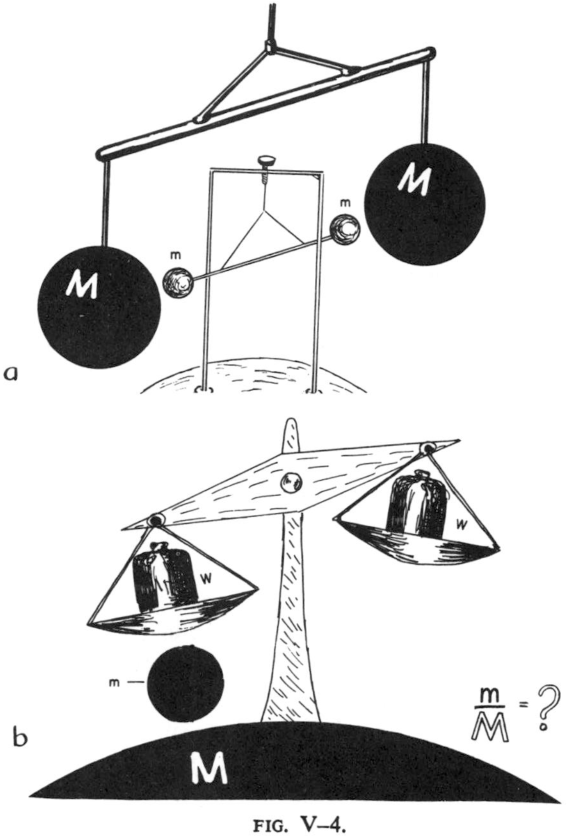


FIG. V–4.

Cavendish apparatus for measuring gravitational attraction was similar to Coulomb's apparatus for measuring electric forces. Changing the position of two large masses *M* suspended from the ceiling (a), one observes the displacement of two smaller masses *m* suspended on a very thin thread. (b) shows modified Cavendish method. Two weights *W*, previously in equilibrium under the action of terrestrial gravity (earth's mass *M*), move when an additional mass *m* is placed under one of them.

the frog's leg. The leg promptly contracted at each touch, and Galvani became certain that this was similar to the electric shock caused by an electric eel. He was, however, quite wrong in this assumption, and his friend the Italian physicist Alessandro Volta soon proved that electric current which caused the contraction of a frog's leg is a purely *inorganic* phenomenon which can always be observed when the ends of a wire, made by soldering together two wires of different metals, are put into a water solution of a salt. Volta called this phenomenon *galvanism* in honor of his physiological friend and constructed what is known as a "Volta pile" by using a large number of alternating copper and iron or zinc discs, separated by layers of cloth soaked in salt solution. Volta's pile was the prototype of the modern electric batteries which we use today in flashlights and in many other devices. In March, 1800, Volta sent a manuscript describing his discoveries for publication to the Royal Society of London, which was at that time the international center for the exchange of scientific ideas. In this paper he says:

> Yes, the apparatus of which I am telling you, and which will doubtless astonish you, is nothing but a collection of good conductors of different kinds, arranged in a certain manner. 30, 40, 60 pieces, or more, of copper, or better of silver, each laid upon a piece of tin, or, what is much better, zinc, and an equal number of layers of water, or of some other humour which is a better conductor than plain water, such as salt water, lye, &c; or pieces of cardboard, leather, &c. well soaked with these humours; such layers interposed between each couple or combination of different metals, such an alternative succession, and always in the same order, of these three kinds of conductors, that is all that constitutes my new instrument; which imitates, as I have said, the effects of Leyden jars, or of electric batteries, giving the same shocks as they do; which, in truth, remains much below the activity of the said batteries charged to a high degree, as regards the force and noise of the explosions, the spark, and the distance over which the discharge can take place, &c., only equalling the effects of a battery charged to a very low degree, of a battery having an immense capacity; but which, besides, infinitely surpasses the virtue and power of these same batteries, inasmuch as it does not need, as they do, to be charged beforehand, by means of outside electricity; and inasmuch as it is capable of giving a shock whenever it is touched, however frequently these contacts are made. . . . I am going to give you here a more detailed description of this apparatus, and of some similar arrangements, as well as the most remarkable experiments concerned with them.

His original drawing of the pile is shown in Figure V-5.

Then a very unfortunate thing happened. Messrs. Carlisle and Nicholson, who were in charge of the Royal Society publications, shelved the manuscript, repeated Volta's experiments, and published the results under their own names. But the trick did not work. The results of Volta's investigations became known through other sources, Carlisle and Nicholson were accused of scientific plagiarism, and they vanished into oblivion. And today *Volta's pile* and the *volt,* a unit of electric potential, commemorate the name of the talented Italian scientist. *Electric potential* characterizes the degree of electrization of charged bodies. Suppose we have a large spherical conductor

FIG. V–5.
Volta's pile.

charged with a certain amount of electricity and want to increase that charge. This can be done by holding with an insulating handle a small metal sphere which is charged with a certain amount of electricity some distance away from the large sphere (theoretically, at an infinite distance) and then brought into contact with it. Because of coulomb repulsion between the two spheres, one will have to do a certain amount of work bringing the two spheres in contact. The work which has to be done in order to increase the charge of the large sphere by one unit of electricity is called its electric potential. If one measures electric charge in *coulombs,* and the work in *joules,* the electric potential will be measured in *volts.*

## ELECTROMAGNETISM

Although the early investigators of electric and magnetic phenomena must have felt that there is some deep relation between them, they could not pin it down. Electric charges did not influence the magnets in any way; neither did the magnets influence the electric charges. The honor of discovering the bridge between electricity and magnetism fell to a Danish physicist, Hans Christian Oersted, who, having heard about Volta's work, constructed an electric pile of his

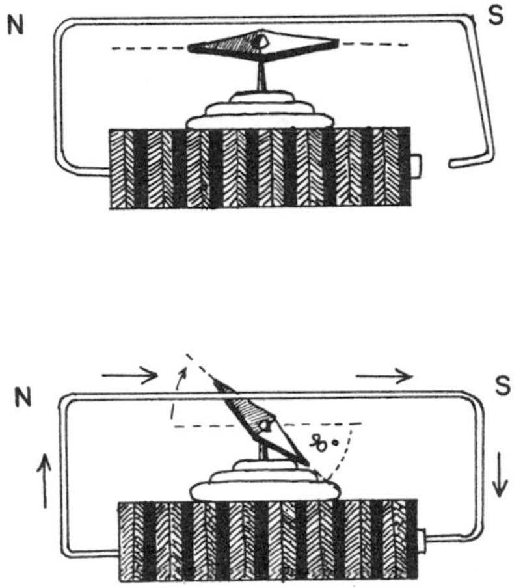


FIG. V–6.

Oersted's discovery of the interaction of an electric current and a magnet.

own, and was carrying on various experiments with it. One day in the year 1820, walking to his lecture at the University of Copenhagen, Oersted got an idea. If static electricity did not affect magnets in any way, maybe things would be different if one tried electricity moving through the wire connecting the two poles of the Volta pile. Arriving at the classroom filled with a crowd of young students, Oersted placed on the lecture table his Volta pile, connected the two opposite ends of it by a platinum wire, and placed a compass needle close to it. The needle, which was supposed to orient itself always in the north-south direction, turned around and came to rest in the direction perpendicular to the wire (Fig. V-6). The audience was not impressed, but Oersted was. After the lecture he remained in the classroom, trying to check the unusual phenomenon he had just discovered. First, he thought that the motion of the compass needle might have been caused by the air currents coming from the wire heated by the electric current. To check this he placed a piece of cardboard between the wire and the compass needle, to stop the air currents. It made no difference. Then he turned the Volta pile by 180 degrees so that the current in the wire was flowing in the opposite direction. The compass needle also turned by 180 degrees, and its north pole was now pointing in the direction in which the south pole previously was. It became quite clear to him that there was an interaction between the magnets and the *moving* electricity, and the direction in which the compass needle was oriented depended on the direction in which the electric current was flowing through the wire. He wrote down all the facts and observations pertaining to that discovery and sent it for publication to the French journal, *Annales de Chimie et de Physique*. The article appeared late in 1820 with the following note by the editors:

> The readers of the *Annales* should have noticed that we do not support too readily announcements of extraordinary discoveries* and up to now we couldn't but congratulate ourselves on this policy. But, in regard to the paper by Mr Oersted, the results obtained by him, however singular they may appear, are accompanied by too many details to give place to any suspicion of error.

Thus *electromagnetism,* as Oersted called it, became a reality!

When the news of Oersted's discovery reached Paris it attracted the attention of the French mathematician and physicist André Marie Ampère, who, in the course of a few short weeks, found that not only does an electric current act on a magnetic needle but two electric currents act on one another. There is an attraction between two parallel wires carrying electric currents if the currents run in the same direction, and a repulsion if the direction of two currents is opposite (Fig. V-7). He showed that a coil made of copper wire and free to rotate around a vertical axis orients itself in the north-south direction if the current is running through in the very same way as a compass needle, and that two such coils interact with one another very much in the same way as two bar magnets. This led him to the

* Presumably because most of them were the work of cranks.

idea that natural magnetism is due to electric current running inside magnetized bodies. He imagined that each molecule of magnetic materials contains inside of itself a circular current, thus representing a tiny electromagnet. When the material is not magnetized, the individual molecular electromagnets are oriented at random in all directions, and the net result is zero. In magnetized bodies molecular magnets are oriented, at least partially, in one direction, thus causing their magnetic attraction or repulsion. These views of Ampère are

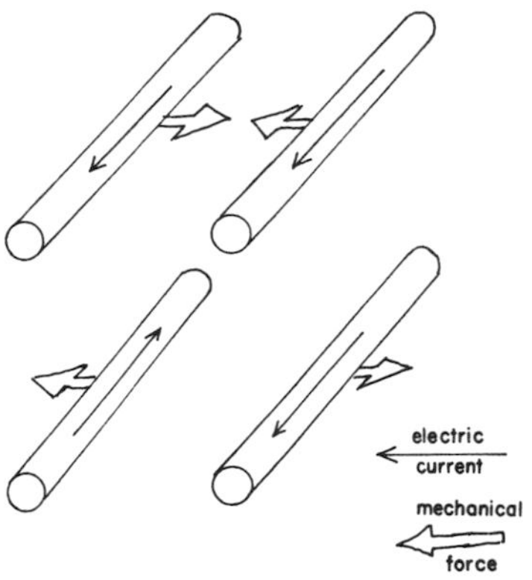


FIG. V–7.
Ampère's laws of interaction between currents.

fully confirmed by modern physics, which considers the magnetic properties of atoms and molecules to be due to electrons revolving around the nuclei, or spinning rapidly around their axes. Because Ampère was the first to formulate clearly the idea of electric current as the motion of electricity along the wire, the unit of electric current carries his name. One *ampere* is defined as a current which carries 1 coulomb per sec through the cross section of a wire.

Being a great scientist in his achievements, he was also a classical example of an absent-minded professor. They say that during his lectures he often used the rag with which chalk is erased from the blackboard to blow his nose, and there is a story that once, walking along the streets of Paris, he mistook the tonneau of a cab parked at the curb for a blackboard and started writing mathematical formulas on it. When the cab moved, he walked and then ran behind it, determined to finish his derivation. Once, when Napoleon Bonaparte visited the Paris Academy, Ampère did not recognize him, and Napoleon remarked, smiling: "You see, Sir, how inconvenient it is not to see one's colleagues frequently. I never see you at the Tuileries, either, but I know how to force you to come, at least to say good-day to me!" And he invited him to dinner at the Palace the next day. But the next day the chair at the dining table of the Palace was vacant; Ampère had forgotten the invitation!

### THE LAWS OF ELECTRIC CIRCUIT

While Ampère was interested mostly in the magnetic effects associated with electric currents, the German physicist George Simon Ohm, who was at that time a schoolteacher in Cologne, wanted to learn how the strength of the electric current depends on the material of the wire through which it flows and the electric potential which keeps it going. He was using a number of Volta piles which, being connected in sequence, produced various degrees of electric tension, and a galvanometer, first constructed by Ampère, in which the strength of electric current was measured by the deflection it caused in the magnetic needle. Using wires of different lengths and different cross sections made of different metals, he found that the strength of the current is directly proportional to the cross section of the wire, inversely proportional to its length, and depends on the material from which it is made. He also found that for any given wire the strength of the current is proportional to the difference of the *electric potentials* between the two ends given by the number of Volta's piles connected in sequence which drive the current through the wire. The situation is quite similar to the case where one pumps water through a pipe filled with some kind of glass fiber which resists free passage of the liquid. In this case, too, the strength of the water current will increase with the pressure supplied by the pump and with the cross section of the pipe, will decrease with the pipe's length, and will depend on the nature and the amount of the material packed into the pipe that resists the free passage of the water.*

Thus, Ohm introduced the notion of the electric *resistance* of different wires, stating that: the strength of the current is directly pro-

* This analogy is quite consistent with modern views on electric current in metal wires; according to the current it is caused by a flux of the so-called free electrons shouldering their way under the action of electric tension through the tightly packed atoms forming the metal.

portional to the difference of the electric potentials producing the current, and inversely proportional to the resistance of the wire which, in its turn, depends on the material of the wire and is directly proportional to its cross section and inversely proportional to its length. He published his findings in 1827 in a paper entitled "The Galvanic Circuit Investigated Mathematically," which laid the foundation for all future studies of electric circuits. Ohm's law can be expressed by two simple formulas:

$$\text{strength of current} = \frac{\text{electric potential difference}}{\text{resistance of wire}}$$

and

$$\text{resistance of wire} = C\frac{\text{wire cross section}}{\text{wire length}},$$

where $C$ is a constant characteristic for the material used. In his honor the unit of electric resistance is called 1 *ohm,* and is the resistance which produces a current of 1 ampere under the electric potential difference of 1 volt. Sometimes, instead of electric resistance, one speaks about electric conductivity, which is just the reverse of it. Appropriately enough, the unit of electric conductivity is called one *mho,* i.e., the reverse of ohm. Fig. V-8 shows various electric gadgets used in the experimental work with electric phenomena.

## DISCOVERIES OF FARADAY

Michael Faraday (Fig. V-9), who brought classical research on electric and magnetic phenomena to its apotheosis and opened a new era which we now call "modern physics" was born in 1791, near London, into the family of a blacksmith. His family was too poor to keep him in school, and, at the age of 13, he took a job as an errand boy in a bookshop run by a certain Mr. Riebau. A year later Mr. Riebau apprenticed him as a bookbinder for a term of seven years. Faraday was not only binding the books which came to the shop but was also reading many of them, from the first to the last page, which excited in him a burning interest in science. Faraday wrote about his youth:

> Whilst an apprentice I loved to read the scientific books which were under my hands, and, among them, delighted in Marcet's "Conversations in Chemistry," and the electrical treatises in the "Encyclopaedia Britannica." I made such simple experiments as could be defrayed in their expense by a few pence per week, and also constructed an electrical ma-

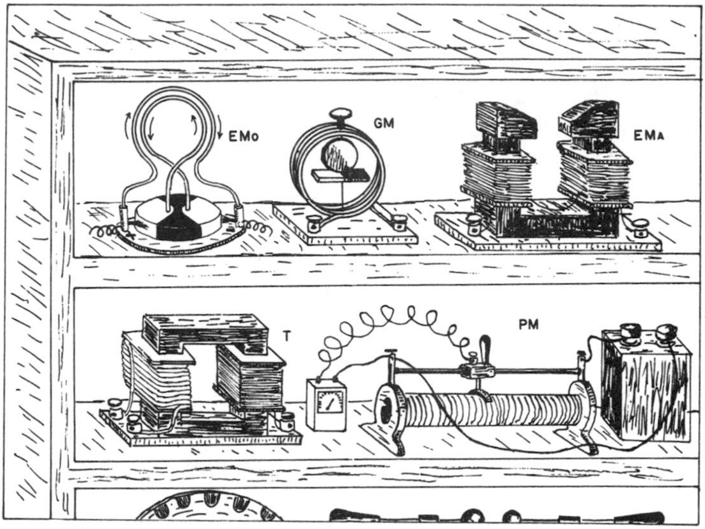


FIG. V–8.

A display of various electric gadgets.

*Electromotor* (E. Mo.). Electric currents passing in opposite directions through two circular wires, one movable and one immovable. The resulting ampere's repulsion forces the movable wire to turn around the axis. However, because of the sliding contact arrangement at the bottom disc, the direction of current reverses and the wire is forced to move on.

*Galvanometer* (*G. M.*). When electric current passes through the coil, a little magnet suspended on a thin wire turns off its normal position. The stronger the current, the larger the angle which is measured by means of a light beam reflected from a little mirror.

*Electromagnet* (E. Ma.) When a DC current passes through the coil, a strong magnetic field appears between two poles.

*Transformer* (*T.*). When AC current of a certain voltage passes through the coil with a small number of windings (*left*), a current of much higher voltage appears in the coil with a large number of windings of thin wire (*right*).

*Potentiometer* (*P. M.*). A current from a battery passes through a rheostat. Varying voltages can be obtained from the wire attached to a gliding contact by moving it to the right or to the left.

chine, first with a glass phial, and afterwards with a real cylinder as well as other electrical apparatus of a corresponding kind.

During the last year of his apprenticeship, when he was just over 20 years old (and when Galvani's and Volta's discoveries were still recent news), he wrote to his old friend Benjamin Abbott:

I have lately made a few simple galvanic experiments, merely to illustrate to myself the first principles of the science. I was going to Knight's to obtain some nickel, and bethought me that they had malleable zinc. I inquired and bought some—have you seen any yet? The first portion I obtained was in the thinnest pieces possible—observe, in a flattened state. It was, they informed me, thin enough for the electric stick, or, as I before called it, De Luc's electric column. I obtained it for the purpose of forming discs, with which and copper to make a little battery. The first

FIG. V–9.
Michael Faraday and his tubes.

I completed contained the immense number of seven pairs of plates!!! and of the immense size of halfpence each!!!!

I, Sir, I my own self, cut out seven discs of the size of halfpennies each. I, Sir, covered them with seven halfpence, and I interposed between, seven, or rather six, pieces of paper soaked in a solution of muriate of soda!!! But laugh no longer, dear A.; rather wonder at the effect this trivial power produced. It was sufficient to produce the decomposition of sulphate of magnesia—an effect which extremely surprised me; for I did not, could not, have any idea that the agent was competent to the purpose. A thought here struck me; I will tell you. I made the communication between the top and bottom of the pile and the solution with copper wire. Do you conceive that it was the copper that decomposed the earthy sulphate—that part, I mean, immersed in the solution? That a galvanic effect took place I am sure; for both wires became covered in a short time with bubbles of some gas, and a continued stream of very minute bubbles, appearing like small particles, ran through the solution from the negative wire. My proof that the sulphate was decomposed was, that in about two hours the clear solution became turbid: magnesia was suspended in it.

This was the discovery of chemical decomposition by means of electric current, or *electrolysis,* as Faraday called it. During the following years of work on this phenomenon Faraday discovered two basic laws which now carry his name. The first law of Faraday states that: *for a given solution the amount of material deposited (or liberated) on the electrodes is proportional to the total amount of electricity* (i.e., current's strength multiplied by time) *which passes through the solution.* This means that the charged molecules (which were later called *ions*) that carry electricity through the liquid solutions have a strictly defined electric charge. (Fig. V-10.)

According to the second law of Faraday: *the monovalent ions of different substances also carry equal amounts of electricity, while two-three-etc.-valent ions carry correspondingly larger charges.* This proves the existence of a universal unit of electric charge which in the days of Faraday was known only as being attached to the various atoms, but later was detected in the form of free electrons flying through space.

But, having discovered electrolysis, Faraday still had to look for a job since his position in the bookshop was expiring in a few months. His biggest aspiration was to work with Sir Humphry Davy, the celebrated chemist, whose lectures Faraday was attending during his apprenticeship. He copied his notes on Davy's lectures in calligraphic fashion, supplied them with masterfully executed drawings, and sent the elegantly bound volume to Sir Humphry with an application for a job in his laboratory. When Davy asked the advice of one of the governors of the Royal Institution of Great Britain, of which he was a director, about the employment of a young bookbinder, the man said: "Let him wash bottles! If he is any good he will accept the work; if he refuses, he is not good for anything."

Faraday accepted, and remained with the Royal Institution for the remaining 45 years of his life, first as Davy's assistant, then as his collaborator, and finally, after Davy's death, as his successor.

Apart from numerous publications in scientific magazines, the most remarkable document pertaining to his studies is his *Diary*, which he kept continuously from the year 1820 to the year 1862. This was recently (1932) published by the Royal Institution in seven

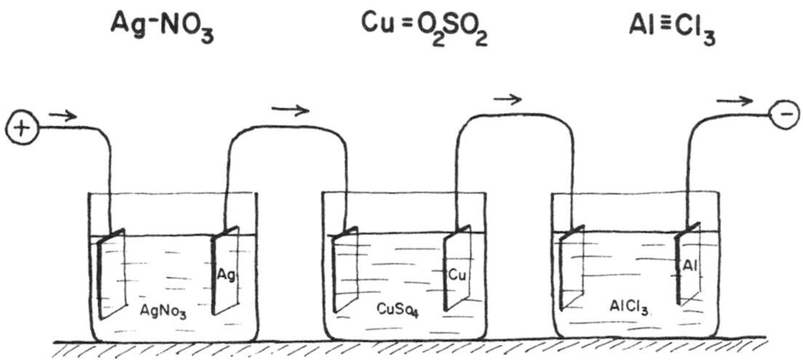


FIG. V–10.

Demonstration of Faraday's laws of electrolysis. If one sends an electric current through the solutions of silver nitrate, copper sulfide and aluminum chloride, the metals will be deposited on the negative electrodes. The amount of deposited metals was found to be proportional to the amount of electricity which passed through (first law of Faraday). It was also found that, if the amount of deposited silver is 108 gm (atomic weight of silver), the amount of deposited copper was only 31.7 gm (a *half* of the atomic weight of copper), and the amount of deposited aluminum only 9 gm (*one third* of the atomic weight of aluminum). Since the same amount of electricity passed through all three containers, one concludes that copper ion carries twice the electric charge carried by silver atom, and the ion of aluminum three times as much. It agrees with chemical valency of these three metals as evidenced by their formulas given at the top of the diagram. This is the second law of Faraday.

thick volumes containing a total of 3,236 pages, with a few thousand marginal drawings. We quote from that *Diary* the description, in Faraday's own words, of his probably most important discovery, that of *electromagnetic induction*:

*Aug. 29th, 1831*

1. Expts. on the production of Electricity from Magnetism etc. etc.

2. Have had an iron ring made (soft iron), iron round and ⅞ inch thick and ring 6 inches in external diameter. [Fig. V-11.] Wound many

coils of copper wire round one half, the coils being separated by twine and calico—there were 3 lengths of wire each about 24 feet long and they could be connected as one length or used as separate lengths. By trial with a trough each was insulated from the other. Will call this side of the ring A. On the other side but separated by an interval was wound wire in two pieces together amounting to about 60 feet in length, the direction being as with the former coils; this side call B.

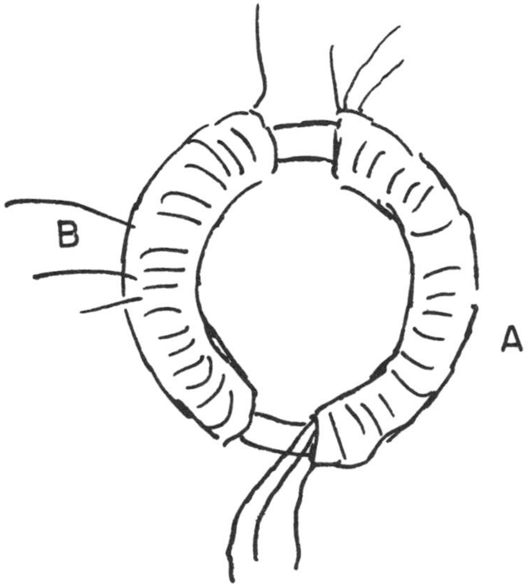


FIG. V–11.

The drawing from Faraday's *Diary* illustrating his discovery of electromagnetic induction. The appearance or disappearance of electric current in coil *A* induces a short-lasting electric current in coil *B*.

3. Charged a battery of 10 pr. plates 4 inches square. Made the coil on B side one coil and connected its extremities by a copper wire passing to a distance and just over a magnetic needle (3 feet from iron ring). Then connected the ends of one of the pieces on A side with battery; immediately a sensible effect on needle. It oscilated and settled at last in original position. On breaking connection of A-side with Battery again a disturbance of the needle.

Thus, electric current in one coil can induce a current in another coil placed nearby, just as an electric charge on one body induces electric polarization of another nearby body. But, whereas in the case of electric polarization the effect is static and lasts as long as the two bodies remain near one another, the induction of electric current is a dynamic process, and the current in the second coil exists only during the periods while the current in the first coil increases from zero to its normal value or decreases from that value back to zero.

Less than three months from the date of this epoch-making discovery, Faraday made another important step in his studies of the relation between electricity and magnetism. Here is how it was done, according to his *Diary*:

*Oct. 17, 1831*

56. Built a cylinder, hollow, of paper, covered with 8 helices of copper wire going in the same direction and containing the following quantities

| | | f. | in. |
|---|---|---|---|
| 1 or outermost | — | 32 | 10 |
| 2 | — | 31 | 6 |
| 3 | — | 30 | |
| 4 | — | 28 | |
| 5 | — | 27 | |
| 6 | — | 25 | 6 |
| 7 | — | 23 | 6 |
| 8 or innermost | — | 22 | |
| | | 220 feet exclusive of projecting ends | |

all separated by twine and calico. The internal diameter of paper cylinder was 13/16 of inch in diameter, the external diameter of whole 1½ inches and the length of copper helices (as a cylinder) 6½ inches.

57. Expts. with 0. The 8 ends of the helices at one end of the cylinder were cleaned and fastened together as a bundle. So were the 8 other ends. [Fig. V-12.] These compound ends were then connected with the Galvanometer by long copper wires—then a cylindrical bar magnet ¾ inch in diameter and 8½ inches in length had one end just inserted into the end of the helix cylinder—then it was quickly thrust in the whole length and the galvanometer needle moved then pulled out and again the needle moved but in the opposite direction. This effect was repeated every time the magnet was put in or out and therefore a wave of Electricity was so produced from mere approximation of a magnet and not from its formation in situ.

58. The needle did not remain deflected but returned to its place each time. The order of motions were inverse as in former expts.—the motions

were in the direction consistent with former expts., i.e. the indicating needle tended to become parallel with the exciting magnet, being on the same side of the wire and poles of the same name in the same direction.
59. When the 8 helices were made one long helix the effect was not so strong on the galvanometer as before, probably not half so strong. So that it is best in pieces and combined at the end.
60. When only one of the 8 helices was used it was least powerful—hardly sensible.

Here again the induction of electric current in the coil was a *dynamic* phenomenon, and the current existed only while the magnet was pushed in or pulled out from the coil. In the days of Faraday the idea that magnetism must produce electric current insofar as electric current produces magnetism was hanging in the air and many physicists were trying to observe this effect. But, being misled by the analogy with electrostatic induction, they tried only the *static* configurations of magnets and wires, such as a magnetized bar with a wire wound around, which stubbornly refused to produce any sparks when the two ends were brought together. It is due to the genius of Faraday, or maybe to the tremendous amount of experimentation which he was carrying on day in and day out, that it became clear that the production of an electric current is a *dynamic* process and requires either a change in the strength of another current or a change in the position of the magnet. The only other physicist who got the same idea was an American named Joseph Henry, but he hesitated so long in announcing it that the priority of that discovery went to the man on the other side of the Atlantic Ocean.

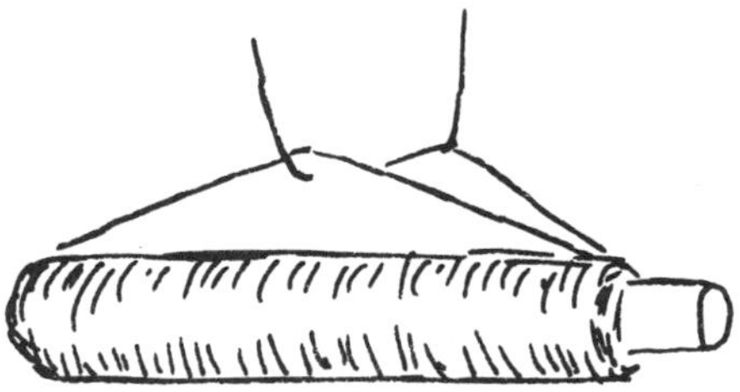

FIG. V–12.
A drawing from Faraday's *Diary*, illustrating an experiment in which a magnet induces an electric current in a coil when the magnet is pushed in or pulled out from that coil.

The inquiring mind of Michael Faraday did not stop with untangling the hidden relation between electricity and magnetism. He also wanted to know if magnets can affect optical phenomena. This culminated in the discovery of the rotation of the plane of polarization of light (cf. p. 87) passing through the transparent materials placed in a magnetic field. Here again we let Faraday himself tell about that discovery:

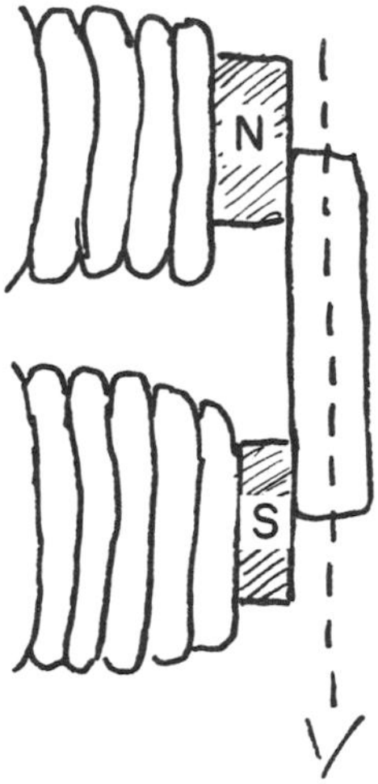


FIG. V–13.

A drawing from Faraday's *Diary,* illustrating his discovery of the effect of the magnetic field on light. When polarized, light propagates along the lines of magnetic force and the plane of polarization turns by an angle proportional to the strength of the field.

*13 Sept. 1845.*

7498. Today worked with lines of magnetic force, passing them across different bodies (transparent in different directions) and at the same time passing a polarized ray of light through them and afterwards examining the ray by a Nichol's Eyepiece or other means. The magnets were Electro magnets, one being our large cylinder Electro magnet and the other a temporary iron core put into the helix on a frame—this was not nearly so strong as the former. The current of 5 cells of Grove's battery was sent through both helices at once, and the magnets were made and unmade by putting on or stopping off the electric current.

After describing several negative results in which the ray of light was passed through air and several other substances, Faraday in the same day's entry:

7504. *Heavy glass.*
A piece of heavy glass which was 2 inches by 1.8 inches, and 0.5 of an inch thick, being a silico borate of lead, and polished on the two shortest edges, was experimented with. It gave no effects when the *same magnetic poles* or the *contrary* poles were on opposite sides (as respects the course of the polarized ray)—nor when the same poles were on the same side, either with the constant or intermitting current—BUT, when contrary magnetic poles were on the same side [Fig. V-13], there *was an effect produced on the polarized ray,* and thus magnetic force and light were proved to have relation to each other. This fact will most likely prove exceedingly fertile and of great value in the investigation of both conditions of natural force.

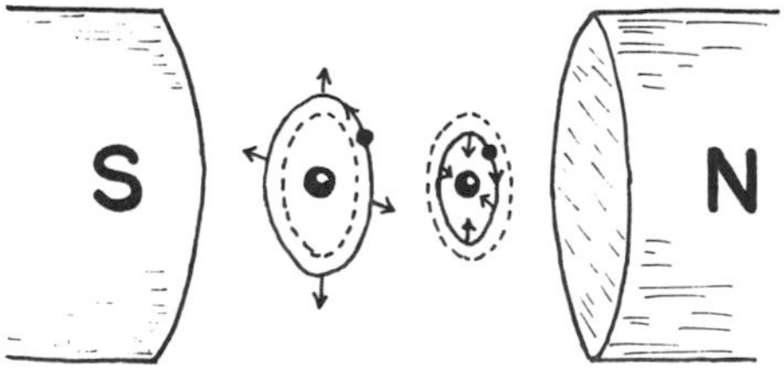


FIG. V–14.
Explanation of the Faraday effect. The force acting on an atomic electron moving in a magnetic field depends on the direction of its motion. In the case of counterclockwise rotation (*left*) the force increases the radius of the orbit and decreases the frequency. In the case of clockwise rotation (*right*) the force decreases the radius and increases the frequency. Interacting with light, these two types of electron motion rotate the plane of polarization.

And it certainly did! The "Faraday effect," or the rotation of the plane of polarization of light propagating along the magnetic lines of force, demonstrates the intimate relation between light waves, which are very short electromagnetic waves, and the electric currents within individual atoms. These tiny electric circuits, the existence of which was first suggested by Ampère (p. 136), are interpreted today as due to rotation of atomic electrons around the central nucleus. Consider two identical atoms placed into a magnetic field in such a way that one electron circles in a clockwise direction and the other in a counterclockwise direction (Fig. V-14). In one case the magnetic field will exert on the moving electron a force directed toward the nucleus, while in the other case the force will be in the opposite direction. Thus, in the first case the diameter of the electron's orbit will thrink and its rotation frequency will increase, while in the second case the opposite will take place. This difference of behavior between clockwise and counterclockwise intra-atomic currents will affect electromagnetic (light) wave propagation through the material and it can be shown that the result will be the rotation of the plane of polarization observed by Faraday.

Being persuaded that all the phenomena observed in the physical world are in one way or another interrelated, Faraday also tried to establish a relation between electromagnetic forces and the forces of Newtonian gravity. In 1849 he wrote in his laboratory *Diary*:

> Gravity. Surely this force must be capable of an experimental relation to electricity, magnetism, and other forces, so as to build it up with them in reciprocal action and equivalent effect. Consider for a moment how to set about touching this matter by facts and trial.

But the numerous experiments he undertook to discover such a relation were fruitless, and he concludes that part of his *Diary* by the words:

> Here end my trials for the present: The results are negative. They do not shake my strong feeling of the existence of a relation between gravity and electricity, though they give no proof that such a relation exists.

A century later another genius was breaking his head for many decades in an attempt to develop the so-called "unified field theory" which would bring together electromagnetic and gravitational phenomena. But, just as did Michael Faraday, Albert Einstein died without being able to accomplish that task.

## ELECTROMAGNETIC FIELD

Impressive as Faraday's experimental discoveries were, they are matched by his theoretical ideas. Having had very little education and having known practically no mathematics, Faraday could not be what is usually called a theoretical physicist. But the fact is that, for conceiving a theoretical picture of a puzzling physical phenomenon, a knowledge of intricate mathematics is often quite unnecessary and sometimes even harmful. The explorer may easily be lost in the jungles of complicated formulas and, as a Russian proverb says, "cannot see the forest for the trees." Before Faraday, electric and magnetic, as well as gravitational, forces were usually considered as acting across empty space, separating the interacting objects. To his simple mind, however, such an "action in distance" did not seem to make physical sense, and seeing a load being moved from place to place he wanted also to see a rope which drags it or a stick with which it is being pushed. Thus, in order to visualize the forces acting

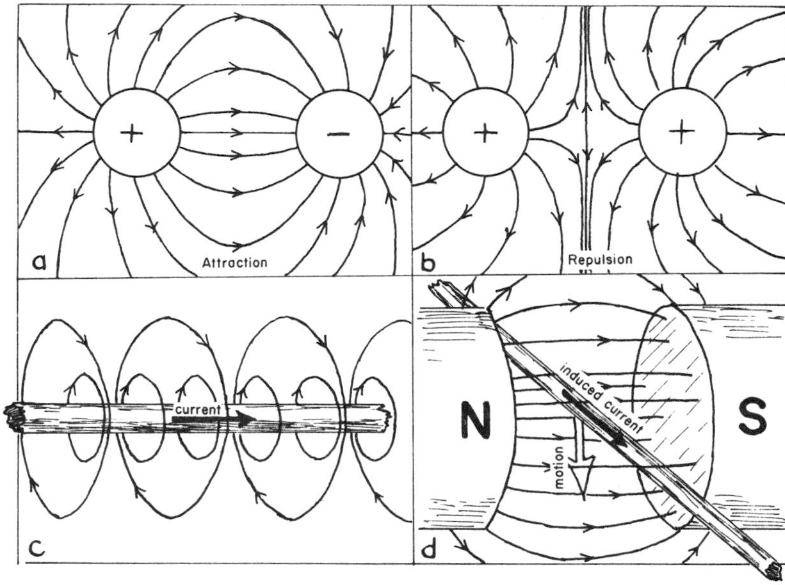


FIG. V–15.

Faraday's lines of force corresponding to different types of electromagnetic interactions: (a) Electric lines of force between two opposite electric charges. (b) Electric lines of force for like electric charges. (c) Magnetic lines of forces around wire carrying electric current. (d) An electric current induced in a wire moving (*white arrow*) through magnetic lines of forces. Small *black arrows* indicate the conventional direction of lines of forces: from positive charge to negative, and from north to south pole.

between electric charges and the magnets, he had to imagine the space between them as being filled by "something" which can pull or push. He spoke about something similar to rubber tubes which stretch between two opposite electric charges or magnetic poles (Fig. V-15a) and pull them together. In case of charges or poles of the same sign the rubberlike tubes run in different ways (Fig. V-15b) and push them apart. The direction of these Faraday tubes in the case of magnets can be detected by spreading fine iron filings on the glass plate on which the magnet is placed. The fillings will be magnetized and orient themselves in the direction of magnetic forces acting along the tubes, producing patterns like those shown in Plate III. In the case of an electric field, similar results can be obtained by using electric polarization, but the experiment is more difficult to perform. According to Faraday, electric and magnetic tubes were also responsible for various electromagnetic phenomena. When a current flows through an electric wire, it becomes surrounded by circular tubes (Fig. V-15c), which exert tension on the magnetic needle orienting it in a proper way. When a conducting wire is moved in respect to a magnet (or vice versa), it crosses magnetic tubes (Fig. V-15d) and, as a result, a current is induced in it.

These ideas of Faraday were in a way rather naïve and to a large extent qualitative, but they spelled a new era in the development of physics. Mysterious forces acting over large distances between bodies were replaced by "something" continuously distributed throughout all the space between and around them, "something" that could be ascribed a definite value at any single point. This introduced into physics the idea of a "field of forces," or simply a "field," be it a case of electric, magnetic, or gravitational interactions. The forces between material objects separated by empty space could now be considered as the result of "close quarter" interactions between the fields surrounding them.

The task of giving to the ideas of Faraday a quantitative mathematical formulation fell on the shoulders of a famous Scotsman named James Clerk Maxwell (Fig. V-16), who was born in Edinburgh just a couple of months after Faraday announced his discovery of electromagnetic induction. In contrast to Faraday, Maxwell was a very good mathematician. At the age of 10, Maxwell went to school in the Edinburgh Academy and was forced to devote much of his time to the study of irregular Greek verbs and other branches of "humanistic sciences." But he would rather have done mathematics and his first success in it was, according to his own words, "making a tetra hedron, a dodeca hedron, and two more hedrons that I don't know the right names for." At the age of 14 he won the Academy's mathematics medal for a paper showing how to construct a perfect oval curve with pins and thread. A few years later Maxwell presented to the Royal Society two papers, one "On the Theory of Rolling Curves" and another "On the Equilibrium of Elastic Solids." Both papers were read before the Society by somebody else because "it was not thought proper for a boy in a round jacket to mount the rostrum there." In 1850, at the age of 19, Maxwell enrolled as a student at Cambridge University, took his degree four years later, and in 1856 was appointed to the chair of Natural Philosophy of Marischal College in Aberdeen, where he remained until he was called back to Cambridge in 1874 as the first director of the then newly erected Cavendish Laboratory.

Although Maxwell's early interests were entirely in the field of pure mathematics, he soon became vividly interested in the application of the mathematical method to various problems of physics. He

$$\operatorname{div} \vec{E} = S \; : \; \operatorname{div} \vec{H} = 0$$

$$\operatorname{curl} \vec{E} = -\frac{1}{c}\,\frac{\delta \vec{H}}{\delta t}$$

$$\operatorname{curl} H = \frac{1}{c}\,\frac{\delta \vec{E}}{\delta t} + S\,\frac{\vec{u}}{c}$$

FIG. V–16.
James Clerk Maxwell and his equations for the electromagnetic field.

made very important contributions to the kinetic theory of heat (cf. Chap. IV), but his undoubtedly most important work was the mathematical formulation of Faraday's ideas concerning the nature and laws of the electromagnetic field. Generalizing the empirical facts that the changing magnetic fields induce electromotive forces and electric currents in the conductors, while the changing electric fields and flowing electric currents produce magnetic fields, he wrote famous equations now carrying his name, which connect the rate of change of the magnetic field with space distribution of the electric field and vice versa. Using Maxwell's equations and knowing the distribution of magnetized bodies, charged conductors, and electric currents, one can calculate in all details the electromagnetic field surrounding them and its change in time. Maxwell has shown that, although electric and magnetic fields are usually "anchored" on electrically charged and magnetized bodies, they can also exist and propagate through space in the form of free electromagnetic waves. To make that point clear let us consider two spherical conductors, one of them charged

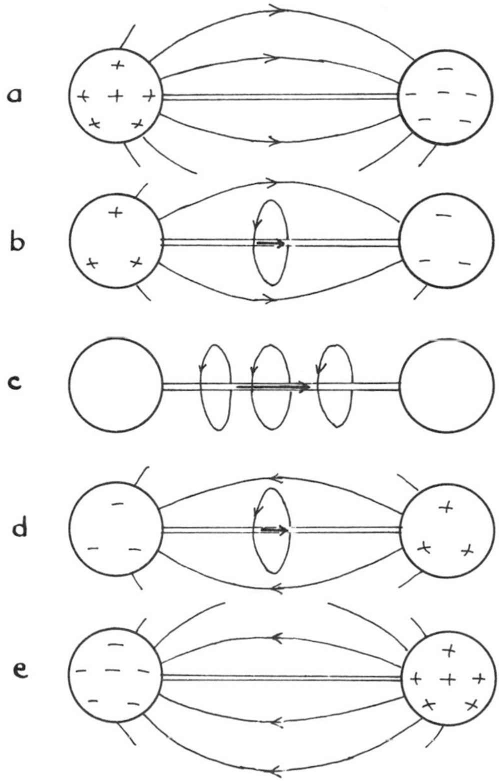


FIG. V–17.

Electromagnetic oscillations between two conductors in which the energy of the electric field (a) periodically transforms into the energy of magnetic field (c), and vice versa (e).

with positive and the other with negative electricity (Fig. V-17a). In the space surrounding these two spheres there exists a static electric field which stores the electric energy of the charges more or less in the same way as a tightly wound spring stores the mechanical energy. If we connect two wires attached to both spheres, electric current will start flowing from one sphere to the other and their electric charges and the electric field surrounding them will begin to diminish rapidly (Fig. V-17b), until they finally vanish (Fig. V-17c). However, electric current flowing through the wire produces a magnetic field around it, and, at the moment when the electric field becomes zero, the entire energy of the system is stored in this magnetic field. But the process does not stop at this, and the electric current in the wire continues to flow, though with decreasing intensity, recharging the two spheres with the electricity of opposite signs (Fig. V-17d). The energy of the magnetic field goes back into the energy of the electric field, and we finally get to the stage with zero current and the two spheres charged with the initial amounts of electricity but of opposite sign (Fig. V-17e). Now the process starts again but in the opposite direction, and electric oscillations proceed to and fro until the gradual loss of energy due to the heating of the wire carrying the current brings them to a stop. The situation is very similar to that of a pendulum where the kinetic energy of motion in the middle of each swing goes over into potential energy at the two extremes.

Using his equations, Maxwell was able to prove that the oscillating electromagnetic field of the above-described type propagates through the space surrounding the oscillator in the form of waves carrying away the energy. Since the electric lines of force lie in the plane passing through the wire, while magnetic lines are perpendicular to it, electric and magnetic vectors in the propagating wave are perpendicular to one another as well as to the direction of propagation (Fig. V-18). The existence of such waves was experimentally confirmed in 1888 by the German physicist Heinrich Hertz soon after Maxwell's paper predicting them, and this led to the development of radio-communication techniques which represent today one of the major branches of industrial civilization.

We want to discuss now in some detail one of the important points of Maxwell's theory, namely the calculation of the propagation velocity of electromagnetic waves. Considering the interaction of electric and magnetic fields, one encounters the question concerning the units to be used for measuring various electromagnetic quantities. We have seen before that the unit of electric charge is defined as a charge repelling an equal charge placed 1 cm away with a force of 1 dyne. Correspondingly, the unit of the electric field must be defined as the field which acts with a force of 1 dyne on a unit of electric charge placed in it. Similar definitions are given to the unit magnetic pole and to the unit of magnetic field. But what happens if we consider the phenomena involving both electricity and magnetism, such as, for example, a magnetic field produced by an electric current?

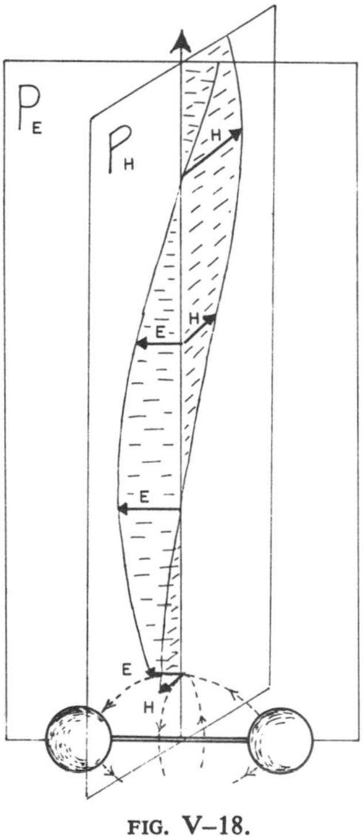


FIG. V–18.
Formation of a propagating electromagnetic wave as the result of the oscillation of charge between two conductors.

Suppose we investigate the action of the electric current on a magnetic pole placed 1 cm from the wire. We can define the unit of electric current as a current carrying the above-defined unit of electric charge per sec. But if we do so, the force with which the magnetic field produced by that current acts on a unit pole placed 1 cm away will not necessarily be 1 dyne, and, indeed, is not by a long shot. Alternatively, we can define a unit current as a current producing a magnetic field which acts with a force of 1 dyne on a unit pole placed 1 cm away. But if we do so, the amount of electricity flowing through the wire carrying a unit electric current will not be equal to the above-defined electrostatic unit of charge. Instead of selecting one possible definition and rejecting the other, physicists prefer to use both, introducing a constant factor for translating one system of units into another, the situation being rather similar to that existing in heat measurements where either 1 calorie or 1 erg (with the ratio of $4.2 \times 10^7$) could be used. The unit of electric charge defined by using the Coulomb law of electric attraction and repulsion (the first of the above two definitions) is known as *electrostatic unit* (esu), whereas the unit of charge defined by Oersted's law of the action of electric current on a magnetic pole is known as *electromagnetic unit* (emu). One electromagnetic unit is equal to $3 \times 10^{10}$ electrostatic units, so that a current carrying 1 electrostatic unit per sec exerts a force of only $\frac{1}{3 \times 10^{10}}$ dynes on a unit pole 1 cm away, while two bodies charged by one electromagnetic unit each, and placed 1 cm apart, will repel one another with a force of $3 \times 10^{10}$ dynes.

Since, in writing his equations, Maxwell had to use electrostatic units for electric fields, and electromagnetic units for magnetic fields, the factor $3 \times 10^{10}$ crept into the formulas containing an electric field on one side of the equation, and a magnetic field on the other. And the application of these equations for describing the propagating electromagnetic waves led to the conclusion that the propagation velocity is numerically equal to the ratio of the two units, i.e., $3 \times 10^{10}$ cm per sec. *And,* lo and behold, this figure coincided exactly with the velocity of light in a vacuum which was measured by various methods long before Maxwell was born! A-ha! probably thought Maxwell, this must mean that light waves are actually electromagnetic waves of very short length, and this thought led to the development of an important branch of physics: *the electromagnetic theory of light.* We now visualize the interaction of light and matter, including the phenomena of emission, propagation, and absorption of light, as the result of forces acting between the propagating short electromagnetic waves and the tiny electrically charged particles, the electrons, busily buzzing around the positively charged atomic nuclei. And, using Maxwell's equations, one can explain in the most minute detail all the phenomena and laws of optics.

The numerical coincidences between seemingly unconnected physical quantities, such as the ratio of electrostatic and electromagnetic units on one side, and the velocity of light on the other, often led to fundamental new discoveries and broad generalizations in physics. Later in this book we will learn that another such coincidence between two physical constants, one pertaining to emission of light and heat waves by hot bodies, and another to emission of electrons from the surfaces illuminated by ultraviolet rays, turned out to be of paramount importance in the development of the quantum theory.

# CHAPTER VI *Relativistic Revolution*

AS DESCRIBED in the previous chapter, the idea of an all-penetrating universal medium filling all the space between and within all material bodies was firmly established in physics by the end of the 19th century. Under the name of Huygens' world ether, this medium served as the substratum for the propagation of light waves; under the name of Faraday's tubes, it was responsible for the forces between electrically charged and magnetized bodies. The work of Maxwell led to a synthesis between these two hypothetical media, showing that light is a propagating electromagnetic wave, and provided an elegant mathematical theory which tied together all the phenomena involving light, electricity, and magnetism. But, in spite of all this success, physicists found it impossible to describe the properties of this mysterious universal medium in terms used for the description of such familiar material media as gases, solids, and liquids, and all attempts in this direction led to violent contradictions.

## THE CRISIS OF CLASSICAL PHYSICS

Indeed, the phenomenon of polarization of light proved beyond any doubt that one deals here with transversal vibrations in which material moves to and fro perpendicularly to the direction of propagation. However, transversal vibrations can exist only in solid materials, which, in contrast to liquids and gases, resist any attempt to change their shapes, so that light ether had to be considered as a solid material. If so, and if world ether fills all the space around us, how can we walk and run on the ground, and how can the planets circle around the sun for billions of years without encountering any resistance whatsoever?

The famous British physicist Lord Kelvin tried to solve this apparent contradiction by ascribing to world ether properties similar to those possessed by shoemaker's glue or sealing wax. These substances possess the property known as *plasticity,* and, while breaking like a piece of glass under the action of a rapidly applied strong force, these substances would flow as liquids under much weaker forces (such as their own weight) acting over a long period of time. He argued that in the case of light waves, where the force changes its direction a million billion times per second, world ether can behave as elastic rigid material, while in the case of the much more leisurely motions of people, birds, planets, or stars, it may give way with practically no resistance. But, if Faraday tubes are the tensions and stresses in the world ether, permanent magnets and static electric charges would not exist for any observable length of time, since the stresses would be rapidly relaxed by the plastic changes in that mysterious material. It is very easy to criticize people for coming to erroneous conclusions after one knows the correct answer, but it is really surprising that the great physicists of the last century did not realize that, if world ether exists, it would have properties entirely different from those of the ordinary material bodies familiar to us. In fact, it was well known that compressibility of gas, fluidity of liquids, elasticity of solids, and all other properties of ordinary material bodies are due to their molecular structure and are the result of the motion of molecules and the forces acting between them. It seems that nobody, except perhaps a Russian chemist, Dmitri Mendeleev, who ascribed to world ether the atomic number zero in his Periodic System of Elements, ever thought that world ether had a molecular structure of its own, and anyway such a hypothesis would only have led to further complications. If the forces between the magnets and charged bodies and the propagation of light through space must be explained by the existence of some kind of substratum, this does not have to resemble in any way the ordinary material substances we are familiar with. But, the human mind is too often much too restricted by traditional thinking, and it took the genius of Einstein to throw the old-fashioned and contradictory world ether out of the window and to substitute for it the extended notion of the electromagnetic field, to which he ascribed physical reality equal to that of any ordinary material body.

### THE VELOCITY OF LIGHT

The first attempt to measure the velocity of light was made by Galileo, who went one evening with his assistant into the countryside carrying two lanterns supplied with shutters. Placing themselves as far apart as they could stand yet still able to see each other, they carried out an experiment in which the assistant flashed his lantern as soon as he saw Galileo's lantern flash. The delay in arrival of the return signal would indicate that light propagates with a final velocity and should permit measurement of it. The result of this experiment was, however, quite negative because, as we now know, light propagates with such a tremendous velocity that the expected delay would be not more than one hundred-thousandth of a second. More than two centuries later, Galileo's experiment in a much improved form was repeated by the French physicist, Armand Hippolyte Fizeau, who used the arrangement shown in Figure VI-1a. It consisted of a pair of cogwheels set at the opposite ends of a long axis. The wheels were positioned in such a way that the cogs of one were opposite the intercog openings of the other so that the light beam from the source on the right could not be seen by the eye on the left, no matter how the axis was turned. However, when the wheels were set in fast rotation and were turning so fast that they moved by half the distance between the neighboring cogs during the interval of time taken by light to propagate from one wheel to the other, the light was expected to pass through without being stopped. The length of the light path between the two wheels was intentionally lengthened by use of three mirrors, one of them placed far away, as shown in the figure. Spinning the wheels at the rate of a few thousand revolutions per minute, Fizeau was gratified to notice that light passed unobtrusively through the system and, putting the observed figures together, concluded that the velocity of light is rather exactly $3 \times 10^{10}$ cm per sec. This figure coincided with that obtained only about three decades after Galileo's death by the Danish astronomer Olaus Roemer, from his observations on the apparent delay of the eclipses of Jupiter's moons when that planet was at different distances from the earth.

Fizeau's method could be used only for the measurement of the velocity of light in air (which is practically the same as in a vacuum) since the mirror used for lengthening the path of light had to be placed at quite a distance away, to make the effect observable. His friend and collaborator, Jean Foucault (both were born in 1819 and were the Castor and Pollux of French science), succeeded in

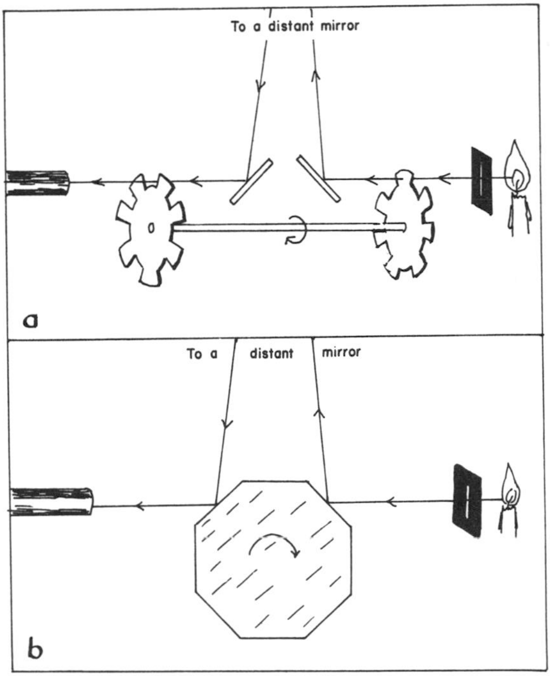


FIG. VI–1.
Fizeau's (a) and Foucault's (b) methods for measurement of the velocity of light.

shortening this distance by substituting a rotating mirror instead of cogwheels. His arrangement, which is shown in Figure VI-1b and is self-explanatory, permitted the shortening of the optical pass to just a few meters, so that he could let light go all the way through water, or any other transparent material. Conducting this experiment, he found that the velocity of light in material bodies is *less* than its velocity in vacuum, thus giving a belated though still ponderous support to Huygens' views as against Newton's. As the wave theory of light predicted, the velocity of light in water, glass, etc., turned out to be exactly its velocity in vacuum divided by the refractive index of the material in question.

### VELOCITY OF LIGHT IN A MOVING MEDIUM

Having the methods for the precise measurement of the velocity of light firmly in hand, physicists of the 19th century were busy conducting various experiments on the propagation of light with the hope of throwing some light on the properties of world ether, this mysterious medium through which light waves were supposed to propagate. A very important experiment, the full significance of which was, however, not realized until after Einstein's first publication, was carried out in 1851 by Fizeau. The idea was to see how the velocity of light is influenced by the motion of the medium through which it propagates. In the case of sound waves traveling through the air, the velocity of propagation is, of course, directly influenced by the motion of the air masses, and the velocity of sound traveling with or against the wind is increased or decreased by the amount equal to the velocity of wind. No doubt about that, but would the same be true in the case of light propagating through a moving medium? To settle this question, Fizeau decided to measure the velocity of light propagating along a tube with water passing swiftly through it. Would the velocity of water be added to or subtracted from the velocity of light in a vacuum in this case? The expected change of the velocity of light in such an experiment is, of course, very small since the highest velocity of water-flow which could technically be achieved is very very small as compared with the velocity of light. Thus, direct measurements of the velocity of light in this case, either by Fizeau's or Foucault's methods described in the previous paragraph, would not show any difference. But, since in this case all one wants to know is only the *difference* between the velocities of light in moving and in still water, one can use here a much more precise method based on the interference of two light rays. The principle involved in that experiment is shown in Figure VI-2.

Monochromatic light from a mercury lamp $L$ falls on a glass plate $P_1$ covered with a very thin layer of silver which is just thick enough to reflect half of it, while another half passes through and is reflected by a mirror $M_1$. Thus one gets two parallel beams of light of equal intensity, the vibrations of which are synchronized just as in the case of Young's experiment described in Chapter III. These two beams of light pass through two tubes $T_1$ and $T_2$ and then are brought together again by means of glass plate $P_2$ and mirror $M_2$. If the water in both pipes is at rest, the two beams come in phase into the eye $E$ of the observer (i.e., crest to crest and hollow to hollow) and add up to the original intensity. If, however, the water in the pipes moves in opposite directions and "drags" light waves with it, the waves of the

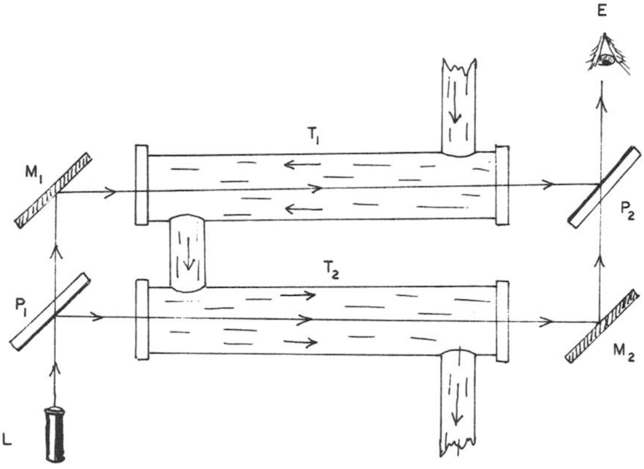


FIG. VI–2.
Fizeau's experiment for observing the change of the velocity of light propagating through a moving medium.

lower beam will arrive at $E$ sooner than the corresponding waves in the upper beam, and, if the difference is just one half of the wave length, destructive interference (crest to hollow, and hollow to crest) will take place. Let us make a rough estimate of how fast water must move in $P_1$ in order to produce this difference of phase. The length of the pipes in Fizeau's experiment was 1.5 meters or 150 cm, and the wave length used was about $0.5\mu$ ($5 \times 10^{-5}$ cm) so there was a train of $3 \times 10^6$ waves in the length of the pipe. In order to change this number by one half of the wave length (from 3 million to 3½ million), the velocity of light in the pipe with moving water must be increased or decreased by a fraction $\frac{0.5}{3 \times 10^6} = 1.7 \times 10^7$. Since the velocity of light in water is about $2 \times 10^{10} \frac{\text{cm}}{\text{sec}}$, the velocity of the water-flow necessary to get this result must be about $\frac{2 \times 10^{10}}{1.7 \times 10^7} = 1{,}000 \frac{\text{cm}}{\text{sec}} = 10 \frac{\text{meters}}{\text{sec}}$, which is rather high but is an achievable flow-velocity of water through pipes. Thus it is possible to notice the expected changes of the velocity of light by observing interference fringes in that experiment.

Carrying out the exact measurements with varying velocity of water-flow, Fizeau arrived at a result which was intermediate between the two expected possibilities. The velocity of light in flowing water *was different* from that in still water, but the difference *was less* than the velocity of water-flow. He found from the observed shift of interference fringes that the velocity of light propagating in the direction of water-flow is increased by 44% of the velocity of water, while the velocity of light propagating in the opposite direction is decreased by the same amount. When other liquids were used, the drag exerted on the light propagating through them was found to have different numerical values, and it turned out that the velocity of light in a moving fluid can generally be expressed by an empirical formula,*

$$V = \frac{c}{n} \pm \left(1 - \frac{1}{n^2}\right)v$$

where $n$ is the refractive index of the fluid in question and $v$ the flow velocity. Neither Fizeau nor anybody else at that time could figure out what it could possibly mean and the case rested until half a century later when Einstein showed that the mysterious empirical formula is a direct result of the theory of relativity.

### THE VELOCITY OF LIGHT ON THE MOVING EARTH

In the year 1887, when Einstein was 8 years old, the American physicist, A. A. Michelson, and his assistant, E. W. Morley, carried out another remarkable experiment. If Fizeau could observe the influence of a fast-moving water stream on the light propagating through it, one should be able to observe the effect of the earth's

* An empirical formula is a formula which is not derived mathematically on the basis of some particular theory, but which is simply adjusted so as to fit empirical data.

PLATE I (Upper) Newton's rings. (Lower) Double refraction in a crystal of Iceland spar. *Courtesy Dalton Kurts, University of Colorado*

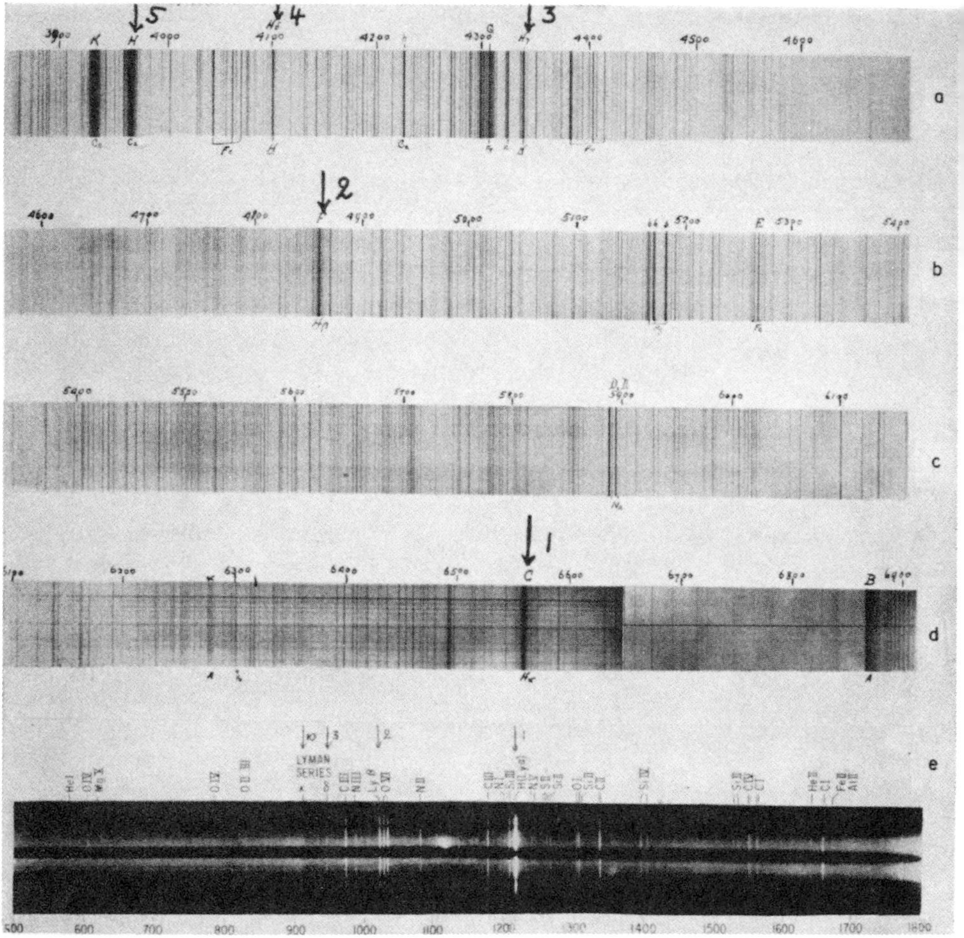

PLATE II (a,b,c,d) The visible part of the solar spectrum taken by 13-foot heliospectrograph. Numbered lines correspond to the Balmer spectrum of hydrogen. *Courtesy Mt. Wilson Observatory.* (e) Far ultraviolet spectrum of the sun taken from a high-flying rocket. Numbered lines correspond to the Lyman spectrum of hydrogen. *Courtesy Naval Research Laboratory*

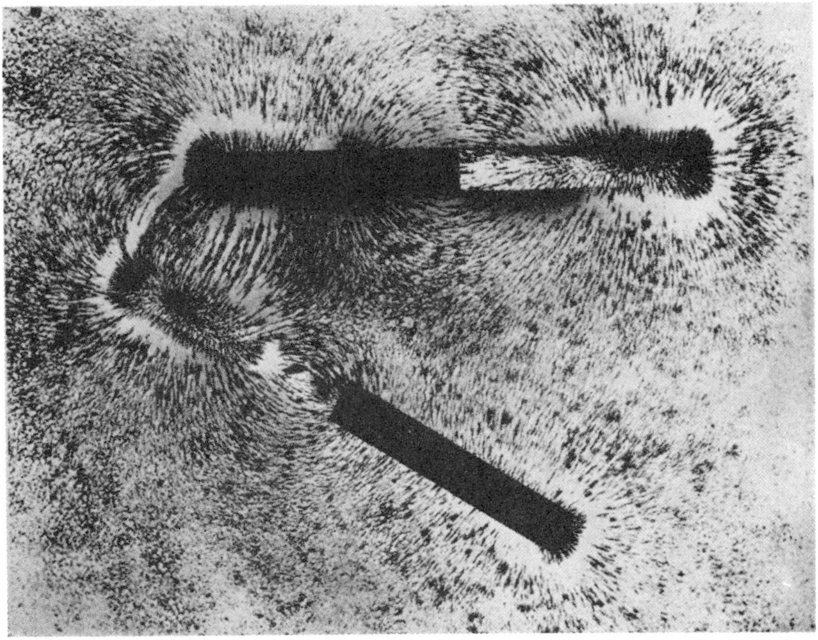

PLATE III Magnetic lines of forces between two magnets oriented in the opposite (upper) and in the same (lower) direction. *Courtesy R. Conklin, formerly at the University of Colorado*

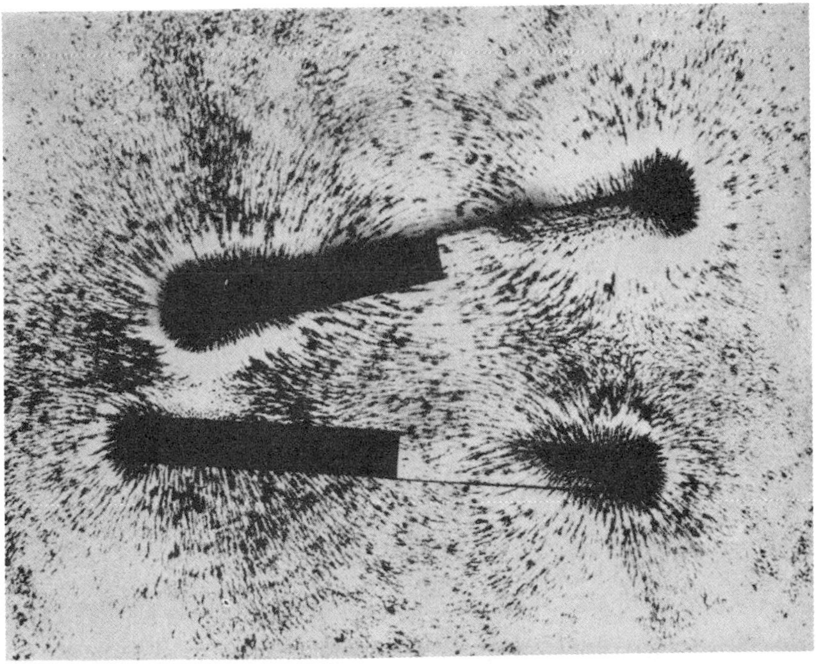

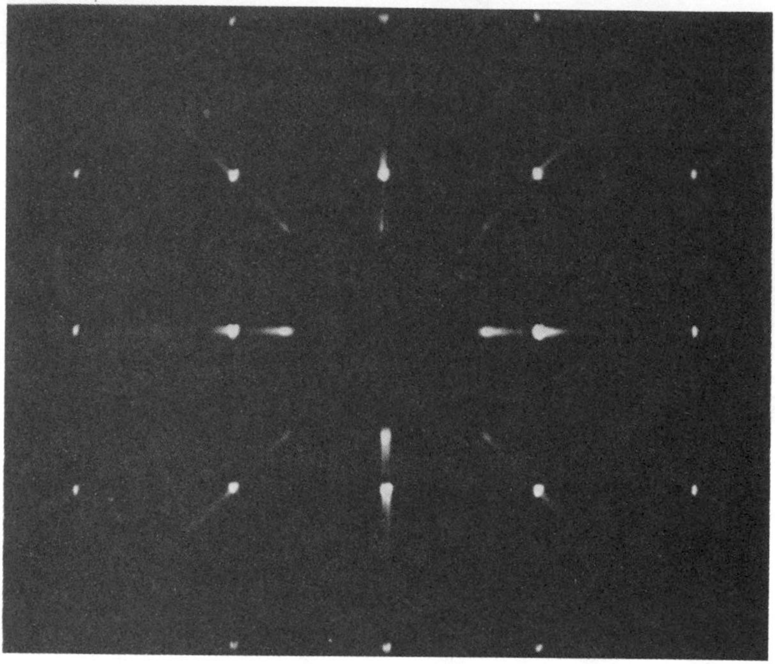

PLATE IV (Upper) Diffraction of X-rays in nickel-iron alloy. (Lower) Diffraction of 100 kv electrons in the same alloy. *Courtesy R. D. Heidenreich, Bell Telephone Laboratories*

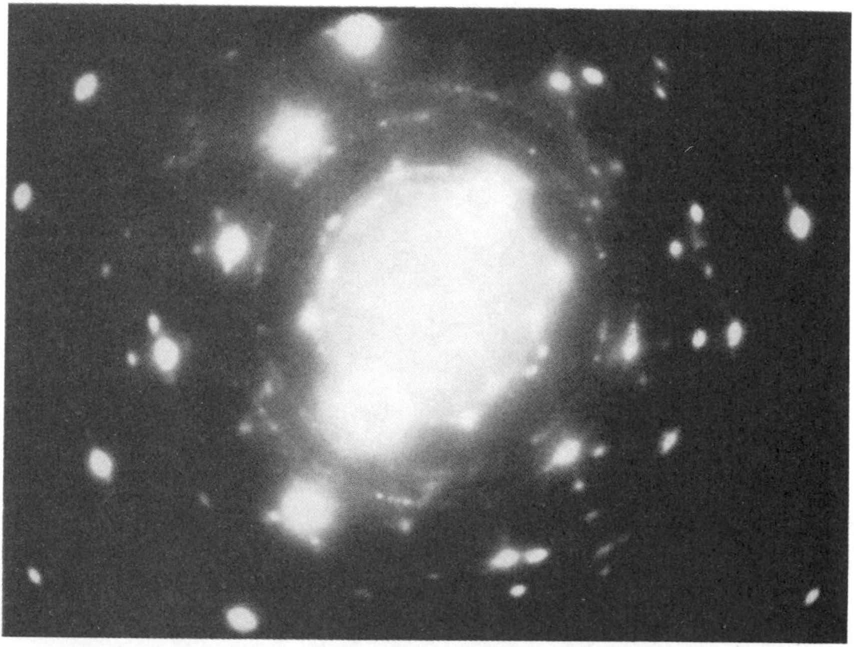

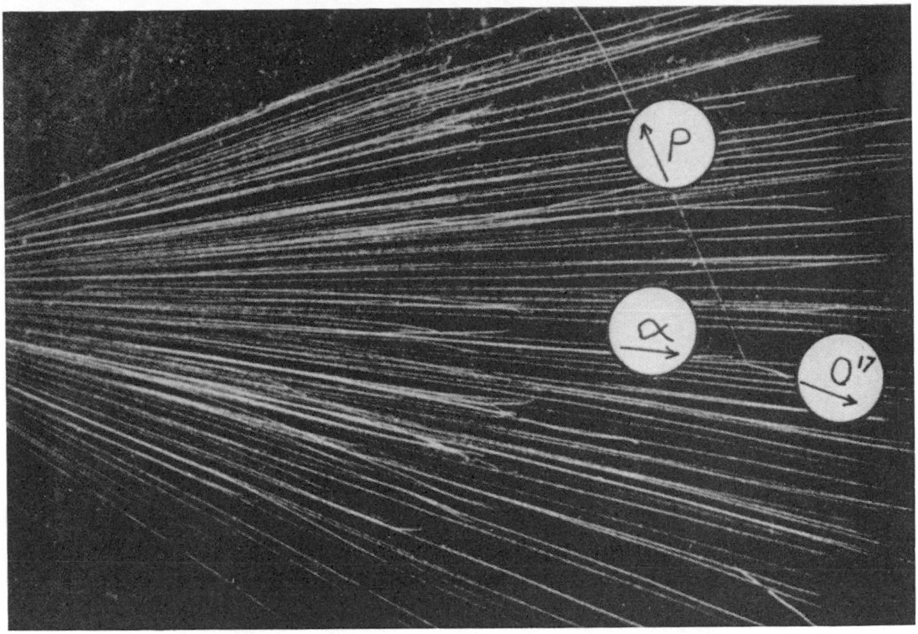


PLATE V (Upper) The first cloud chamber photograph of artificial nuclear transformation. *Courtesy P. M. S. Blackett, formerly at Cambridge University.* (Lower) The breakup of boron nucleus into three alpha particles. *Courtesy P. Dee and C. Gilbert, formerly at Cambridge University*

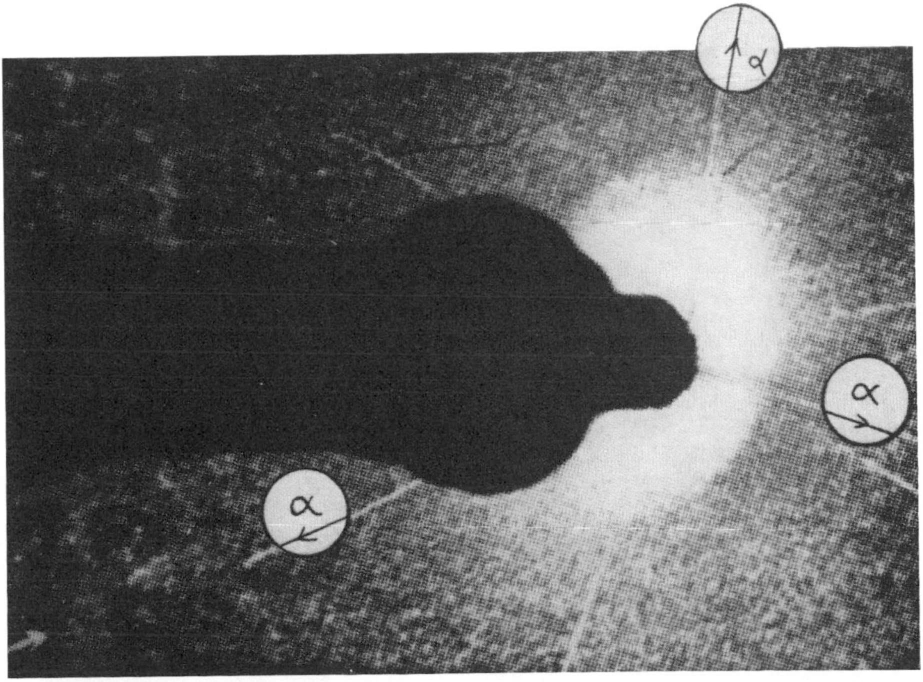

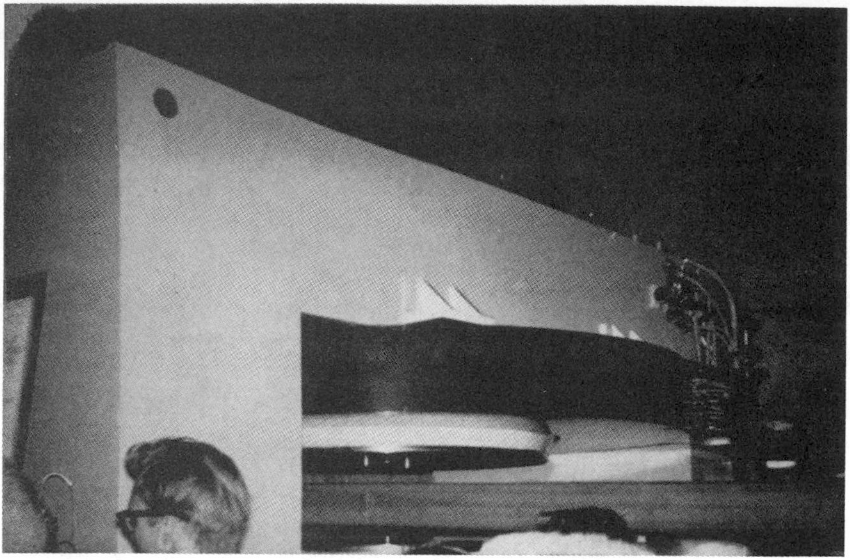

PLATE VI (Upper) University of Colorado cyclotron, showing a pole of the electromagnet and the beam. *Courtesy Nuclear Research Laboratory*. (Lower) A section of the University of California bevatron. *Courtesy Lawrence Radiation Laboratory*

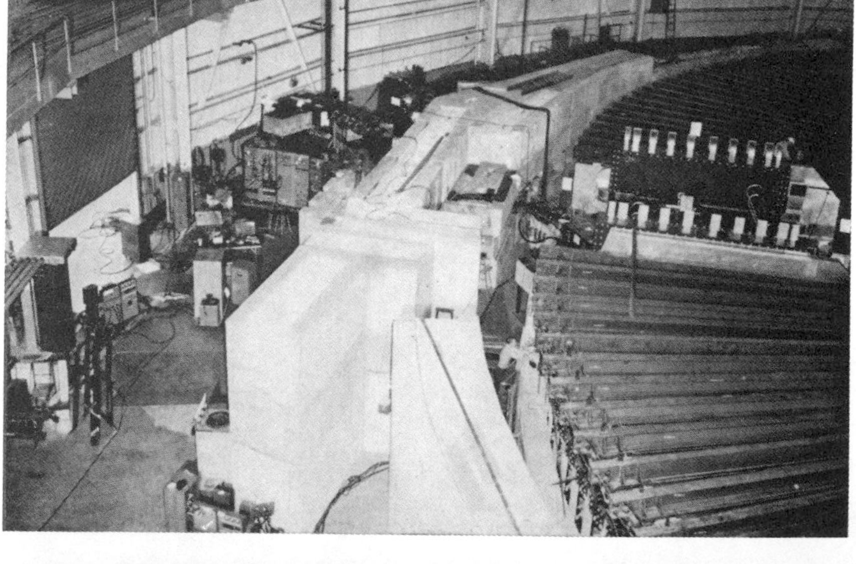

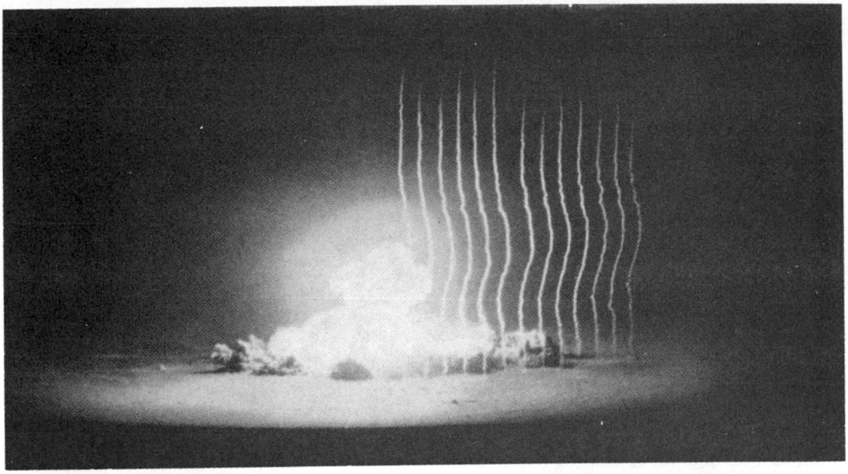

PLATE VII (Upper) Atomic bomb test in Nevada. (Lower) Swimming pool reactor in Oak Ridge. *Courtesy Atomic Energy Commission*

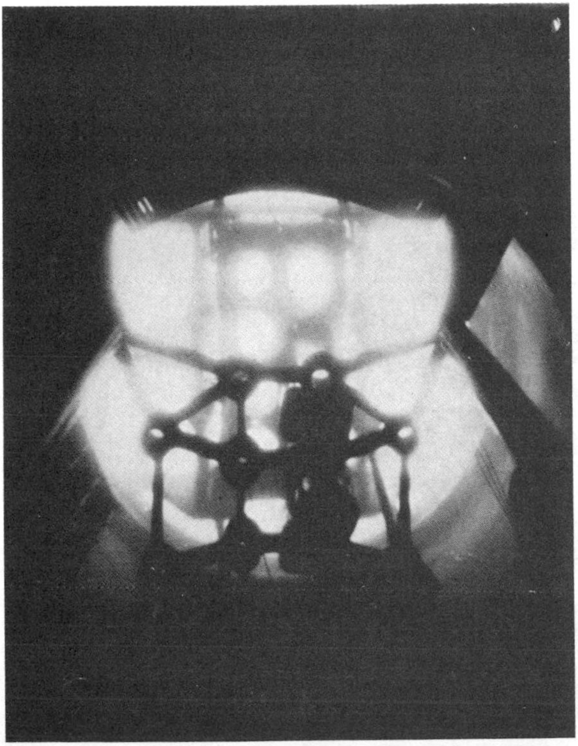

PLATE VIII (Upper) Formation of a pion and its subsequent decay into a muon and electron. Thick emulsion photograph. *Courtesy E. Pickup, formerly at Canadian National Research Council.* (Lower) A series of nuclear events in a bubble chamber. *Courtesy L. Alvarez, University of California*

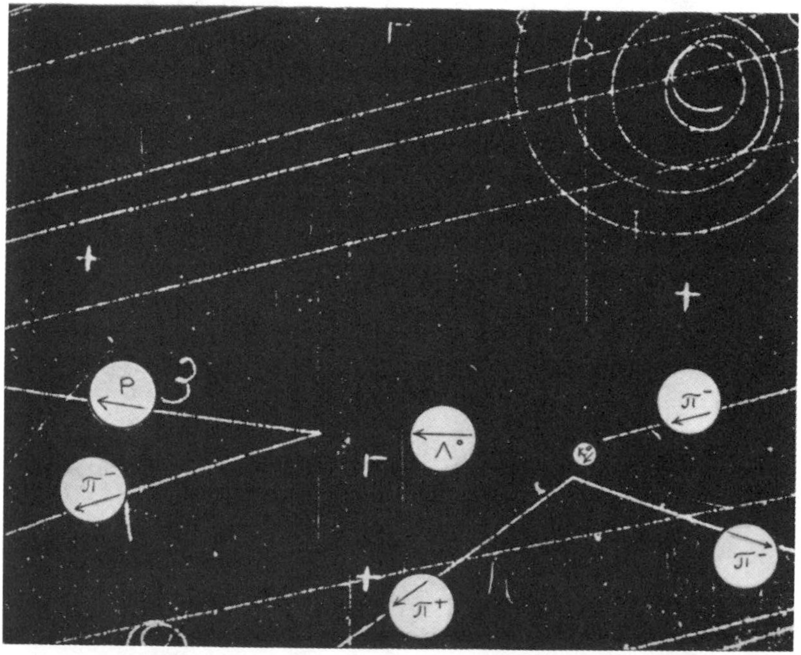

motion through space on the velocity of light as measured on its surface. Indeed the earth moves on its orbit around the sun at the velocity of about 30 km/sec, and there must be ether wind blowing over the surface, and probably also through the body of the earth, just as in the case of an automobilist driving an open convertible on a windless day. The experiment of Michelson and Morley followed the same principle as that of Fizeau, but had to be modified because in this case one apparently could not have the equivalent of two parallel pipes, through which ether wind was blowing in opposite directions. Instead, they undertook to measure the time for the round trip of light, in one case propagating in the direction of the expected ether wind, and in another case propagating perpendicularly to it. In order to understand the principle of that experiment, let us consider a motor launch which makes round trips, in one case along a broad river and in another case across it. In the first case, during one part of the journey the launch will sail with the stream, and its velocity will be $V + v$ where $V$ is the velocity of the launch in respect to water, and $v$ is the velocity of the river. On the way back, the launch will be sailing against the stream and will move with the velocity $V - v$. If $L$ is the distance between two landing places along the river, the time for the return trip will be:

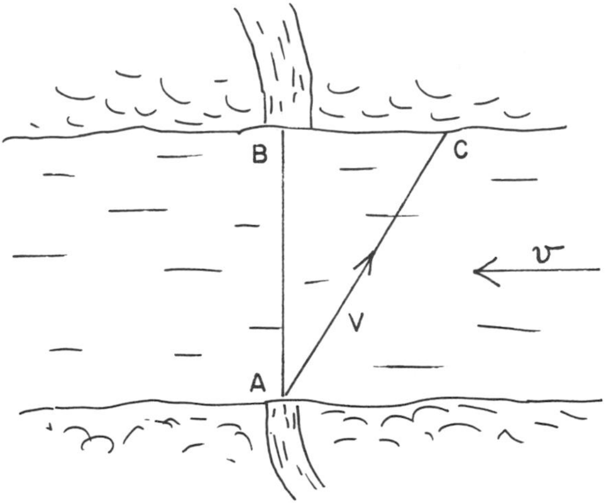


FIG. VI–3.
The problem of a launch crossing a river.

$$t_{\rightleftarrows} = \frac{L}{V+v} + \frac{L}{V-v} = \frac{2LV}{V^2 - v^2} = \frac{2L/V}{1 - \dfrac{v^2}{V^2}}$$

Since $\frac{2L}{V}$ would be the time of the return trip in nonmoving water, we see that the presence of the flow will always make that time longer. In particular, if $v$ is equal to or larger than $V$, the boat will never return, and $t_{\rightleftarrows}$ becomes infinite.

Let us now consider the case of a launch sailing across the river (Fig. VI-3). If it starts from the point $A$ and must land at point $B$ straight across the river, the launch must keep its course slightly upstream to compensate for the drift. Thus while it covers the distance $AC$ in respect to water, it drifts downstream by the distance $CB$. Clearly, the ratio $BC/AC$ is equal to the ratio of stream and launch velocities. Applying the Pythagorean theorem to the rectangle $ABC$, we get:

$$\overline{AB}^2 + \left(\overline{AC} \times \frac{V}{v}\right)^2 = \overline{AC}^2$$

or

$$\overline{AB}^2 = \overline{AC}^2 \times \left(1 - \frac{V^2}{v^2}\right)$$

or

$$\overline{AC} = \frac{\overline{AB}}{\sqrt{1 - \dfrac{v^2}{V^2}}}$$

If $AB = L$, we find for the time of crossing and coming back:

$$t_{\uparrow\rightarrow} = \frac{2\overline{AC}}{V} = \frac{2L/V}{\sqrt{1 - \dfrac{v^2}{V^2}}}$$

Just as in the previous case, time is longer than it would be in still water, but the correction factor $\sqrt{1 - \frac{v^2}{V^2}}$ is smaller than the previously obtained factor $1 - \frac{v^2}{V^2}$.

Now substitute, for the flowing river, ether wind and, for the launch, light wave, and you have the Michelson-Morley experiment. The arrangement used is shown schematically in Figure VI-4. It was mounted on a solid marble slab floating in mercury so it could be rotated around its axis without much difficulty or shaking. A light beam from the lamp $L$ fell on a glass plate located in the center of the slab. The glass plate was covered with a thin layer of silver which reflected one half of the incident beam, and let through the other

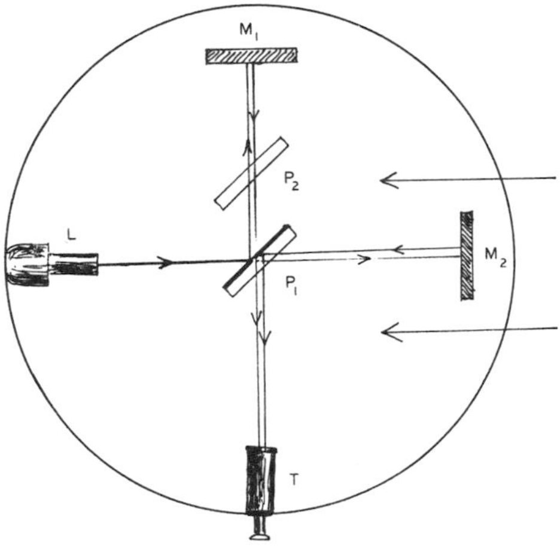


FIG. VI–4.

Michelson-Morley apparatus, showing the paths of light rays. Rays falling and reflected from mirrors $M_1$ and $M_2$ are shown somewhat shifted in respect to one another for convenience in drawing. Plate $P_2$ is introduced to compensate for additional path in the plate $P_1$ of the ray directed toward $M_2$.

half. The two beams were further reflected by two mirrors $M_1$ and $M_2$, located equal distances from the center. Returning to the silvered glass plate, the ray reflected from $M_1$ was partially coming through (nobody cared what happened to the other part of it) while the ray returning from $M_2$ was partially reflected (nobody cared what happened to its other part), and these two rays were entering the telescope $T$. If there were no ether wind, both rays would come in phase and produce the maximum illumination of the field of the telescope. If there were ether wind blowing, say, from right to left, the ray going across the wind would be delayed less than that going with and against the wind, and there would be at least a partial destructive interference. Here is a rough numerical estimate of the situation. The ratio of two time periods, $t_1$ and $t_2$, for light traveling in two mutually perpendicular directions is, according to the previously described formulas:

$$\frac{1 - \frac{v^2}{c^2}}{\sqrt{1 - \frac{v^2}{c^2}}} = \sqrt{1 - \frac{v^2}{c^2}}$$

where $V$ is substituted by the velocity of light $c$. The ratio $\left(\frac{v}{c}\right)^2$ is in this case equal $\left(\frac{3 \times 10^6}{3 \times 10^{10}}\right)^2 = 10^{-8}$ or 0.00000001. It can be shown* that, for such a small value of $v^2/c^2$, the radical $\sqrt{1 - \frac{v^2}{c^2}}$ is very closely represented by $1 - \frac{1}{2}\frac{v^2}{c^2} = 1 - 0.000000005 = 0.999$-999995. Thus, the expected difference in arrival of two waves is only 5 ten-millionths of a percent. But, it is sufficiently large to be noticed by sensitive optical instruments. In fact, if the diameter of the marble slab was 3 meters (and that is about right), the total travel time (plate to mirror and back) was $\frac{300}{3 \times 10^{10}} = 10^{-8}$ sec. Thus the difference of the arrival time of two waves to the telescope was:

$$5 \times 10^{-9} \times 10^{-8} = 5 \times 10^{-17} \text{ sec.}$$

For the wave length of $6 \times 10^{-5}$ cm the vibration period was

$$\frac{6 \times 10^{-5}}{3 \times 10^{10}} = 2 \times 10^{-15} \text{ sec.}$$

Thus the difference of the arrival time was $\frac{5 \times 10^{17}}{2 \times 10^{-15}} = 2.5 \times 10^{-2}$ or 2.5% of the vibration period, and should produce a noticeable amount of destructive interference. In the actual experiment, the effect

* In fact, it was the content of the first mathematical paper by Sir Isaac Newton.

was observed not by the decrease of the intensity, but by a shift of the series of interference fringes by 2.5% of the distance between them. Turning their apparatus by 90 degrees (this is why it had to float on mercury), and thus exchanging the role of mirrors $M_1$ and $M_2$, one would expect the same shift in the opposite direction, so that the total shift of fringes should be 5% of the distance between them, and, if observed, that shift would show that the earth's velocity through space is 30 km per sec.

Well, the experiment was performed, and there was no shift whatsoever. How could it be? Was light ether dragged 100% along with the body of the moving earth? The repetition of Michelson's experiment on a balloon floating high above the ground disproved that possibility. Physicists were breaking their heads and could not make head nor tail of it. A very revolutionary proposal was made by the British (and very Irish) physicist G. F. Fitzgerald, who suggested that all material bodies moving with the velocity $v$ through the world ether shrink in the direction of motion by a factor $\sqrt{1 - \frac{v^2}{c^2}}$. Such shrinking, which had to be assumed to be the same for all bodies independent of their physical structure, would reduce the distance between the central plate and the ether windward mirror in the Michelson-Morley experiment by just a proper amount to make the arrival times equal, and to eliminate any shift of interference fringes. Numerous attempts were made to explain the hypothetical "Fitzgerald contraction" by the interplay of electric and magnetic forces between the atoms forming material bodies, but all was in vain. This brave and ingenious proposal led to a limerick which runs as follows:

> There was a young fellow named Fiske
> Whose fencing was strikingly brisk;
> So fast was his action
> Fitzgerald contraction
> Reduced his rapier to a disc.

But it was only a half truth, and not all the truth.

## AN INTERMEZZO

Before we come to grips with Einstein's explanation of the negative result of the Michelson-Morley experiment, it is nice to discuss a problem which, having nothing to do with the theory of relativity, has, however, a relativistic flavor. A man in a boat travels upstream (Fig. VI-5) on a river, and there is a half-empty bottle of whisky standing on the stern of the boat. While the boat was passing under the bridge, a wave reflected from the bridge's pillars shook the boat, and the bottle fell into the water without the man's noticing it. Now for 20 minutes the boat continues upstream while

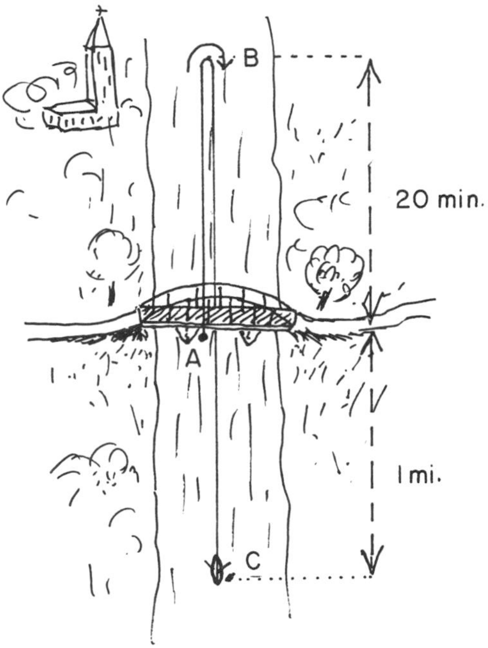


FIG. VI–5.
Recovery problem of a dropped bottle.

the bottle floats downstream. At the end of 20 minutes, the man notices that the bottle is gone, turns the boat (neglect the time necessary for that operation), and moves downstream with the same velocity in respect to water as before. He picks up the bottle a mile below the bridge. The question is: what is the velocity of the river? Try to solve that problem before reading further, and you will find how difficult it seems. In fact, several good mathematicians were completely stumped by it.

But things become very simple if, instead of considering the events described relative to the shore line, as it is natural to do, one describes them relative to river water. Suppose we sit on a raft floating down the stream, and look around. In respect to us, water will be at rest, but the shores and the bridge will be moving along at a certain velocity. A boat passes by and the whisky bottle falls into the water. The boat continues on its course while the bottle floats motionless at the spot where it fell out. (Remember: water is *not* moving relative to us.) Twenty minutes later we see the boat turning around, and coming back to recover the bottle. Naturally, it would take the boat another 20 minutes to come back. Thus, the bottle was in the water for 40 minutes, and during that time the shore line and the bridge moved 1 mile. Thus, the velocity of the bridge in respect to water, or, what is the same, the velocity of water in respect to the bridge and shore line is one mile in 40 minutes, or 1½ miles per hour. Simple, isn't it?

### BIOGRAPHICAL FRAGMENT

Just for the record, it should be stated that Albert Einstein (Fig. VI-6) was born on March 14, 1879, in the little, but famous (for the Meistersingers) German town of Ulm near Munich where his father owned an electrotechnical works. Spending his boyhood in Munich, he then moved to Switzerland, studying in the Polytechnical School in Zurich and earning his living by tutoring less gifted students in mathematics and physics. In 1901 he was married and secured a quiet but not too well-paid position as examiner of patents in the Swiss Patent Office in Berne. In 1905, at the age of 26, he published in a German magazine, *Annalen der Physick,* three articles which shook the scientific world. These three articles pertained to the three broad fields of physics: heat, electricity, and light. One, which was already mentioned in Chapter IV, contained the detailed theory of Brownian motion and was of fundamental importance for the development of the mechanical interpretation of the heat phenomena. Another explained the laws of photoelectric effect on the basis of at that time juvenile quantum hypothesis, and introduced the notion of individual packages of radiant energy or photons. This will be described in the next chapter. The most important of the three in the development of physics carried a rather dull title, "On the Electrodynamics of Moving Bodies," and was devoted to the

FIG. VI–6.
Albert Einstein.

paradoxes in the measurements of the velocity of light. It was the first paper on the theory of relativity.

RELATIVITY OF MOTION

The accumulated difficulties and contradictions concerning the nature of the hypothetical substratum responsible for electromagnetic interactions and the propagation of light waves became entangled into an undisentanglable etherial knot, very similar to the legendary knot of cornel bark which bound the yoke to the pole in the chariot of the ancient Greek peasant king Gordius. The prediction of an oracle that the person who would disentangle the Gordian knot would rule over all Asia was fulfilled by Alexander the Great, who cut the knot by a stroke of his sword. Similarly, Albert Einstein became the ruler of modern physics by cutting the etherial knot with the sharpness of his logic, and throwing the twisted pieces of world ether out of the window of the temple of physical science.

But, if there is no world ether filling the entire space of the universe, there cannot be any absolute motion, since one cannot move in respect to nothing. Thus, said Einstein, one can speak only about the relative motion of one material body in respect to another, or one system of reference in respect to another system of reference, and the two observers located in these two systems of reference have equal right to say: "I am at rest, and this other fellow is moving." If there is no world ether providing a universal reference system for motion through space, there cannot be any methods to detect such a motion, and, in fact, any statement concerning such a motion must be branded as physically nonsensical. It is no wonder, therefore, that Michelson and Morley, measuring the velocity of light in different directions in their laboratory, could not detect whether their laboratory and the earth itself were or were not moving through space.

Let us remember the words of Galileo (p. 44-45 of this book):

> Shut yourself up with some friend in the largest room below deck of some large ship, and there procure gnats, flies, and such other small winged creatures. Also get a great tub full of water and within it put certain fishes; let also a certain bottle be hung up, which drop by drop lets forth its water into another narrow-necked bottle placed underneath. Then, the ship lying still, observe how those small winged animals fly with like velocity towards all parts of the room; how the fishes swim indifferently towards all sides; and how the distilling drops all fall into the bottle placed underneath. And, casting anything towards your friend, you need not throw it with more force one way than the other, provided the distances be equal; and jumping broad you will reach as far one way as another.
>
> Having observed all these particulars, though no man doubts that, so long as the vessel stands still, they ought to take place in this manner, make the ship move with what velocity you please, so long as the motion is uniform and not fluctuating this way and that. You shall not be able to discern the least alteration in all the forenamed effects, nor can you gather by any of them whether the ship moves or stands still.

We can paraphrase Galileo's words in respect to the Michelson-Morley experiment in the following way: Shut yourself up with an assistant in a large laboratory on the earth and there procure light sources, mirrors, and all other kinds of optical equipment. And also get all kinds of gadgets which can measure electric and magnetic forces, currents, and other things. Then persuade yourself by a logical argument that, if the earth were standing still, the propagation of light rays, the interaction of charges, magnets, and electric currents, would not depend on their relative locations and directions in respect to the walls of your laboratory. Then assume, as is true, that the earth moves around the sun and, with the sun, around the center of the stellar system of the Milky Way. You shall not be able to discern the least alteration in all forenamed effects, nor can you gather by any of them whether the earth moves or stands still.

Thus, what was true for gnats, fishes, water drops, and thrown objects on Galileo's hypothetical ship sailing across the blue Mediterranean waters turns out to be also true for light waves and other electromagnetic phenomena on the earth moving through space. Of course, Galileo could easily find out whether his ship moved in respect to the earth or not, by walking out from his inside cabin to the deck and looking at the water or the shore line. In the same way, we can establish the motion of the earth around the sun, and of the sun in respect to stars, by looking at the stars and observing the change in their apparent positions (paralactic displacement) and the wave lengths of light coming from them (Doppler effect). But, without looking out, it is just as impossible to detect motion through space by observing electromagnetic phenomena as it is impossible to do so by observing the mechanical ones.

## THE UNION OF SPACE AND TIME

Einstein realized that this broadened form of the Galilean principle of relativity of motion required a drastic change in our fundamental ideas about space and time. From time immemorial, space and time were considered as two entirely independent entities, and in his *Principia* the great Newton wrote:

> Absolute space, in its own nature, without relation to anything external, remains always similar and immovable.
>
> Absolute, true, and mathematical time, in itself, and from its own nature, flows equably without relation to anything external.

While Newton's definition of space implies the existence of an absolute reference system for the motion through it, his definition of time involves the existence of an absolute timing system such as could be provided by a large number of synchronized chronometers, or simply clocks, located in different parts of universal space and all showing standard universal time. While the experimentally proved constancy of the velocity of light undermined the idea of absolute space, it also dropped a monkey wrench into the universal timing system. To understand this universal-time catastrophe, let us ask what is the best way to synchronize two clocks located a certain distance away from one another. Of course, an employee of a Universal Time Company could travel from one location to another, carrying with him a standard-time chronometer for setting local clocks; this is just what was done by old-time navigators carrying chronometers on their ships. But who can guarantee that a chronometer will not go wrong in transit? The modern timing system is of course based on radio signals which carry time-information with the velocity of light. For any practical purpose of timing on the earth, a slight delay caused by the finite velocity of light can easily be neglected, but it will certainly be of importance in case of interplanetary time-setting where the delay may be many hours. However, this difficulty can easily be circumvented by making the signal travel to and fro, being reflected (without any time delay) at the signal recipient station. Thus, if the time signal was sent at time $t_1$ and returned at time $t_2$, the correct clock-setting at the recipient station at the moment of its arrival should be $(t_1+t_2)/2$. Since, according to the Michelson-Morley experiment, the velocity of light in a vacuum is always the same no matter what the conditions of motion are, the above-described method must be considered as absolutely precise and objectionless. Its alternative would be to send two light signals in opposite directions from a point located exactly halfway between two stations, and to consider the two clocks synchronized if they show the same time when the signals arrive.

The next step is to correlate the clock on two systems moving uniformly in respect to one another as, for example, two railroad trains passing each other in opposite directions. We have chosen the example of trains because railroad employees are extremely proud of their large, often gold-plated watches which always show the exact time. To carry out the above-described synchronization method, the brakeman should swing his lantern from a point in the middle of the train, while the engineer and the conductor, leaning out from locomotive and caboose respectively, should check on their watches the exact time they saw the light arriving.

The above-described procedure reminds us of old Galileo's attempt at measuring the velocity of light by flashing a lantern, but of course we do not mean here that such an experiment must really

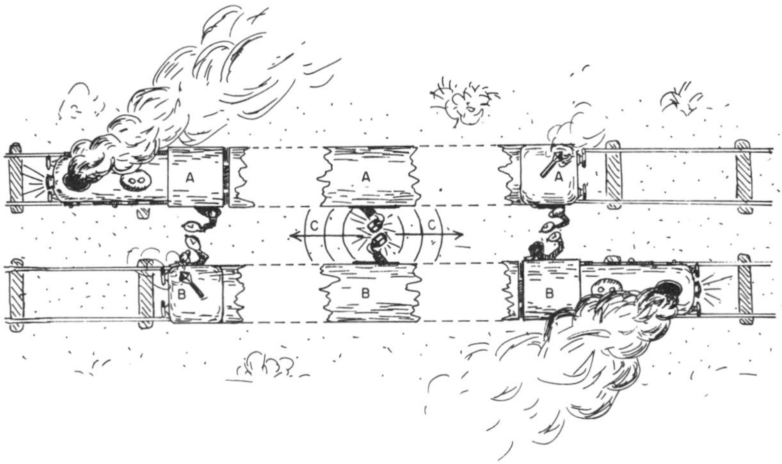


FIG. VI–7.
Synchronizing watches on two trains moving in respect to each other.

be carried out by the crews of two railroad trains. It is rather what Einstein liked to call "a thought-experiment" (Eine Gedankenexperiment), in which one only visualizes the situation and tries to conclude what will happen on the basis of the known results of experiments (such as that of Michelson and Morley).

Using this method on both trains *A* and *B*, one can synchronize the clocks on each of the trains, and now one is facing a problem of comparing the timing on one train with the timing on another. That can be done at the moment that the two trains are in such a position that the locomotive *A* just passes the caboose *B*, and the caboose *A* passes the locomotive *B* (Fig. VI-7). Indeed, at that moment, the engineer of *A* and the conductor of *B* can compare their watches directly by leaning out of windows and placing their watches side by side. The same goes for the conductor of *A* and the engineer of *B*.

We can correlate these direct clock comparisons with the previously described light wave comparison by assuming that two brakeman, on *A* and *B*, swing their lanterns when in passing they are exactly opposite each other. In this case, of course, there will be only one light wave, since both lanterns practically coincide.

Let us now consider the result of such a procedure. Since light propagates with final velocity, it will take some time before it reaches the two ends of the train, and, when it finally arrives, locomotive *A* will be to the left of caboose *B*, while caboose *A* will be to the left of locomotive *B*. Thus, after passing caboose *B*, the light wave will take some time until it reaches locomotive *A*. Therefore, if, because of the agreement concerning the light-signal method for watch-setting, the engineer *A* and conductor *B* had their watches set in such a way that they show equal time when the engineer and conductor see light, the watch of engineer *A* must have been behind the watch of conductor *B* at the moment of passing. By the same argument, the watch of conductor *A* must have been *ahead* of the watch of engineer *B* at the moment of their encounter. Now, since the people on train *B* are sure that their watches are correctly synchronized because they used the light-signal method, they will insist that the watch-setting on train *A* is wrong, and that the watch on the locomotive *A* is running *behind* the watch in the caboose of the same train. Similarly, the people on train *A*, considering their own watch-setting to be correct, will have their doubts about *B*'s watch-setting. The engineer of *A* will say that *B*'s conductor's watch is *ahead* of the correct time, while the conductor of *A* will insist that the watch of the engineer on *B* is *behind*. Both will agree that the watch-setting on train *B* is definitely wrong and that the watch on the locomotive *B* is running *behind* the watch in the caboose *B*. The argument can never be settled, since trains *A* and *B* have no priority over each other, and we must conclude that *the clocks synchronized in one system will appear as not being synchronized when observed from another system moving in respect to it, and vice versa.* In other words, *two events which occur simultaneously in one system some distance apart* (train length) *will be observed as not simultaneous when observed from another system moving in respect to it.* Thus, space is, at least partially, interchangeable with time, and a purely spatial separation of two events in one system leads to a certain time difference between them when viewed from another moving system.

To illustrate this statement, let us consider a man eating his dinner in a dining car of a moving train. He first eats his soup, then the steak, and then the dessert. These events all occur at the same place (same table) in respect to the train but at different times. However, from the point of view of an observer on the ground the man eats his soup and his dessert miles apart. This trivial fact can be formulated as follows: *Events which occur at the same place but at different times in one system occur at different places when observed from another system moving in respect to it.* Now replace in the above sentence the word "place" by "time" and vice versa, and it will read: *Events which occur at the same time* (i.e., simultaneously) *but at different places in one system, occur at different times when observed from another system moving in respect to it.* And this is exactly the result we have arrived at above.

If a zero time interval becomes larger than zero when viewed from a moving system, then *a finite time difference between two events must increase when viewed from the same system.* This is the famous *time-dilation,* or slowing of the clock (along with all other physical, chemical, and biological processes) when observed from a moving system. Like all relativistic phenomena, time dilation is symmetrical in respect to two systems moving relative to one another, and, while the clocks on train *A* will be observed to lose time by the crew of a bypassing train *B,* the crew in train *A* will insist that it is the clocks in train *B* which are slowed down. It can be shown that the expected relativistic slowing down of the clock is given by the formula

$$t = \frac{t_0}{\sqrt{1 - \frac{v^2}{c^2}}}$$

which is similar to that of Fitzgerald contraction except that the radical stands in the denominator.

Slowing down of all physical processes in fast-moving systems was observed directly in the case of the decay of "mesons," the unstable elementary particles which constitute an essential fraction of cosmic rays coming down to the surface of the earth at extremely high speed. It will be discussed in more detail in the last chapter of this book. The idea is to place into a satellite circling the earth an atomic clock, a very precise timepiece in which the motion of hands is synchronized with vibrations of the molecules of a gas placed inside it. Comparing by means of radio signals the rate of the clock on the satellite, and an identical clock resting on the ground, it will be possible to prove the validity of the time dilation effect on a large scale.

## RELATIVISTIC MECHANICS

The shrinkage of distances and the dilation of time intervals when viewed from a moving system necessitates a drastic change in the formulas which connect space and time measurements in one system of coordinates with the measurements of the same quantities carried out from another system moving in respect to it. Let us consider two coordinate systems $(x, y)$ and $(x', y')$ moving in respect to each other with a velocity $v$ and let us count the time in both systems from the moment when their origins $O$ and $O'$ coincided with each other. Let us consider an object $P$ located immovably in the primed system of coordinates at the distance $x'$ from its origin $O'$. What is the $x$-coordinate of that object in the unprimed system at time $t$, i.e., what is its distance from the origin $O$? The answer is very simple if we use the classical Newtonian point of view. During time interval $t$ the origin of two coordinate systems got separated by distance $vt$, so that

$$x' = x + vt.$$

One could also add the formula

$$t' = t$$

which simply restates Newton's definition of absolute time.

Before Einstein, these two formulas, which are today referred to as "Galilean transformations of coordinates," were considered to be a matter of common sense, and the second one was never even written down. But the possibility of partial transformation of space distances into time differences demands the replacement of these seemingly trivial formulas by more sophisticated ones, and it can be shown that in order to satisfy the condition of the constancy of the velocity of light, and other relativistic effects discussed above, the old Galilean transformations must be replaced by a new set:

$$x' = \frac{x + vt}{\sqrt{1 - \frac{v^2}{c^2}}}$$

$$t' = \frac{t + \frac{v}{c^2}x}{\sqrt{1 - \frac{v^2}{c^2}}}.$$

These expressions, known as Lorentz transformations, were derived by the Dutch physicist H. A. Lorentz, soon after the publication of results of the Michelson-Morley experiments, but were considered by their author and other physicists of the time more or less as an amusing purely mathematical trick. It was Einstein who first realized that Lorentz transformations actually correspond to physical reality and require a drastic change in the old-fashioned common-sense notions concerning space, time, and motion.

We notice that, whereas Galileo transformations were not symmetrical in respect to space and time coordinates, Lorentz transformations are. In calculating the new time $t'$ one has to add to $t$ an additional term depending on the relative velocity $v$, which is similar to the term added to the old space coordinate $x$ in order to get the new space coordinate $x'$. In all the cases which we encounter in everyday life, where all the velocities involved are much smaller than the velocity of light ($v \ll c$), the second term in the nominator of the time transformation becomes practically zero, and the factor in the denominator of both formulas becomes practically equal to one. This brings us back to the old Galilean transformations. But, if the velocities involved are comparable to that of light, the additional term in time transformation results in the breakup of the notion of absolute simultaneity, while the square root factors lead to contraction of distances and dilation of time.

It is necessary at this point to discuss a misunderstanding concerning the relativistic contraction of length. This misunderstanding existed among physicists for 54 years, from the time of publication of Einstein's original paper in 1905 until it was straightened out in a short critical paper published by a young American physicist, J. Terrell, in 1959. It was always believed that the contraction of length by a factor $\sqrt{1 - \frac{v^2}{c^2}}$ could really be observed by looking at a moving object if the velocities close to that of light could be achieved. Thus, a man riding on a Pan-American plane would see a TWA plane passing close by in the opposite direction (against all the regulations of FAA!), shrunk from cockpit to tailfin, while a TWA passenger would observe the same thing happening to a Pan-Am plane. Terrell has shown that this concept was incorrect and that, from the standpoint of *visual observation* of a fast-moving object, it will not look any shorter than when it is at rest. This result is due to the fact that, because of the finite velocity of light, we will see light coming from the nose and tail of the passing plane with different time delays, and that this time difference will cancel the effect of relativistic contraction of length. If light would propagate with an infinite velocity, this error in observation would not exist, but then, of course, for $c = \infty$, the relativistic contraction of length would anyway be zero for any value of the relative velocity of two systems.

According to Terrell's argument, while relativistic contraction of length cannot be seen by an individual observer, it still can be photographed provided the size of the lens is larger than the length of the moving object. We can imagine a special photographic plane equipped with a camera which extends all the way from its nose to its tail. This camera must have a long cylindrical lens and a "simultaneous shutter," i.e., a shutter the nose end of which closes simultaneously with its tail end (in the time-synchronization system valid on the plane). If such a plane passes an unidentified object moving fast in the opposite direction, and a photograph of that object is taken, the picture will show all the features of relativistic length contraction. It goes without saying that, if the unidentified object could photograph the camera plane, it would radio to the pilot, "You, too, are shorter!"

It is not the place in this book to develop mathematical consequences of Lorentz transformations, and we will only indicate the most important results to which they lead. One of the most important results pertains to the addition of two velocities. Suppose an aircraft carrier speeds across the ocean at 35 knots, i.e., about 40 miles per hour, and a motorcyclist drives along its deck from stern to bow, at a speed of 60 miles per hour (Fig. VI-8). What is the velocity of the motorcycle in respect to the water? In classical mechanics the answer is simple: $40 + 60$, i.e., 100 miles per hour. This simple rule for the addition of velocities cannot, however, be true in relativistic mechanics. In fact, if the velocity of the flat top

FIG. VI–8.
**Relativistic addition of two velocities.**

and of the motorcycle are both, say, 75% of the velocity of light (which is possible, at least, in principle), the velocity of the motorcycle in respect to the water would be 50% more than the velocity of light. The relativistic formula for the addition of two velocities, $v_1$ and $v_2$ is:

$$V = \frac{v_1 + v_2}{1 + \frac{v_1 v_2}{c^2}}$$

where $V$ is the resulting velocity. It can easily be seen that, if both $v_1$ and $v_2$ are smaller than $c$, $V$ is also smaller than $c$. In fact, even if we put $v_1 = c$ we get:

$$V = \frac{c + v_2}{1 + \frac{cv_2}{c^2}} = \frac{c + v_2}{1 + \frac{v_2}{c}} = \frac{c\,(c + v_2)}{c + v_2} = c$$

meaning that any velocity added to the velocity of light does not increase the latter by any amount. If we put $v_1 = c$ and $v_2 = c$ we have again:

$$V = \frac{c + c}{1 + \frac{c \times c}{c^2}} = \frac{2c}{1 + 1} = c$$

The relativistic formula for addition of velocities explained the experiment of Fizeau, described earlier, which occurred about half a century before. Substituting for $v_1$, the value of the velocity of light $c/n$ in water, and writing for $v_2$ simply $v$ for the velocity of water in the pipe, we get:

$$V = \frac{c/n + v}{1 + \frac{cv}{nc^2}} = \frac{c/n + v}{1 + \frac{v}{nc}}$$

Multiplying nominator and denominator by $\left(1 - \frac{v}{nc}\right)$, we get further:

$$V = \frac{(c/n + v)\,(c/n - v)}{1 - v^2/n^2c^2} = \frac{c/n + v - v/n^2 + v^2/nc^2}{1 - v^2/n^2c^2}$$

Now, since $v$ is much smaller than $c$, $\left(\frac{v}{c}\right)$ is a very small number, and $\left(\frac{v}{c}\right)^2$ is still smaller. Thus, neglecting the terms containing $v^2/c^2$ in the above formula, we obtain:

$$\mathrm{V} = \frac{c}{n} + v - \frac{v}{n^2} = \frac{c}{n} + v\left(1 - \frac{1}{n^2}\right)$$

which is exactly Fizeau's empirical formula. Thus, there is no such thing as the "drag of ether" by moving fluid, and the resulting velocity is simply the relativistic sum of the velocity of light in the liquid and the velocity of the liquid's flow through the pipe.

Another important consequence of relativistic mechanics is that the mass of a moving particle does not remain constant as in the

Newtonian system, but increases with the increasing velocity. The factor affecting the mass of moving bodies is the same as that affecting the shrinkage of length and dilation of time, and the mass of a body moving with the velocity $v$ is given by the expression:

$$m = \frac{m_0}{\sqrt{1 - \frac{v^2}{c^2}}}$$

where $m_0$ is the so-called "rest mass," i.e., the inertial resistance to the force which tends to move the particle initially at rest. As the velocity of the particle increases, approaching the velocity of light, it becomes progressively more difficult to increase its velocity, and at $v = c$ the resistance to further acceleration becomes infinitely large. This gives us another aspect of the basic statement of the theory of relativity, to the effect that no material body can move faster than light; in fact, because of the increased inertial resistance, the energy which would be needed to accelerate a material body and to make it move with the velocity of light becomes infinite.

## MASS-ENERGY EQUIVALENCE

Rejecting the notion of world ether and returning the interstellar space to its previous state of emptiness, Einstein had to do something in order to retain the physical reality of light waves and electromagnetic fields in general. If there is no ether, *what* surrounds the electric charges and magnets, and *what* propagates through a vacuum carrying to us the light of sun and stars? This could be done only by considering the electromagnetic field as some kind of material medium even though quite different from ordinary material media we are well familiar with. In physics, the adjective "material" is equivalent to "ponderable," i.e., possessing some weight or mass. Thus, electric charges and magnets must be surrounded by some ponderable substance even though it may be very light, which is comparatively dense in their neighborhood, and is thinning out to zero at the distance where the electric and magnetic forces disappear. Similarly, light rays should be pictured as vibrating streams of this material ejected from luminous bodies (just as streams of water are ejected from garden hoses) rushing through a completely empty space. The difference between the new and old views is illustrated schematically in Figure VI-9. Whereas world ether was previously assumed to be uniformly distributed throughout space, while the electric and magnetic fields were considered only as certain deformations in it, the new "etherial" material is assumed to exist only in the places where electric and magnetic forces are present and is not so much the carrier of these forces as the materialized forces themselves. The physical properties of that

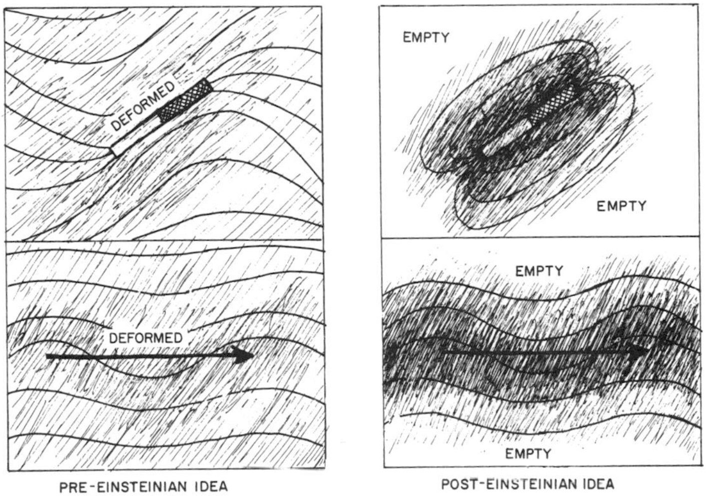


FIG. VI–9.

The old and the new concept of electromagnetic field. Before Einstein it was believed that all penetrating world ether was deformed in the region of electromagnetic field. Now we believe that electromagnetic field is a physical (ponderable) entity existing for itself in an empty space.

material are not to be described by old-fashioned terms such as rigidity, elasticity, etc., applicable only to material bodies made up of atoms and molecules, but by Maxwell's equations which describe in all details electromagnetic interactions. This new point of view requires some time and effort to be assimilated, but it frees the mind from the old "matteromorphical" (as in anthropomorphical) point of view about the light.

But what are the arguments for ascribing ponderable mass to this new "etherial" substance, and how much mass should be ascribed to it? The easiest way to answer this question is to consider what happens when a beam of light falls on a mirror and is reflected by it. It was known in physics for a long time that light reflected by a mirror exerts on it a certain pressure, which, being not strong enough to push over a mirror standing in front of a candle, succeeds in pushing gas molecules from the bodies of comets approaching close to the sun. These gas molecules form brilliant tails spread-

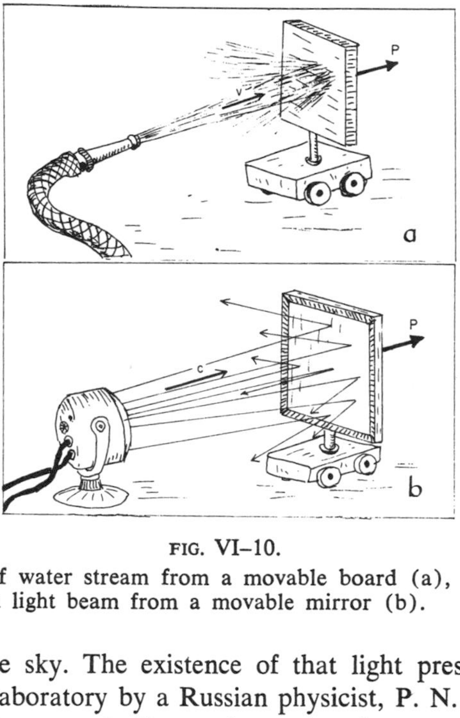


FIG. VI–10.
Reflection of water stream from a movable board (a), and the reflection of a light beam from a movable mirror (b).

ing through the sky. The existence of that light pressure was first proved in the laboratory by a Russian physicist, P. N. Lebedev, and was shown to be numerically equal to twice the amount of reflected energy divided by the velocity of light.

A close mechanical analogy to the pressure exerted by a light beam reflected from a mirror is the pressure exerted by a stream of water from a garden hose directed at a board placed in its way (Fig. VI-10). According to the laws of classical mechanics, the pressure exerted by a stream of material particles on the wall from which they are reflected is equal to the rate of change of their momentum, or the "amount of motion" in Newtonian terminology (cf. Chap.

IV). If $m$ is the mass of water carried by the stream per unit time, and $v$ is the velocity of the stream, the change of momentum is $2\,mv$, since it changes from $+\,mv$ to $-\,mv$. (Indeed, $mv - (-\,mv) = mv + mv = 2\,mv$.)

If we apply a similar argument to a beam of light reflected from a mirror, we will have to ascribe to it a mechanical momentum equal to the product of the "mass of light" $m$ falling on the mirror per unit time by its velocity $c$. Thus we write for the pressure of light:

$$P_{\text{light}} = 2mc$$

Comparing this expression with the empirical relation

$$P_{\text{light}} = \frac{2E}{c},$$

quoted before, we come to the conclusion that:

$$m = \frac{E}{c^2} \text{ or } E = mc^2$$

This is Einstein's famous "mass-energy equivalence law," which grants to the "imponderable" radiant energy of classical physics an equality with ordinary ponderable matter. Since $c^2$ is a very large number, $9 \times 10^{20}$, the mass of appreciable amounts of radiant energy is very small if expressed in customary units. Thus a flashlight with a 10-watt bulb emitting $6 \times 10^9$ ergs of light per minute becomes lighter by $\frac{6 \times 10^9}{9 \times 10^{20}} = 7 \times 10^{-12}$ gm. The sun, on the other hand, loses $4 \times 10^{11}$ tons per day by pouring its radiation into the surrounding space.

The relation between mass and energy must of course be generalized for all other kinds of energy. The fields surrounding electrically charged conductors, and magnets, become a ponderable physical reality even though the mass of the field surrounding a copper sphere 1 meter in diameter and charged to the potential of 1 kv weighs $2 \times 10^{-22}$ gm, whereas the field of an ordinary laboratory magnet would tip the scales by only $10^{-15}$ gm.

Heat energy must also possess ponderable mass, and 1 liter of water at 100° C weighs $10^{-20}$ gm more than the same amount of cold water, while the total energy liberated by a 20-kiloton atomic bomb weighs about 1 gm.

A few words must be said concerning the statement which infests newspapers and popular magazine articles to the effect that Einstein's energy-mass relation was the basis for the invention of the atomic bomb. It is quite incorrect, and one can say with the same right that the relation serves as a basis for Nobel's discovery of nitroglycerine, or Watt's invention of the steam engine. In all cases when a physical or chemical transformation takes place, with liberation of a certain amount of energy, the mass of the products is less than the mass of the initial ingredients by the mass of liberated energy. Thus, the gases resulting in the explosion of nitroglycerine weigh less than the original amount of explosive; the steam coming from a steam engine weighs less than hot water in the boiler; and the weight of liberated gases and ashes in burning wood weigh less than the original wood log. But in all these cases the weight of liberated energy is so small as compared with the weight of original material that it cannot be measured by using even the most precise scales. No physicist can notice the difference in weight between a glass of hot and cold water, and no chemist has ever detected the difference between the weight of water and the weight of gaseous hydrogen and oxygen which unite to produce it.

In the case of nuclear reactions, the amounts of energy produced are much greater, and although it would be an impossible task to collect all fission products of a bomb and to prove that they weigh just 1 gm less than the original plutonium core, one can determine by delicate methods of nuclear experimentation the exact mass values of individual atoms and the difference between the combined masses of atoms entering a nuclear reaction and those produced in it. But it is all just the difference in precision. Thus, Einstein's role in atomic bomb development was not the formulation of the $E = mc^2$ law, but a letter which he wrote to President Roosevelt which, by the weight of Einstein's authority, started the Manhattan Project rolling on its way.

A material body moving with a certain velocity carries within itself the kinetic energy of motion, and the additional mass of that energy accounts for the relativistic increase of mass. Einstein's equivalence law also applies to the transformation of elementary particles. To create a pair of an electron and an antielectron (or a proton and an antiprotron), an amount of energy equivalent to their combined mass has to be supplied, and the same amount of energy is liberated as high-frequency radiation when two particles are mutually annihilated.

## THE WORLD OF FOUR DIMENSIONS

Relativistic space contraction is mathematically equivalent to Fitzgerald's contraction of moving objects, but, whereas Fitzgerald thought about that contraction as a real physical effect caused by the motion of material bodies through ether, the theory of relativity considers it as an apparent shrinking of distances when they are viewed from a moving system. Both space contraction and time dilation are symmetrical in respect to both systems in a state of relative motion. Whenever space distances shrink, time intervals lengthen, which is in a way analogous to the case of vertical and horizontal projections of a stick with a given length $L$. If the stick is placed vertically its vertical projection is $O$ and its horizontal projection is $L$. If the stick is placed horizontally its vertical projection is $L$ and horizontal projection $O$. If the stick is placed at a certain angle $\theta$, both vertical and horizontal projections are different from zero.

But, no matter what the angle $\theta$ is, we have by the Pythagorean theorem:

$$\Delta x^2 + \Delta y^2 = L^2$$

This analogy led the German mathematician H. Minkowski (whose work closely followed Einstein's early publications) to the conclusion that time can be considered, in a way, as the fourth coordinate complementary to three space coordinates, and that the motion of one system in respect to another can be treated as a rotation of this four-dimensional coordinate cross.

In everyday life we specify different events by giving place and time information; we say the meeting will take place on the 15th floor at the corner of Sixth Avenue and 32nd street at 8:00 P.M. And it is customary to make diagrams in which positions are plotted against time. But such diagrams, which are not much different from diagrams which show the change of stock market prices from month to month, are nothing more than a graphical presentation of the dependence between two interrelated quantities, and cannot be considered in any sense as subject to any standard geometrical rules and operations. If time is to be considered as a legitimate fourth coordinate, it must first of all be measured in the same units as the three space coordinates. This can be done by multiplying time, originally given in seconds, by some standard velocity which leads to the distance expressed in centimeters, i.e., the same as the three space coordinates. It would be irrational to choose for that purpose some arbitrary unit such as the speed limit on highways (which depends upon local legislature), or even the velocity of sound (which depends on material and temperature). Clearly, the best choice will be the velocity of light in a vacuum, which is apparently connected with basic laws of nature, and was proved by the Michelson-Morley experiment to be invariable. Thus, by using $x$, $y$, and $z$ for the three first (space) coordinates, we will use $ct$ for the fourth (time) coordinate. But this is only the very beginning of what has to be done. In the case of space coordinates $x$, $y$, and $z$, all three are freely interchangeable, and the length of a wooden box becomes its height if we turn it on its side. It is clear that such a complete interchange cannot exist in the case of time and space coordinates. Otherwise, one would be able to turn a clock into a yardstick, and vice versa! Thus, if time is to be considered as the fourth coordinate, we should not only multiply it by $c$, but also by some other factor which, without spoiling the harmony of the four-dimensional coordinate system, would make the time coordinate physically *different* from the three space coordinates. Mathematics provides us with just such a factor known as an "imaginary unit" and designated by the symbol $i$. An "imaginary unit" is defined as the square root of minus one:

$$i = \sqrt{-1}$$

Since, according to elementary algebra, $(+1)^2 = +1$ and also $(-1)^2 = +1$ the number $i$ has no place among ordinary positive or negative numbers, and is therefore called an imaginary unit. It does not have any use in ordinary counting, and whereas "having \$1.00" means that there is one dollar in your bank account while "having \$(−1)" means that you are one dollar in the red, \$$i$ means nothing whatsoever in banking operations.

But mathematicians and theoretical physicists find the use of $i$ very convenient in their calculations provided it falls out in the final results which must have a physical interpretation. And it always happens when the final results contain only the squares of $i$, since

$i^2 = -1$ and is an ordinary negative number. Thus, let us use the "imaginary unit" as the additional cofactor, and write the fourth coordinate as *ict*. Since it is impossible to draw four mutually perpendicular axes, we skip the third space coordinate $z$ and use in its stead the new time coordinate *ict*. The result is a diagram shown in

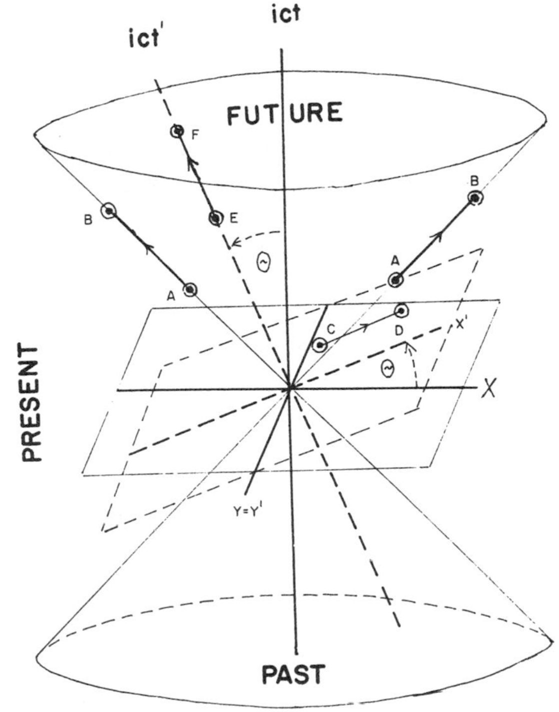


FIG. VI–11.

Space-time continuum containing two space coordinates ($x$ and $y$) and the time coordinate (*ict*). The conical surfaces representing the propagation of light ($x^2 \times y^2 - c^2t^2 = 0$) divide the continuum into "present," "past," and "future."

Figure VI-11 where space coordinates $x$ and $y$ axes are in horizontal plane (in respect to the reader) and the imaginary time axis runs vertically. Each point on this diagram represents an *event*, i.e., something that happened in a definite place at a definite time. Events which are simultaneous (in the particular reference system for which the diagram is drawn) are represented by the points in planes perpendicular to the time axis. Events which occur at different times but in the same place (again in this particular reference system) are on straight lines parallel to the time axes. The conical surface with a 90 degree opening, known as a "light cone," corresponds to events which can be connected by a light signal. If, for example, the point (event) *A* represents a flash which emitted a light wave, the point *B* corresponds to the illumination by that light of an object located somewhere else in space.

As was discussed earlier, observations of space and time intervals from a moving system can be interpreted geometrically as the rotation of a four-dimensional coordinate cross in which the time axis is turned by a certain angle (dashed lines and letters in Fig. VI-11). Since, however, the velocity of motion can never exceed the velocity of light *c,* the angle $\theta$ by which the *ict*-axis is turned can never be larger than 90 degrees. Thus, we can distinguish between two different kinds of event-pairs.

1) Events, like *E* and *F,* for which the angle between the line *EF* connecting them with the time axis makes an angle *smaller than 90 degrees.* In this case we can find a coordinate system moving in respect to the original one with such a velocity that both events will be on the new time axis, *ict'*, and their space distance becomes zero. This kind of rotation of space-time axes is a trivial one, and we all encounter it in everyday life. If, for example, we want to watch a football game on Monday in one city, and on Tuesday in another city, located a couple of hundred miles away, we just drive away after the end of the first game to be there before the beginning of the second. Although the positions of the two stadiums are different in respect to the equator and to the Greenwich meridian, both stadiums will be practically in the same place in respect to the coordinate system bound to the automobile. The space-time separation of the two athletic events described above is called *timelike* separation, because, by moving over with an appropriate speed, we can reduce to zero their space separation, and observe them from the same place (the seat of the car) at different times (a day apart).

2) Events, like *C* and *D,* for which the angle between the line *CD* and the time axis is *larger than 90 degrees.* In this case we cannot get from the first show to the second unless we move faster than light. Thus, for example, since light takes some 5 hours and 20

minutes to cover the distance from Mercury to Pluto, we cannot possibly attend a 1 o'clock luncheon on Mercury and a 5 o'clock cocktail party given the same day on Pluto. On the other hand, we can always select an appropriate traveling velocity, to reduce to zero the time difference between the two events and to make them simultaneous in the chosen space-time coordinate system. The space-time separation of such event-pairs is called *spacelike,* since by moving in an appropriate way, we can reduce the time difference to zero.

We can now give a new definition to the old notions of "past, present, and future." If we consider ourselves to be in the beginning of coordinates shown in Figure VI-11, and say: "I am here ($x = 0$, $y = 0$, $z = 0$), and now ($t = 0$)," all the events located in the upper part of the cone (positive $t$) will be the future since, no matter how we move, some time will pass before we see them. We can influence these future events by doing something about it, but we cannot be influenced by them. Similarly, all the events located in the lower part of the cone (negative $t$) will be the past, since no matter how fast we move we cannot see them. For example, it is impossible to fly out into space so fast as to catch the light waves of the explosion of the first atomic bomb or the burning of Rome. These past events can influence us, but we cannot influence them! Between the upper and the lower parts of the light cone lies the "no man's land" of what we ordinarily call "present." This includes the events which are either simultaneous from our point of view, or can be made simultaneous if we observe them from a system of reference moving with lesser velocity than that of light. The fact that the "present" in Figure VI-11 occupies such a large space is due, of course, to our decision to use $ct$, instead of just $t$, for counting time intervals. If we plot $t$ instead of $ct$, the vertical scale would shrink by a factor of $3 \times 10^{10}$, the upper and lower parts of the light cone would widen up, as the space between them would shrink to practically nothing. This is what we observe in everyday life with the velocities negligibly small compared with that of light.

Now, returning to three-dimensional space, and introducing $z$-coordinate, we can make some mathematical tricks involving the "imaginary unit" in the expression for the fourth coordinate. Suppose we send a light signal from the beginning of coordinates $x = 0$,

$y = 0$, $z = 0$ at the time $t = 0$. At the time $t$ the light signal reaches some position with space coordinates $x$, $y$, and $z$, and its distance from the origin will be, according to the Pythagorean theorem:

$$\sqrt{x^2 + y^2 + z^2}$$

Since light always propagates with the velocity $c$, this distance must be equal to $ct$, and we can write:

$$\sqrt{x^2 + y^2 + z^2} = ct$$

or

$$x^2 + y^2 + z^2 = (ct)^2$$

or

$$x^2 + y^2 + z^2 - (ct)^2 = 0$$

But, since $-1 = i^2$ we can rewrite the above as:

$$x^2 + y^2 + z^2 + (ict)^2 = 0$$

the left side of which is the Pythagorean sum of squares for four-dimensional space. In the primed coordinate system, moving in respect to the original one, we will have:

$$x'^2 + y'^2 + z'^2 + (ict')^2 = 0$$

so that the sum of *four* squares does not change because of the rotation of the four-dimensional coordinate system. One can show, by using Lorentz transformations, that the same is true for the space and time separation of any two points in $(x, y, z, ict)$ space representing two events. Thus the expression:

$$x^2 + y^2 + z^2 + (ict)^2$$

is invariant (i.e., unchangeable) no matter from which system of reference the two events are observed. Their three-dimensional space and one-dimensional time separations will change, but their four-dimensional separation given by the above expression always remains the same. Thus, using $ict$ as the fourth coordinate, we achieve a mathematical union of space and time, and can consider all physical events as taking place in the four-dimensional space-time world. We should not forget, however, that this can be achieved only by using an "imaginary unit," which is a treacherous helper, and that when the cards are down and the real values are called for, space and time are not exactly the same thing.

## RELATIVISTIC THEORY OF GRAVITATION

As we discussed above, Einstein's theory of relativity can be considered as a brilliant culmination of Galileo's arguments concerning mechanical experiments carried out in an inside cabin of a smoothly sailing ship. The generalization of that theory to the case of non-uniform motion, which is often called the general theory of relativity but is better described as the relativistic theory of gravitation, also has its roots in Galileo's experiment in which a light and a heavy body were dropped from the top of the Leaning Tower of Pisa. The empirical fact that light and heavy material bodies fall with exactly the same acceleration remained a complete mystery through the ages until Einstein's article on the relation between accelerated motion and the forces of gravity was published in 1914.

In this article Einstein describes imaginary experiments which can be carried out within a closed chamber freely floating in interstellar space. Because of the absence of gravity, all objects within have no tendency to move in any direction. If, however, the chamber is accelerated, say, by a couple of rocket motors attached to its bottom, the situation inside will be quite different; all the objects will be pressed to its floor as if there were a gravitational force pulling them down. Consider a man standing on the floor of such a space laboratory moving with a uniform acceleration $a$, in his hands two spheres, one light and one heavy. Because of acceleration of the entire system, the feet of the man will be firmly pressed to the floor, and the two spheres will press against the man's palms. What happens now after he releases simultaneously both spheres? Being disconnected from the body of the rocket, both spheres will continue to move with the velocity they had at the moment of release and will therefore remain side by side. On the other hand, since the motion of the rocket is accelerated, it will continuously gain speed and the floor of the chamber will soon overtake the two spheres and hit them at the same time. After that impact the spheres will remain pressed to the floor, being accelerated along with the rest of the system. The observer within the chamber, however, will observe that the two spheres he had released started falling with equal accelerations and thus hit the floor at the same time. This is the equivalence between gravity and acceleration, which is a matter of common knowledge in the "space age" in which we live.

But is this similarity of the mechanical phenomena taking place within an accelerated rocket ship and in the field of gravity produced by the large mass of the earth purely coincidental, or does it have a deeper connection with the nature of gravitational forces? Einstein felt sure that the latter was the case, and asked himself how, in that case, a light ray would behave within an accelerated chamber. Imagine that a flashlight is attached to the wall of the chamber, sending the beam of light across it. To observe the passing of the beam one can place on its way a number of equidistantly located plates of fluorescent glass (Fig. VI-12). If the chamber

FIG. VI–12.

Optical experiment in an accelerated rocket ship which suggests that light rays must be deflected by gravitational field.

is not accelerated, the points at which the beam crosses the glass plates will of course be on a straight line and it will be impossible to tell whether the rocket is at rest or in a state of uniform motion, let us say in respect to the fixed stars. However, the situation will be different if the chamber moves with a uniform acceleration $a$. Time needed for the light to reach the first, the second, the third, etc., plates of glass increases as arithmetical progression 1, 2, 3, etc., while the displacements of the rocket moving with a constant acceleration increases as geometrical progression 1, 4, 9, etc. Thus, the traces of the light beam at the plates of fluorescent glass will form a parabola similar to the trajectory of a horizontally thrown stone. Hence, if the equivalence of acceleration and gravity extends to electromagnetic phenomena, light rays must be bent by the gravitational field. However, because of the high speed of light, its bending in the gravitational field of the earth is too small to be observed. In fact, if a horizontal light beam travels, let us say, 30 meters before falling on the screen, it covers the entire distance $\frac{3 \times 10^3}{3 \times 10^{10}} = 10^{-7}$ seconds. Since the acceleration of gravity on the surface of the earth amounts to about $10^3 \frac{\text{cm}}{\text{sec}^2}$, vertical displacement of the light beam on the screen is expected to be:

$$\frac{1}{2} \times 10^3 \times (10^{-7})^2 = 5 \times 10^{-12} \text{ cm},$$

being comparable with the diameter of the atomic nucleus!

Einstein realized, however, that noticeable deflection of light rays may be expected when they pass close to the surface of the sun. Here is a rough estimate of expected deflection. Acceleration of gravity near the surface of the sun is the product of the gravitational constant ($6.7 \times 10^{-8}$) and mass of the sun ($2 \times 10^{33}$) divided by the square of the sun's radius ($7 \times 10^{10}$ cm) will be:

$$\frac{6.7 \times 10^{-8} \times 2 \times 10^{33}}{(7 \times 10^{10})^2} = 3 \times 10^4 \frac{\text{cm}}{\text{sec}^2}.$$

The distance traveled in the gravitational field of the sun is comparable to solar diameter ($1.4 \times 10^{11}$) and time taken to cover it is $\frac{1.4 \times 10^{11}}{3 \times 10^{10}} = 5$ sec. During that time the light beam will "fall":

$$\frac{1}{2} \times 3 \times 10^4 \times 25 = 3.7 \times 10^5 \text{ cm},$$

and the deflection angle will be:

$$\frac{3.7 \times 10^5}{7 \times 10^{10}} = 6 \times 10^{-6} \text{ radians}$$

or about 1 angular sec.

More exact calculations for the deflection of the light ray grazing the solar disc give the value 1.75 angular sec. Because stars near the sun can only be seen during a total solar eclipse, a British astronomical expedition went to Africa in 1919 where an eclipse was taking place. (German astronomers could not go because of the war blockade.) The results fully conformed to Einstein's predictions. When Einstein was told about these results he simply smiled and said that he would be very much surprised if the results were negative. This and other confirmations of the theory proved beyond any doubt the correlation between phenomena taking place in gravitational fields and in accelerated systems.

## GRAVITATION AND THE CURVATURE OF SPACE

Everybody knows what is meant by a curved line or a curved surface, but it takes some imagination to understand the meaning of curved space in three dimensions. The difficulty of forming a concept of curved space is that, while we can look at a surface from the outside and see if it is flat or curved, we live inside of the space and cannot get out of it to take a look. The best way to discuss the properties of curved space is to use an analogy of imaginary two-dimensional creatures who live on a surface, and have no idea that there is a direction perpendicular to their surface. How can they tell whether or not the surface they live on is a plane, a sphere, or still something else, without getting out of it? The answer is of course that they should study geometry on their surface by drawing various figures, measuring angles, etc. In Figure VI-13 we give an example of such two-dimensional geometers studying a triangle drawn on a plane, on a sphere, and on a so-called "saddle surface."

If it is a plane surface (a), the rules of Euclidean plane geometry apply, and the sum of three angles of a triangle will always be found to be equal to 180 degrees. On a spherical surface (b) the sum of three angles will always be larger than 180 degrees, as one can easily see by drawing on a globe a triangle formed by two half meridians, and a section of equator enclosed by them. Since meridians intersect the equator under a straight angle, the sum of the two angles at the base of our spherical triangle is already 180 degrees. To this we have to add an angle at the pole which can also be quite large. For smaller spherical triangles the sum of three angles will be closer to 180 degrees, but the difference disappears only when the triangle is infinitely smaller than the sphere on which it is drawn. On a saddle surface (c) the situation is different, and the sum of three angles is smaller than 180 degrees. It is customary

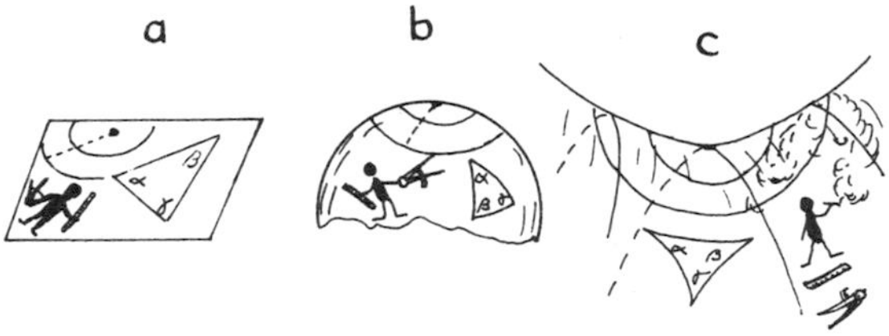


FIG. VI–13.

Three different types of curved (two-dimensional) surfaces. (a) Plane surface: zero curvature. (b) Spherical surface: positive curvature. (c) Saddle surface: negative curvature. The difference between the three cases can be discovered by two-dimensional intelligent beings if they study the geometry of circles or triangles.

to ascribe to a spherical surface positive curvature and to a saddle surface a negative curvature.

We can extend these conclusions to the case of the space of three dimensions, and say that the space is flat or possesses a positive or a negative curvature, depending on whether the sum of the angles of the triangles drawn between any three points in this space is equal to, larger than, or smaller than 180 degrees. Let us consider a large scale triangulation experiment in which three astronomers equipped with theodolites place themselves on the earth, on Venus, and on Mars, and measure the three angles of the triangle EVM. Since, as was discussed in the previous chapter, light rays are deflected by the gravitational field of the sun (being curved *toward* the gravitating body) the three rays forming that triangle will look as shown in Figure VI-14, and the sum of three angles will be found by astronomers to be *larger* than 180 degrees. Thus, our astronomers will conclude that the space around the sun is curved, possessing a positive curvature. If the measurement is repeated using the planets Jupiter, Saturn, and Uranus, which are farther away from the sun, the deflection of light rays by the sun's gravity will be smaller and the sum of the three angles will be closer to 180 degrees, indicating that the curvature of space around the sun decreases with the distance from it. One could object to the above-given interpretation of such measurement, saying that what the astronomers were measuring is not actually a regular triangle, since its sides are not straight lines. But, what *is* a

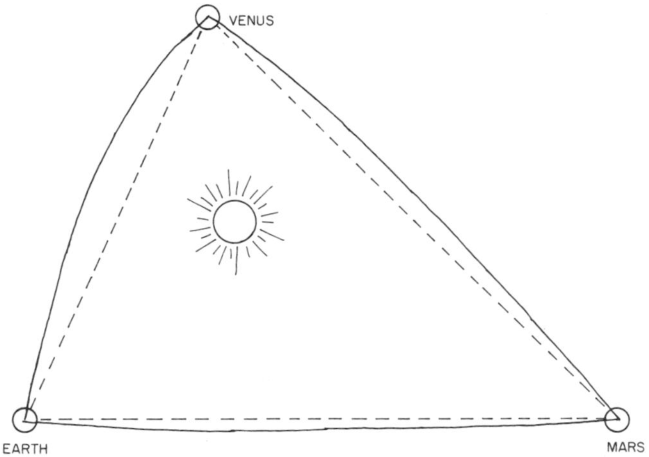


FIG. VI–14.
Triangulation of space around the sun.

straight line? The only reasonable definition of it is: "the line of sight," but the line of sight *is* the line of propagation of light through empty space! One can also define a straight line as "the shortest distance between two points," but all the science of optics is based on the postulate that light always takes the shortest route. If one thinks about the situation seriously, one finds that there is no other rational way to define a straight line, that the solid lines in Figure VI-14 must be considered as "straight" lines in "curved" space and that the dotted lines shown in that figure have no physical meaning at all. To avoid confusion in terminology, the term "straight line" is reserved only for the shortest distances in plane geometry, while on a curved surface and in curved space we speak instead of "geodesical lines." Thus, on the surface of a sphere the equivalents of straight lines are the arcs of great circles, and we use them to build spherical triangles. We may notice here that in spherical geometry the old statement of Euclidean geometry that "parallel lines never meet" does not hold any more, since *any* two great circles always intersect in two points, and two airplanes starting from two points on the equator in parallel directions perpendicular to the equator and flying without changing their course will collide upon reaching the pole.

The equivalence between the gravitational field and the curvature of space can be further clarified by the following two-dimensional example. If we roll a billiard ball along a flat horizontal table it will move, of course, along the conventional straight line. But if, for some reason, the table has a shallow depression and slightly bulges up on the path of the ball, the ball will be deflected from its "straight" path, turning toward the center of the depression and away from the top of the bulge. If we observe the motion of the ball from above the table (through a hole in the ceiling) we will not notice the defect of the table's surface, and will be inclined to believe that there is a certain force attracting or repelling the ball from a certain point on the table's surface. Similarly, the deflection of light rays and moving material bodies in the vicinity of the sun can be interpreted either as a force acting on them, or as the result of the curvature of space in the neighborhood of large masses.

Let us now look at our problem from another angle, and consider physical phenomena as seen by observers riding on a large rotating platform (Fig. VI-15). This thought experiment is similar to Einstein's box case discussed in the previous section, with the difference, however, that, instead of linear acceleration (i.e., the change of the numerical value of the velocity without the change of direction), we have a circular acceleration (the change of the direction of the velocity without the change of its numerical value). We may add to this rotating platform a semispherical dome rotating with it which prevents the people inside from noticing the trees and houses going around. As everybody knows, the people on a rotating platform will experience a centrifugal force pushing them away from the center, and they may interpret it as a special force of gravity which is repulsive rather than attractive. The analogy with gravity will be strengthened by the fact that, if one of these men, planting himself firmly on the platform, puts two spheres on the platform, one heavy and one light, the spheres will roll side by side very much in the same way as two objects dropped from a tower. Since the men on the platform are trained physicists and know all the arguments presented earlier in this chapter, they may correlate this "pseudogravitational field" with the geometry of space and try to carry out some geometrical measurements. First, they may try to build a triangle with the

FIG. VI–15.
**Geometrical studies on a rotating platform.**

vortices as *A, B,* and *C,* and to measure the sum of its angles. Using the definition of a straight line as the shortest distance between two points, physicist No. 2 (physicist No. 1 is just a Big Boss and supervises the work) gets a box of wooden sticks, all of exactly the same standard length, and tries to nail them in a line between the points *A* and *B* so as to use the smallest number of them. If the platform were not rotating, the best way to do it would be a broken line shown in the figure. But, in case of rotation the situation changes. The sticks are now moving in the direction of their length and are subject to Fitzgerald contraction; in fact, physicist No. 5, standing on the ground, will be sure they do. The middle stick moves exactly along its length and suffers the Fitzgerald contraction in full measure, while the sticks located closer to the periphery have at least a component of velocity along their length. Because of the shrinkage, there will appear spacings between the sticks, and No. 2 will have to add more sticks to make the line continuous. But there is at least a partial remedy for that trouble: if sticks are moved somewhat closer to the center of the platform, their linear velocities and also their shrinkage will be somewhat reduced and fewer additional sticks will be necessary. Thus, No. 2 will place his sticks as shown in the figure, and will be forced to do the same for the other two sides of the triangle. The sum of the three angles will now be *less* than 180 degrees, and the physicists on the platform will conclude that their space has negative curvature.

We may add that in case these physicists decide to check the above-described results by optical methods, they will come to the same result. Indeed, since the field of centrifugal forces is in all respects similar to a repulsive gravitational field, light rays connecting the vortices *A, B,* and *C* will be deflected *away* from the center of the platform and will follow the path laid by wooden sticks.

Now there are two more people on the platform, No. 3 and No. 4, doing something different. They are trying to measure the ratio of the circumference to the diameter which, in plane geometry, is designated by the Greek letter $\pi$. Here the rotation of the platform will also play havoc: while No. 3 will have no trouble because the sticks he uses move perpendicularly to their length, becoming thinner without changing their length, the sticks used by No. 4 will be subject to the maximum Fitzgerald contraction and he will have to use a whole lot of them. Thus the ratio of the circumference to the diameter as measured on the platform will come out larger than the figure 3.1416 . . . used in plane geometry. This result again confirms the conclusion concerning the negative curvature of space.

Let us return for a minute to the two-dimensional curved surfaces and see what happens if we draw circles on them. On the globe the circles with the pole as the center are known as "parallels," and it is apparent that the ratio of the length of a parallel to its diameter (as measured along the meridian) is less than the number $\pi$. In fact, the length of the Equator (0th parallel) divided by the length of the meridian is only 2. The length of the parallels increases more slowly than their radii as measured along the meridian, and for the 80th, 70th, 60th, etc. parallels (with radii 10, 20, 30 degrees, etc.) the lengths increase more slowly than 1, 2, 3, etc. Similarly the surface area within these parallels increases more slowly than 1, 4, 9, etc. The opposite situation exists on a saddle-surface, where the length of the circles increases *faster* than their radii, and the areas faster than the square of the radii. If we cut a circular piece of leather from a football, and put it on the table, it will bulge in the middle and we must stretch its periphery if we want to flatten it. On the contrary, a piece of leather cut from a western saddle will have too much leather on the rim, which must be shrunk to make it flat. By this analogy again, one must ascribe a negative curvature to the space within a rotating laboratory.

In the case of three-dimensional space, the surface of the sphere increases more slowly than $r^2$ and its volume more slowly than $r^3$ in case of positive curvature, the opposite being true for the space of negative curvature. This mathematical result provides a basis for a very interesting piece of work in the field of astronomy which was carried out by Edwin Hubble at Mt. Wilson Observatory quite a number of years ago. Hubble, being a great expert on stellar galaxies, billions of which are scattered through the space of the universe within the range of big telescopes, decided to investigate whether the number of galaxies within various distances from us increases in direct proportion, more slowly, or faster than the cubes of these distances. If the first possibility is true, we must conclude that the space of the universe is Euclidean. In the second case, the space has positive curvature, and must eventually close on itself. In the third case, the space has a negative curvature, being wide open in all directions. Unfortunately the observational technique of measuring intergalactic distances was, at that time, not sufficiently developed, and Hubble's results were self-contradictory and inconclusive. It is hoped that repetition of Hubble's "galactic counts" with better observational means will lead to an answer to that important problem of cosmology.

Being led by the considerations described above, Einstein developed a theory according to which all gravitational interactions should be interpreted as due to the curvature of space. Fortunately for Einstein, a detailed mathematical theory of curved spaces of any number of dimensions was developed many decades earlier by a German mathematician, Bernhard Riemann, so all Einstein had to do was to apply the existing mathematical formula to the physically real curved space. Of course it was a four-dimensional space, with the coordinates *x, y, z,* and *ict* discussed earlier in this chapter. Correlating the so-called "curvature tensor" of space-time continuum, with the distribution and motion of masses (this basic formula is shown under Einstein's picture in Fig. VI-6), Einstein was able to get all the results of Newton's theory of gravity as the first approximation. However, more exact calculations indicated that there must be some small deviations from Newton's original theory of gravity, and the discovery of these deviations would have to prove the superiority of Albert's views over Isaac's. One of the conclusions of Einstein's theory of gravity, the deviation of the path of light in the gravitational field, has already been discussed. Another important point concerns the motion of planets around the sun. Newton has shown that, according to his law of gravity, planets must move along the elliptical orbits around the sun in agreement with the empirical laws discovered by Kepler. In Einstein's theory, all motions should be considered in the four-dimensional world (*x, y, z, ict*), which is curved if gravitational fields are present. The lines representing the "history of motion" of any material body in the four-dimensional world, known as "world lines" of that body, must be the "geodesic," i.e., the shortest, lines, and can be calculated on the basis of the relativistic theory of the gravitational field.

In Figure VI-16 we give a graphical presentation of the world line of the earth in its motion around the sun. The two space coordinates *x* and *y* are taken in the plane of elliptics while the third one is time coordinate *ict*. The space-time continuum in the vicinity of the sun is curved and the world line of the earth corresponds to the straightest (i.e., geodesic) line in this curved space. Thus, the line *ABCD* is the shortest distance between the points (events) *A* and *D* in the three-dimensional, space-time continuum, and its projection on the plane (*x, y*) is the orbit of the earth around the sun. The exact calculations have shown, however, that this ellipse does not remain stationary in space as Newton's theory indicated, but is slowly rotating with its major axis turning by a small angle in the course of each revolution. The effect must be most noticeable in case of the orbit of Mercury, which is more elongated than the orbits of other planets and which is closest to the sun. Einstein calculated that the orbit of Mercury must turn by 43 angular seconds per century, and solved herewith the old riddle of celestial mechanics. It was calculated by mathematical astronomers long before Einstein was born that the major axis of Mercury's orbit must slowly turn around because of the perturbations, i.e., gravitational disturbances, of the other planets of the solar system. But, there was a discrepancy between the calculations and the observations amounting to 43 angular seconds per century which could not possibly be explained. Einstein's relativistic theory of gravitation

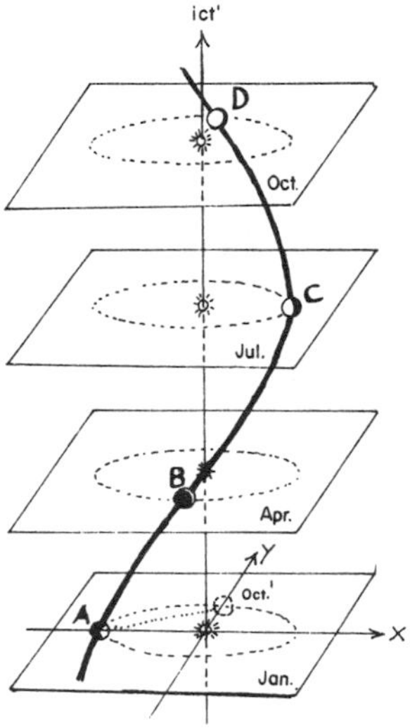


FIG. VI–16.

The world line of the earth in its motion around the sun plotted in *x, y, ict,* coordinate system. The space-time dimensional distance between earth's position in January and in October is the shortest distance. But the distance between January position and the *projection* of October position on January plane (*Oct.'*) is, of course, not the shortest.

filled this gap, and became an indisputable conqueror over the old Newtonian theory.

## THE UNIFIED FIELD THEORY

The life work of Albert Einstein resulted in geometrization of a large part of physics: time became a legitimate fourth companion of three space coordinates (except for *i*-factor), and the forces of gravity were interpreted as due to the curvature of this four-dimensional world. But electric and magnetic forces still remain outside this geometrical conquest, and, having progressed so far, Einstein directed all his energy to putting on the stubborn electromagnetic field a rigid geometrical bridle. What is the so far unexplored geometrical property of four-dimensional space which can account for electric and magnetic interactions? Einstein himself, and many other "interested bystanders" such as the famous German mathematician Hermann Weyl did their best to give to the electromagnetic field a purely geometrical interpretation. But, with typical Scottish stubbornness, William Clerk Maxwell's child, the electromagnetic field, refused to be geometrized. For almost four decades, until his death in 1955, Einstein worked on the so-called "unified field theory," i.e., the theory which would unify the gravitational and electromagnetic fields on a single geometrical basis. But, as the years dragged by, the task became more and more hopeless. Every so often Einstein came out with new sets of formulas which, he claimed, would be destined to solve the riddle of the unified field theory, and these complicated tensor expressions were printed on the front pages of the *New York Times* and other papers all over the world. But the formulas always turned out to be unfit for the job, and silence fell again until the next revelation. Theoretical physicists, old and young, gradually lost confidence in the possibility of giving to the electromagnetic field a purely geometrical interpretation. It would have been fine if this could have been done, but Nature cannot be forced to do what is not in her nature. On the other hand, physics was making rapid progress in the newly discovered fields, and in addition to the classical gravitational and electromagnetic fields, new fields introduced by wave mechanics took a firm position in science. If a purely geometrical interpretation is given to the electromagnetic field, we would have to subdue meson fields, hyperon fields, and many other new fields, in order to be able to say: physics is nothing but geometry. Einstein himself became more and more touchy about his thesis, and more and more reluctant to discuss these problems with other physicists. During one of his visits to Great Britain in the early thirties, he gave a lecture on the unified field theory at a girls' school in northern England (the blackboard with the complicated tensor formulas has been preserved by the school authorities), but refused to talk in Cambridge University. His attention became more and more occupied by the problems of Zionism and world peace, but his scientific agility remained as sharp as ever. When the author of this book used to visit Einstein during the World War II years in his quiet home in Princeton, he found him to be his same old charming self, and remembers many informative and interesting conversations on various branches of modern physics. There were pieces of paper scattered over Einstein's desk, covered with complicated tensor formulas apparently pertaining to the unified field theory. But about that Einstein would never talk. Now he is certainly in paradise, and must know whether he was right or wrong in his attempt to geometrize all physics.

# CHAPTER VII *The Law of Quantum*

## DIVISIBILITY OF MATTER

AS EVERYBODY KNOWS, the *atom* (which means "indivisible" in Greek) is a brain child of Democritus, who lived and taught in Athens some 23 centuries ago. He considered it inconceivable that material bodies could be divided into smaller and smaller parts without any limit, and had postulated that there must be ultimate particles, so small that no further division into still smaller parts is possible. Democritus was recognizing four different kinds of atoms—those of stone, of water, of air, and of fire—and believed that all the variety of known materials results from the various combinations of these four elements. His views, which were adopted and put on a firm experimental foundation early in the 19th century by the British chemist John Dalton, form the basis of all modern chemistry even though we know now that atoms are not at all indivisible and, in fact, possess a rather complicated internal structure. But Democritus' idea of ultimate elementarity is now transferred to much smaller particles which constitute the internal structure of atoms, and one may hope that electrons, protons and other so-called "elementary particles" are really and truly elementary and indivisible in the good old Democritean sense of the word. It may be that this impression results from our comparatively weak familiarity with those comparatively recently discovered particles, and that we are making here the same mistake as the 19th-century physicists and chemists who believed that the divisibility of matter stopped at atoms. And it may also happen, of course, that if the elementary particles of today are in the future found to be complex structures with new names invented for their constituent parts, this will prove not the end of the road, and years later a next step to still smaller particles will be made. There is no way to predict scientific developments of the future, and the question of whether Democritus' original philosophical concept of indivisibility was right or wrong will never be answered by empirical means. But somehow, many scientists, including the author, feel happier with the thought that in the study of matter "things will come to an end," and that the physicists of the future will know all there is to know about the inner structure of matter. And it also seems quite plausible that the elementary particles of modern physics deserve their name one hundred percent, because their properties and behavior seem to be much simpler than could ever be said about the atoms.

### A CHIP OFF THE OLD ATOM

Toward the end of the 19th century, physicists turned their attention to the passage of electricity through gases. It was known for centuries that gases, being ordinarily rather good electric insulators, sometimes can be broken through by high electric tensions. The intensities of discharge range from tiny sparks between the doorknob and the hand of a man who walks across a carpeted floor in rubber shoes, to the mighty lightning during thunderstorms. But Sir William Crookes, whose contributions to science are only partially obscured by his beliefs in spiritism and the supernatural, has shown that the passage of electricity through gases takes place in a much more peaceful way if the gas pressure is reduced to a small fraction of 1 atmosphere. Crookes' tubes were glowing with quiet light of a color depending on the nature of the gas, and they are still glowing on the streets of cities today advertising hotels, night clubs, and thousands of other things. When the gas pressure in a tube to which a high electric tension is applied is sufficiently low, there appears a sharply defined beam shooting from a cathode to an anode, and hitting the far end of the tube if the mischievous physicist moves the anode out of the way of the beam. Hitting the glass wall, the mysterious beam emanating from the cathode plate would make it glow with diffuse green light, and any object placed in its way would cast well-defined shadows. Placing a magnet near the tube, Crookes observed the deflection of the beam as it would be in the case of electric current or a swarm of negatively charged particles flying away from the cathode. About the same time, Jean Perrin in France found that a metal plate placed in the way of that beam acquired a negative electric charge. This all seemed to indicate that these must be negatively charged particles flying through rarefied gas in very much the same way as Faraday's ions moved through liquids in the process of electrolysis. The essential difference was, of course, that whereas in the case of electrolysis the ions had to shoulder their way slowly through tightly packed molecules of the liquid and would never miss their way to the opposite electrode, the *cathode rays* (as they were called) in rarefied gases were flying straight and were hitting whatever was in their way.

These views were opposed by the German physicist, Philipp Lenard, who had found that the cathode ray could pass quite easily through various screens placed along the way without making holes in them, as any material particles would surely do. Only waves, and not a beam of material particles, could do it, argued Lenard. Of course today when we know that concrete walls many feet thick should be built around atomic piles to stop neutrons from coming out and causing radiation sickness in the atomic plant personnel, Lenard's argument sounds rather weak. But it was a very strong argument at the time it was given.

The task of resolving the experimental contradictions proving that cathode rays *are* streams of particles, and finding the physical characteristics of these particles, was given by the Supreme Council of the Progress of Science to Joseph John Thomson, later Sir Joseph (Fig. VII-1, *right*), a Manchester-born physicist, who was by that time 40 years old and the director of the famous Cavendish Laboratory at Cambridge, one of the major centers of contemporary physics. Assuming that cathode rays are formed by fast-flying particles, Thomson decided to measure their mass and electric charge. One piece of information concerning these quantities was the observed deflection of cathode rays in the magnetic field (Fig. VII-2b). This deflection depends not only on the charge and mass of the flying particles but also on their velocity, and by measuring it one can find only the product $\frac{\textit{mass} \times \textit{velocity}}{\textit{charge}}$ or $\frac{mv}{e}$ in conventional notations. It followed from the theory, however, that the deflection caused by the electric field (Fig. VII-2a) depended on the other combination of the same quantities, namely on the product $\frac{mv^2}{e}$. Thus, measuring both deflections and combining the results, Thomson could find separately the velocity of their motion $v$ and their charge-mass ratio $\frac{e}{m}$. While $v$ depended on electric potential applied to the tube, $\frac{e}{m}$ was always the same, being equal to $5.28 \times 10^{17} \frac{\text{esu}}{\text{gr}}$.*

FIG. VII–1.
Lord Rutherford (*left*) and Sir J. J. Thomson.

Although it was almost certain that *e* must have the same numerical value as the elementary electric charge found by Faraday in his experiments with the electrolysis of liquids, Thomson undertook a special experiment to measure that value for gas ions. His method was based on the discovery of another Cavendish physicist, C. T. R. Wilson ("A brilliant star is C. T. R.!" goes the old Cambridge song), who found out that, if the dustless air saturated with water vapor is suddenly cooled by expansion, tiny water droplets are formed on any

* esu is the electrostatic unit as described in Chapter V; gr is a unit of mass.

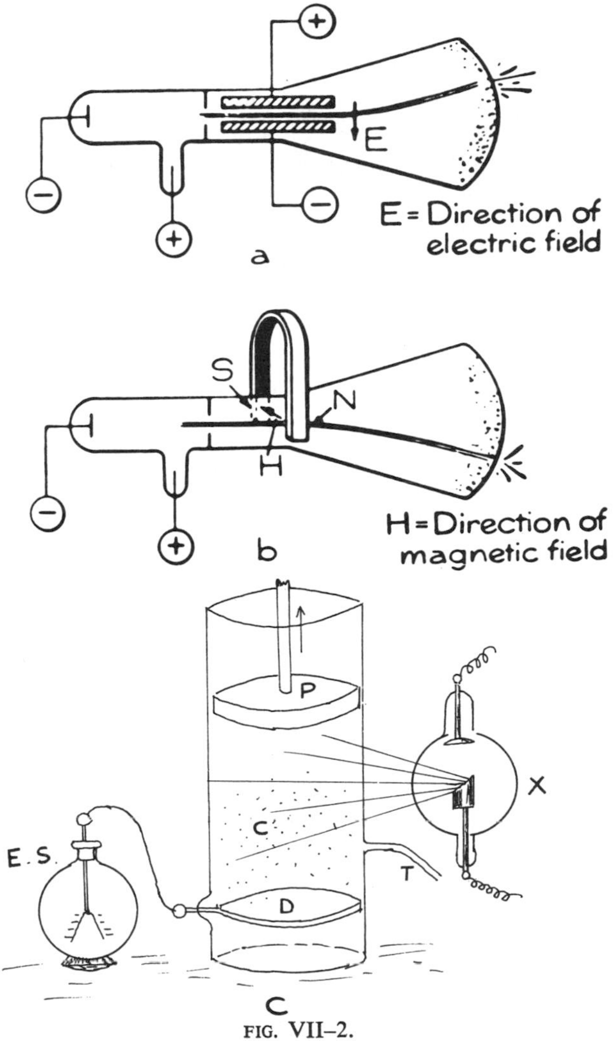


FIG. VII–2.

Sir J. J. Thomson's measurement of the mass of an electron. (a) Deflection in an electric field permits the measurement of $\frac{m}{e}.\ v_2$. (b) Deflection in a magnetic field permits the measurement of $\frac{m}{e}.\ v$. Combining the two results, one finds $\frac{m}{e}$. (c) The rate of fall of droplets formed on gas ions permits the measurement of $e$. Knowing $\frac{e}{m}$, and $e$, $m$ is easily found.

ions that may be present in it.* For small expansions (below 30%) only negative ions serve as condensation centers, whereas for larger expansions, water vapor condenses both on positive and on negative ions. The scheme of Thomson's experiment is shown in Figure VII-2c. It consists of a glass cylinder *C,* with a piston *P,* and a metal disc *D* connected with an electroscope. The cylinder is filled with moist air through the tube *T* and illuminated by the X-rays. When the piston is suddenly pulled up, producing the expansion of the air (smaller than 30%), a cloud of fog formed by water condensation on negative ions appears in the chamber. The fog slowly settles down on the disc *D,* and the total electric charge of the ions formed is measured by the electroscope. Knowing the initial amount of water vapor in the cylinder and the average size of fog droplets, one can find the total number of droplets produced or, what is the same, the total number of ions. Since the droplets were too small to see, Thomson decided to find their size from the velocity with which the fog settles down to the disc. The smaller the droplet, the more slowly it settles down, and there exists a formula, first derived by Stokes, which gives the relation between the fall velocity, the radius of the droplet, and the viscosity of the air. Using this method, and dividing the total charge received by the electroscope by the number of droplets, Thomson found that the electric charge of each droplet has the value $4.77 \times 10^{-10}$ esu, i.e., the same value as in the case of the electrolysis of liquids.

Now Thomson could find the value of $m$ from the previously measured $\frac{e}{m}$ ratio, and it turned out to be $0.9 \times 10^{-29}$ gm, i.e., 1,840 times smaller than the mass of the hydrogen atom.

Here was a great discovery: a particle almost two thousand times lighter than the lightest of all atoms! Thomson concluded that, while Faraday's ions were the atoms carrying electric charges, the particles forming cathode rays were just free electric charges themselves, and gave them the name of *electrons.* He visualized an atom as a sphere of positively charged massive material with a multitude of tiny electrons spread through its body as are black seeds in the red flesh of a watermelon. It was, as one says, a "static model," i.e., electrons within the atom were supposed to be at rest at certain equilibrium

* If the air contains some dust, saturation of vapor will first take place on the dust particle, confusing the experiment.

positions determined by the balance of the forces of electrostatic repulsion between the negatively charged electrons, and the forces of electrostatic attraction between the electrons and the center of the positively charged body of an atom. When an atom becomes excited, i.e., gets some surplus energy from outside, electrons in its interior were supposed to oscillate about their equilibrium positions, emitting electromagnetic (light) waves of various lengths. Laborious calculations were carried out in the attempt to correlate the vibration frequencies of different electronic configurations with the observed line spectra of various chemical elements, but this work was in vain, and the problem rested until the appearance of Rutherford's atomic model.

### MYSTERIOUS X-RAYS

A number of important discoveries which occurred toward the end of the 19th century and rapidly transformed physics from its "classical" to its "modern" form occurred accidentally. But these discoveries always involved keen-minded scientists who were attentive enough to notice unusual things and to pursue their studies until the important facts came out. On November 10, 1895, the German physicist Wilhelm Konrad Roentgen (Fig. VII-3), carrying out some experiments with cathode rays in a Crookes tube, noticed that a fluorescent screen standing by chance nearby on the table became brilliantly luminous when the electric current was passing through the tube. Roentgen covered the tube with a piece of black paper, but the fluorescence did not vanish. A metal sheet, on the other hand, definitely stopped the effect. Thus, here was a new radiation emanating from the tube which could easily pass through materials opaque to ordinary light. The first photograph which Roentgen took by means of the newly discovered radiation, which he called X-rays, was the hand of his wife, which clearly showed the bone structure and her wedding ring. Further studies indicated that this penetrating radiation came from the far end of the glass tube which was struck by the beam of the cathode rays. The intensity of X-rays could be considerably increased by placing in the path of the cathode ray a plate of heavy metal known as "anticathode" (Fig. VII-3). The emission of X-rays is ascribed to the impact of the fast-moving electrons which form cathode rays (as the reader remembers, electrons were discovered by Sir J. J. Thomson just two years later) against the target placed in their way. Being suddenly stopped in their tracks, electrons spit out their kinetic energy in the form of very short electromagnetic waves, similar to sound waves resulting from the impact of bullets against an armor plate. And, just as in the case of bullets, the emitted sound contains all possible frequencies and is described as "noise" rather than as a pure musical tone. X-rays represent a mixture of a continuum of wave lengths. Germans call them "Bremsstrahlung" ("Bremse" means brake; "Strahlung" means radi-

FIG. VII–3.
Max von Laue (*left*) and Wilhelm Roentgen.

ation) and this term is customarily used, as are many other German terms, in English. (Indeed, it would sound odd to say "brake radiation"!)

Since X-rays were not deflected by the magnetic field, Roentgen had assumed from the very beginning that there are vibrations similar to ordinary light. If so, they should show the phenomenon of diffraction, and Roentgen spent years in the attempt to prove it experimentally but without any positive result. Twelve years after his great discovery, Roentgen, who was at that time professor of experimental physics at the University of Munich, was invited by a young (33 at that time) theoretical physicist, Max van Laue, of the same university, to inspect some photographs which had just been taken by assistants W. Friedrich and P. Knipping. At his first glance he realized that this was exactly what he had been looking for for years: the beautiful diffraction pictures produced by X-rays passing through a crystal (Plate IV, *upper*). Von Laue came to the idea of using a crystal as a diffracting grating on the basis of purely theoretical consideration. Since X-rays did not show any diffraction phenomena when ordinary optical gratings were used, they must have a much smaller wave length. Now, the crystalline lattice is formed by regular layers of atoms or molecules located about $10^{-8}$ cm apart. When a beam of X-rays falls on a crystalline surface, it penetrates deep into the crystal, being partially reflected from each layer it passes through (Fig. VII-4). If the angle of incidence is such (b) that the reflected wavelets are in phase, the intensity of the reflected beam will be increased. For the other angle (a), in which the wavelets are out of phase, darkness is to be expected. Just as in the case of optical gratings the diffraction pattern can be observed either in the reflected or transmitted beam. The situation is complicated by the fact that crystals have many systems of parallel molecular layers so that the expected picture looks more complicated than in the case of ordinary light. Plate IV, *upper,* taken in the Bell Telephone Laboratories shows the diffraction of X-rays in nickel-iron alloy.

It was later discovered that, apart from the continuous "Bremsstrahlung," the X-rays contain also the sequences of sharp lines which are quite similar to optical spectra, and originate from electron transitions deep inside the atoms. Much work on X-ray line spectra was carried on by W. Bragg (father) and W. L. Bragg (son) who developed the precise mcthods of X-ray spectrography.

## ISOTOPES

In the beginning of the 19th century a British chemist, W. Prout, was impressed by the fact that atomic weights of various elements, expressed in terms of the atomic weight of hydrogen, are very closely represented by integer numbers. This observation led him to the hypothesis that the atoms of different chemical elements are nothing else but agglomerates of different numbers of hydrogen atoms: helium = 4 hydrogens; carbon = 12 hydrogens; oxygen = 16 hydrogens, etc. Prout's contemporaries did not share his views, and were quick to point out a number of facts which contradicted his brave hypothesis. Thus, for example, the atomic weight of chlorine and cadmium were found to be 35.457 and 112.41 respectively, which is just about halfway between two integers. Also for the elements with atomic weights close to the integer, the values were always some-

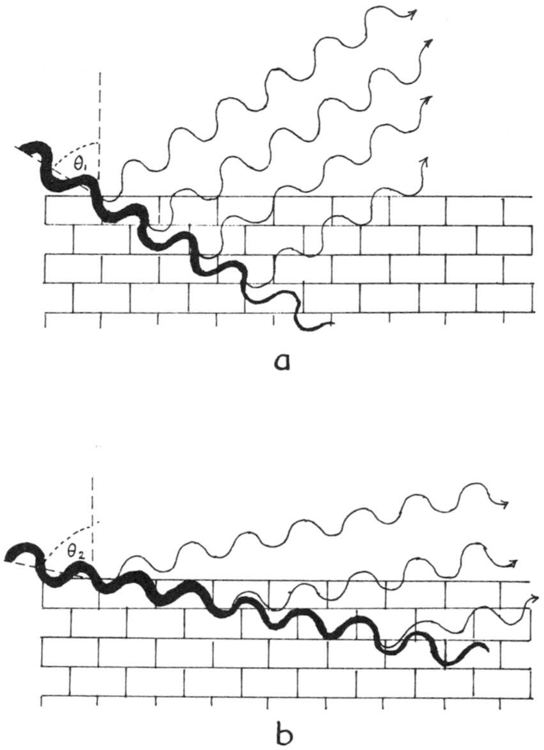


FIG. VII–4.

Reflection of X-ray or de Broglie waves from the surface of a crystal. In (a) the wavelets reflected from successive layers of crystal lattice (symbolized by bricks) are out of phase and cancel one another. In (b) the wavelets are in phase, which results in an increased intensity.

what smaller than would be expected if their atoms were formed by aggregation of the atoms of hydrogen. Since the atomic weight of hydrogen is 1.0080,* the figure for helium should be 4 × 1.0080 = 4.0320, whereas it is actually 4.003, i.e., 0.8% lower. Similarly,

* Chemists adjust atomic weight so as to have atomic weight of oxygen equal to 16.0000000000.

twelve hydrogens put together would weigh 12 × 1.0080 = 12.096, whereas the chemically estimated atomic weight of carbon is only 12.010. Because of these "apparent" discrepancies, Prout's hypothesis was rejected and forgotten for almost half a century until its glorious resurrection in 1907 as a result of J. J. Thomson's studies.

Having established the existence of the electron, and having measured its mass and charge by the deflection of electron beams in the electric and the magnetic fields, J. J. turned his attention to the particles moving in the opposite direction through the electric discharge tubes. The beams of these positively charged particles were

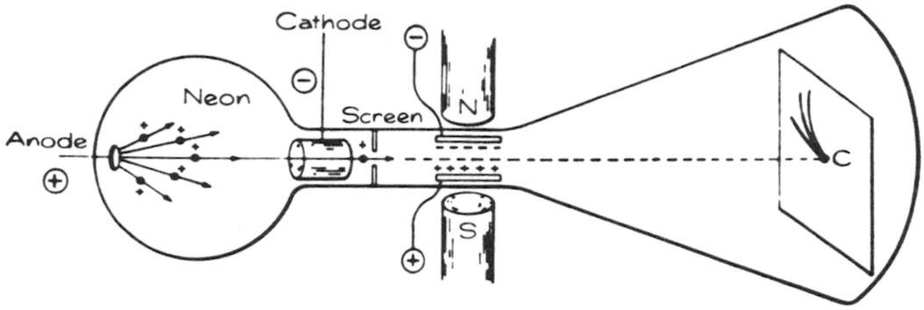


FIG. VII–5.

Thomson's apparatus for the study of canal rays. Positive ions moving from anode toward cathode pass through channels driven in the cathode, and after passing through a screen enter into the region of electric and magnetic field oriented in the same direction. Since magnetic deflection (in horizontal direction) depends on the velocity of particles, whereas magnetic deflection (in vertical direction) depends on the square of that velocity, particles of the same mass but moving with different velocities will be distributed along a parabola on the screen *C*.

known as "canal ways" since they were first observed by drilling holes (canals) in the cathode plate, which permitted the particles to pass through into the space beyond it. The apparatus used by Thomson for studying canal rays is shown in Figure VII-5, and is based on the same principle used in his studies of electron beams. Positively charged particles, originating in the gas discharge between the anode and the cathode, passed through a hole (canal) drilled in the cathode, and entered the region of electric and magnetic fields oriented in the same direction. As was discussed before, the vertical deflection of the beam caused by the electric field is proportional to $\frac{e}{m}v^2$, while the horizontal deflection due to the magnetic field varies as

$\frac{e}{m}v$. Thus, for the particles with the same charge to mass ratio but different velocities, vertical deflections are proportional to the squares of the horizontal deflections, and the curves observed on the fluorescent screen *S* are expected to be parabolas.

This was exactly what J. J. observed, but instead of just one parabola (for any given chemical element) there were two or more, indicating the presence of atoms with different mass. In chlorine for example, one parabola for chlorine atoms with mass 34.98 and another for chlorine atoms with mass 36.98 are obtained, both figures being very close to integers. The atoms of the same element having, however, different atomic weights were christened "isotopes," i.e., occupants of the same place in Mendeleev's table. The relative number of the chlorine atoms of these two different weights (estimated from the blackening of the photographic plate) was found to be 75.4% and 24.6% respectively. Thus, the average atomic weight would be $34.98 \times 0.754 + 36.98 \times 0.946 = 35.457$, which coincides exactly with the chemically estimated atomic weight of chlorine. Further studies by F. W. Aston have shown that the same is true for other chemical elements. Thus, for example, cadmium consists of eight different kinds of atoms with the weights 106, 108, 110, 111, 112, 113, 114, 116, and the relative amounts 1.4, 1.0, 12.8, 13.0, 24.2, 12.3, 28.0, 7.3%; the average weight is 112.41 in perfect agreement with chemical measurements. Thus, Prout's old idea came back to prominence.

But even with the discovery of isotopes some discrepancy remained since, for example, the exact atomic weights of the two isotopes of chlorine were 34.98 and 36.98 instead of 35.280 ($= 35 \times 1.008$) and 37.296 ($37 \times 1.008$). This time, however, it was not a trouble but a pleasure, since, according to Einstein's energy-mass equivalence law, the combination of several particles should weigh less than the original particles by the amount of their mutual binding energy divided by $c^2$. Thus, the difference between the mass of a composite atom and the combined mass of its constituents tells us about the energy involved in the formation process. Let us take, for example, the carbon atom ${}_6C^{12}$, which consists of 6 protons and 6 neutrons. The exact mass of a hydrogen atom is 1.008131, whereas for a neutron it is 1.008945. Thus the total mass should be $6 \times 1.008131 + 6 \times 1.008945 = 12.102456$. The exact measurements, however, give for the mass of the carbon atom the value 12.003882, i.e., 0.098546 units less. This so-called *mass defect* must represent the mass of energy liberated in the formation of the carbon nucleus from neutrons and protons. According to Einstein, this corresponds to energy $0.0986 \times 1.66 \times 10^{-24}$* $\times 9 \times 10^{20} = 1.48 \times 10^{-4}$ erg or 92.5 mev.

## RUTHERFORD'S ATOMIC MODEL

Ernest Rutherford (Fig. VII-1) was born in 1871 near the city of Nelson on the southern island of New Zealand, and when many years later he was given a British title for his scientific merits he became Lord Rutherford of Nelson. At the age of 24 he came to Cambridge to study under J. J. Thomson at the Cavendish Laboratory, and after graduation received a professorship at McGill University in Montreal, where he made his first important contributions to the study of the newly-discovered phenomena of radioactivity. Later he moved to the University of Manchester and in 1919, after the retirement of J. J. Thomson, became Director of Cavendish. He was known among his colleagues by the name of "Crocodile," given to him by one of his favorite students, a Russian physicist, Peter Kapitza. It must be noted that, while to the English who often go (or rather went) to Egypt and are bitten or eaten by crocodiles this nickname sounds rather derogatory, for the Russians, who never see crocodiles in their native country, it stands as a symbol of robust power. While nobody dared to mention that nickname in his presence, Rutherford knew about it and was secretly proud of it, and the wall of the new building constructed for Kapitza's studies on very strong magnetic fields carries, for a reason never made official, a large bas relief of a crocodile.

Recalling Cavendish days reminds the author of a Cambridge incident which took place in connection with "Crocodile". . .

> . . . that handsome, hearty British lord
> We knew as Ernest Rutherford.
> New Zealand farmer's son by birth,
> He never lost the touch of earth;
> His booming voice and jolly roar
> Could penetrate the thickest door,
> But if to anger he inclined

* $1.66 \times 10^{-24}$ gm is 1/16 of the mass of an oxygen atom.

You should have heard him speak his mind
In living language of the land
That anyone could understand!

One day George Gamow, as his guest,
By Rutherford was so addressed
At tea in honour of Niels Bohr
(Of whom you may have heard before).
The men talked golf, and cricket too;
The ladies gushed, as ladies do,
About a blouse, a sash, a shawl—
And Bohr grew weary of it all.
"Gamow," he said, "I see below
Your motorcycle. Will you show
Me how it works? Come on, let's run!
This party isn't any fun."
So to the motorcycle Bohr,
With Gamow running after, tore.
Gamow explained the this and that
And Bohr, who on the saddle sat,
Took off to skim along the Backs,
A threat to humans, beasts and hacks,
But though he started full and strong
He didn't sit it out for long.
No less than fifty yards ahead
He killed the nervous engine dead
And, turning wildly as he slowed,
Stopped traffic up and down Queen's Road.

While Gamow, rushing to the fore,
Was doing what he could for Bohr
Who should like Jove himself appear
But Rutherford. In Gamow's ear
He thundered: "Gamow! If once more
You give that buggy to Niels Bohr
To snarl up traffic with, or wreck,
*I swear I'll break your bloody neck!*"

But let us return to Rutherford's years in Manchester. He did not like Thomson's watermelon model of the atom and decided to probe the atomic interior by shooting through it new kinds of projectiles which fell into the hands of physicists after the discovery of radioactivity. During his earlier years in McGill, Rutherford had shown that the so-called alpha particles emitted by various radioactive elements are actually the beams of positively charged helium ions ejected with tremendously high energy from the unstable atoms. Interacting with the charged parts of the atoms, alpha particles must be deflected from their original paths, and the resultant scattering of the beam should disclose information about the distribution of electric charges in the atomic interior. Thus Rutherford was directing the beams of

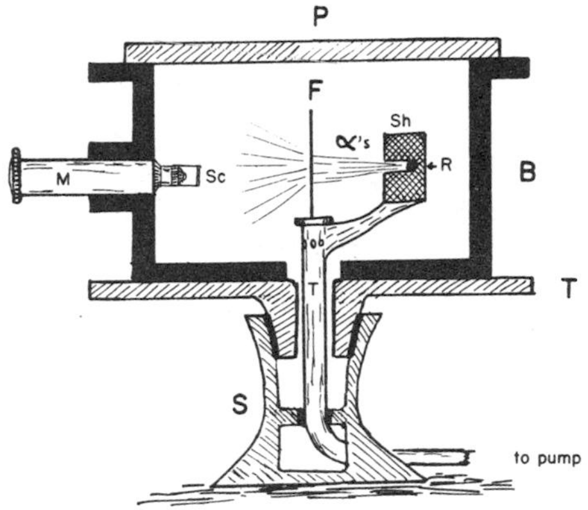


FIG. VII-6.

The first apparatus for the study of alpha-ray scattering. An evacuated box *B* with removable upper plate *P* is placed on a revolving table *T*. Radioactive source *R* placed in a lead shield *Sh* and the scattering filament *F* are attached immovably to the stand *S*. Microscope with the scintillation screen *Sc* is attached to the box and can rotate around horizontal axis.

alpha particles at thin foils made of different metals (Fig. VII-6) and counting the number of particles scattered in different directions after passing through the foil. In those days particle counting was a laborious procedure. Whereas today a physicist can set an automatic Geiger counter and go for a walk or to the movies, Rutherford had to look through a microscope at a fluorescent screen placed in the way of the beam, and count on his fingers the scintillations, i.e., the tiny sparks occurring when a high-energy particle hits the screen. Some nuclear physicists at that time were even swallowing belladonna in order to open the pupils of their eyes wider. As a result of these studies, Rutherford found that the scattering of alpha particles passing through metal foils was considerable. Although most of the particles from the incident beam retained their original direction of motion, a number of them were deflected by many degrees, and some were even thrown practically backward. This result did not fit at all with what would be expected from Thomson's model, in which the mass and the positive charge were distributed almost uniformly throughout the entire body of the atom. In fact, in that case, the

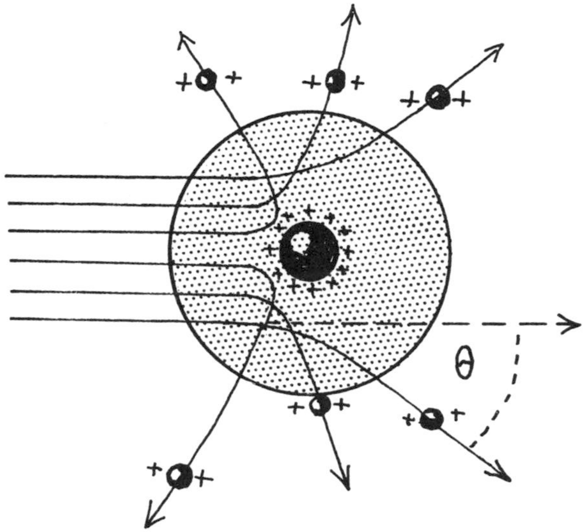


FIG. VII–7.
Nuclear model of an atom.

interaction between the charge of the incident particle, and the internal charges of the atom would never be strong enough to deflect an alpha particle by a large angle from its original direction, not to mention the possibility of throwing it backward. The only possible explanation was that the positive charge and the mass of an atom were concentrated in a very small region, practically in a point, in the very center of the atom (Fig. VII-7). To see if such an assumption was in agreement with the observed scattering, a formula based on the laws of mechanics had to be developed, for the deflection of particles passing at various distances from the center of repulsion.

Like many great experimentalists, Rutherford disliked mathematics, and, at least according to gossip, this formula was derived for him by a young mathematician, R. H. Fowler, who later married Rutherford's daughter. According to Rutherford's formula, the number of alpha particles deflected by an angle $\theta$ from the original direction of their motion must be inversely proportional to the inverse fourth power of $\sin \frac{\theta}{2}$, and this conclusion agreed very nicely with the observed scattering curves. Thus, there emerged an entirely new picture of an atom with a tiny but massive and heavily charged central core, which Rutherford called the *atomic nucleus,* and a swarm of electrons circling around it under the action of coulomb attraction. It looked more or less like our system of planets circling around the sun and held in their orbits by the forces of Newtonian gravity. It was later established by the work of Rutherford's students, H. Geiger and E. Marsden, that the positive charge of the atomic nucleus, and, which is the same, the number of electrons circling around it, is equal to the position number, or *atomic number,* of the element in question in Mendeleev's Periodic System of Elements. Thus the present picture of atomic structure came into being.

## ULTRAVIOLET CATASTROPHE

We must now go back a little way in history to the last decade of the 19th century when physics was going through the pains of metamorphosis from the classical larva to the modern butterfly. At that time, the kinetic theory of heat was well developed by the works of Boltzmann, Maxwell, and others, and there was no doubt that what one calls "heat" is the result of irregular random motion of innumerable molecules which form all material bodies. In the simplest case, that of gases in which the molecules fly freely through space, one could derive simple mathematical expressions from the distribution of velocities, the number of intermolecular collisions, and other molecular characteristics of thermal phenomena. At this stage, a renowned British physicist, astronomer, and popular book writer, Sir James Jeans, decided to apply statistical methods, which proved to be very successful in the study of the thermal motion of molecules, to the problem of thermal radiation. We have seen in Chapter IV that hot bodies emit continuous light spectra with vibrations of all frequencies and wave lengths. We have seen also that, for each given temperature, there exists a certain distribution of available energy between different wave lengths, and that the wave length corresponding to the maximum concentration of energy changes with the change of the temperature (Fig. IV-13). Jeans asked himself whether the distribution of energy between different wave lengths in the case of radiation is subject to the same statistical laws as the distribution of energy between the molecules of gas. Let us consider a so-called "Jeans' cube," which is a box inlaid by "ideal mirrors," i.e., mirrors which reflect 100% of the light which falls on them. Of course every mirror in existence absorbs some of the incident light before reflecting it, but we are speaking here about a "thought experiment" similar to Einstein's box in the relativistic theory of gravitation. Having such a Jeans' cube with a small window and a shutter on its side, we can open the shutter, shine in some light from a lamp, and imprison the light by closing the shutter again. Since the light cannot be absorbed by the walls of the container, it will undergo innumerable reflections, and if, an hour or two later, we open the shutter again, the light will rush out of the container, as gas rushes out from the open valve of an automobile tire.

In Figure VII-8 we give a comparison between two containers, one filled with molecules in thermal motion, another with thermal radiation of various wave lengths. In the first case, the molecules rush through space in all possible directions with all possible velocities, being reflected from the walls of the container, and occasionally colliding between themselves in flight. In the second case, we have light waves of various lengths propagating in all possible directions, being reflected from the mirror walls.

Lacking in the second picture are the "collisions between the waves" which would permit the exchange of energy between them. Indeed, the basic property of any kinds of waves, be they waves in the ocean, sound waves, or radio and light waves, is that they do not influence each other when they meet. Bow waves of two ships sailing side by side, sound waves carrying the conversation of several people in a room, radio waves from two broadcasting stations located in the same city, or the two beams of light from searchlights crossing in the sky, pass through each other as if they were two good old medieval ghosts. To remove this lack of analogy, let us imagine inside of the Jeans' cube a few tiny particles of coal dust which may absorb some energy of one wave length, and remit it in another wave length. We speak figuratively about coal dust because it is black, and it is known that black bodies (or rather, *ideal* black bodies to match the *ideal* mirrors of Jeans' cube) would absorb and emit the radiation of any wave length. The coal dust particles are introduced

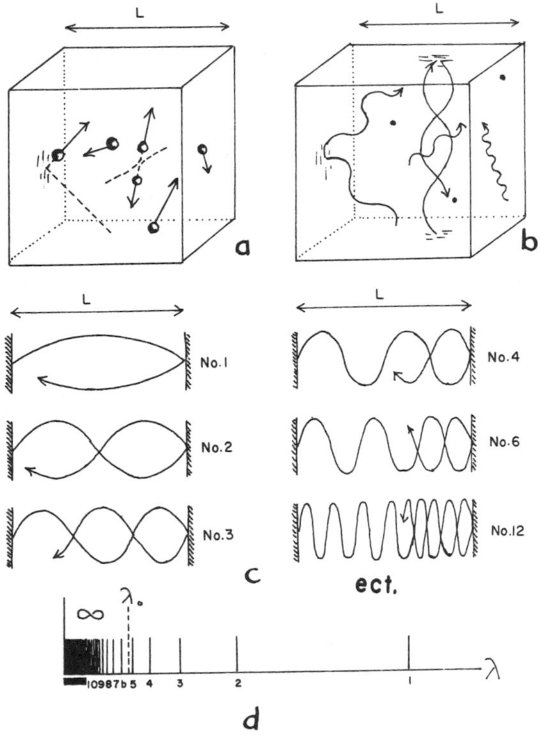


FIG. VII–8.

Comparison between the random motion of gas molecules in a closed container (a), and random motion of waves in Jeans' cube (b). Black dots in (b) represent tiny particles of coal dust serving as energy exchangers between the waves. (c) shows various modes of vibrations in Jeans' cube (for simplified one-dimensional case), while (d) gives the corresponding spectrum.

in this thought experiment just to permit the exchange of energy between the light vibrations of different wave lengths. They can do it without taking energy away from the system because of their very small size and correspondingly small heat capacity.

Now let us see how the available energy will be distributed between different vibrations which can exist in the Jeans' cube. In statistical physics there is one basic rule known as "the law of equipartition of energy." It states that, if we have a very large number of systems (such as individual gas molecules) which are in statistical interaction among themselves, the available energy will be on the average distributed equally among all of them. Thus, if there are altogether $N$ gas molecules in the container, and the total available energy is $E$, each molecule will have an average energy:

$$\mathcal{E} = \frac{E}{N}$$

The same simple law must be applicable to the multitude of waves which can exist inside of the Jeans' box. But how many such waves can there be? Considering, for simplicity, only the waves running horizontally between the right and left wall of the box (Fig. VII-8c), we find that the situation is similar to that of a violin string fastened at the two ends (compare the work of Pythagoras described in Chap. I). The longest wave possible will be No. 1, with a wave length twice the distance $L$ between walls. The next shorter one will be No. 2, with a wave length $L$ or, what is the same, $\frac{2L}{2}$. The still shorter waves will be $\frac{2L}{3}, \frac{2L}{4}\left(\text{or } \frac{L}{2}\right), \frac{2L}{5}, \frac{2L}{6}\left(\text{or } \frac{L}{3}\right), \ldots \frac{2L}{100}, \frac{2L}{101}, \ldots \frac{2L}{1{,}000{,}000} \ldots$. There is no lower limit to the possible wave length of electromagnetic vibrations, and, continuing the above sequence, we will pass through visible light, the ultraviolet, X-rays, gamma rays, and so on. Thus, the number of possible vibrations is infinite, and, generalizing this argument for the waves propagating in all three directions, we obtain, of course, the same result. Thus, following the classical equipartition law, and dividing the available energy, no matter how large it is, among all possible waves, we get:

$$\mathcal{E} = \frac{E}{\infty} = 0$$

Physically, it means the following. If we divide all possible wave lengths shown in Figure VII-8d into two groups, by a vertical line marked $\lambda_0$, there will always be a finite number of possible vibrations on the right of $\lambda_0$, but an infinite number between $\lambda_0$ and zero point.

The equipartition principle will demand that all available energy should be given to the vibrations with the wave length shorter than $\lambda_0$, no matter how small $\lambda_0$ is. Therefore, if we fill Jeans' cube with red light, this light will begin to turn (through the absorption and re-emission by coal dust) into ultraviolet rays, X-rays, gamma rays, etc. . . . What is true for the hypothetical Jeans' cube must also be true in general, and opening an oven door in the kitchen, or shutter in a locomotive furnace, we would be hit by lethal short wave radiation, and be dead then and there. This conclusion is clearly nonsensical, but, on the other hand, it results from the application to radiant energy the most fundamental laws of classical physics.

For a number of years after the publication of Jeans' paper, neither Jeans nor anybody else knew how to explain this paradoxical result. Then, during the last week of the last year of the last century, a German physicist, Max Planck (Fig. VII-9), stepped to the blackboard at the Christmas meeting of the German Physical Society, and made an extraordinary proposal. The idea was that light and all other kinds of electromagnetic radiation, which were always considered as continuous trains of waves, actually consist of individual energy packages with well-defined amounts of energy per package. The amount of energy per package depends on its vibration frequency $\nu$, and is directly proportional to it, so that one can write:

$$\mathcal{E} = h\nu$$

where $h$ is a universal constant. Planck called these energy packages *light quanta* (or, more generally, *radiation quanta*), and the constant $h$ is known as the *quantum constant.*

Now, how does this revolutionary idea of Max Planck remove the danger of Jeans' ultraviolet catastrophe? To give to the reader a glimpse of how this can be done, let us consider the case of a man who died, leaving an estate of, say, \$600. He had no heirs, but only five creditors: a bartender, a butcher, a druggist, a grocer, and a tailor, each of whom wanted to collect his money, the total debts being considerably larger than the available money. A simple solution would be to use the "equipartition law" and to give \$100 to each. But things are complicated by the fact that each of the creditors would take only all money due him or nothing. The bartender wants to collect all his \$600, the butcher and druggist demand \$300 each, while the grocer asks for \$200 and the tailor for \$100. Since there is not enough money to pay all these debts, the judge must revert to what is known among lawyers as "equity," i.e., the common-sense solution. It would clearly be irrational to give all $600 to the bartender, and to deprive the rest of the creditors of any money at all. A more reasonable solution would be to spend more money to satisfy the creditors with lower demands, and refuse the demands

FIG. VII–9.
Niels Bohr (*left*) and Max Planck, and quantum transitions in a hydrogen atom.

of those who asked too much. Thus, for example, the tailor could be given $100, the grocer $200, either the butcher or druggist $300 (by tossing a coin), and nothing to the bartender. (It may be noticed here that this principle of distributing money is actually used by the National Science Foundation, which has very little money and tries to distribute it fairly among various contract-seekers.) It is doubtful that equity provides a unique solution for problems of that kind, but statistical physics does. Once Planck's hypothesis concerning the minimum amount of energy in the light quanta of different wave lengths was introduced, the exact laws of mathematical statistics came to work, depriving many of the short-wave vibrations of any energy at all, because of their unreasonably high demands. As a result, one obtains a formula for the distribution of energy in thermal radiation in which most of the energy is given to the average wave length, while the high-demanding short wave vibrations get very little or nothing.

The formula derived by Planck on the basis of his light quanta hypothesis turned out to be in perfect agreement with all the known laws of thermal radiation. But introducing the idea of individual energy packages into the classical picture of wave propagation of light has produced a revolution of ideas comparable only with that which resulted from the Michelson-Morley experiment.

### THE REALITY OF LIGHT QUANTA

While the original Planck notion of the packages of radiant energy was rather vague, and served only as the basis for statistical energy distribution among different wave lengths in the spectra, it took a more definite shape in the hands of Albert Einstein five years later. In one of three articles published by him in 1905,* Einstein applies the idea of light quanta to the explanation of the so-called "photoelectric effect." It was known for some time that light (particularly ultraviolet light) which falls on metallic surfaces communicates to them a positive electric charge. After the discovery of electrons, it was proved that this effect is due to the ejection of electrons from the illuminated surfaces.

A standard arrangement for the study of photoelectric effect is shown in Figure VII-10. Light from an electric arc *A* (which contains a lot of ultraviolet rays) passes through a system of two quartz lenses and a prism ("monochromotor") which separates the different wave lengths. The selected beam (which can be changed by rotation of the prism) enters through a quartz window *W* into an evacuated tube *T*, passes through a hole in the bottom of a copper cylinder *C*, and falls on a metal plate *Pl* which can be made of different materials. A variable electric potential between the plate and the cylinder slows down the motion of emitted photoelectrons. (Battery *B* and

* As already mentioned, the other two were on Brownian motion and on the theory of relativity.

variable resistance *R* supply electric tension, while the galvanometer *G* measures the current.) When the applied electric potential multiplied by the charge of electron becomes equal to their kinetic energy, the current in the circuit stops. Thus, varying the intensity and wave length of incident light and measuring the potential at which the current stops, one finds the relation between the intensity and frequency of light and the velocity of photoelectrons. Experimental studies of the photoelectric effect resulted in two laws:

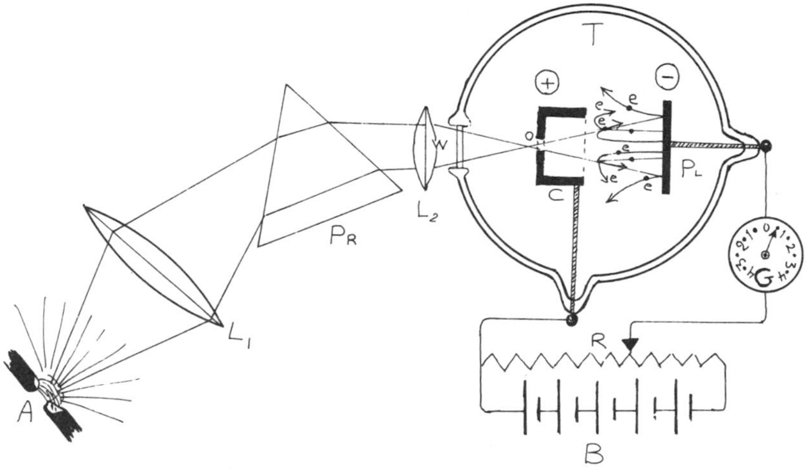


FIG. VII–10.

Arrangement for the study of the photoelectric effect. Photoelectrons ejected from the plate *Pl* toward the cylinder *C* are stopped by the electric field if the potential difference between *Pl* and *C* is large enough.

I. For the given frequency of incident light, the energy of ejected photoelectrons does not change, but their number increases in direct proportion to light intensity.

II. When the frequency of incident light changes (increases) no electrons are emitted until a certain threshold frequency (depending on the metal) is reached. For higher frequencies, the energy of photoelectrons increases in direct proportion to the difference of the frequency used and the threshold frequencies.

These two laws are shown graphically in Figure VII-11. These very simple laws, however, did not fit at all with the predictions of the classical electromagnetic theory of light. According to that theory, the increase of *the intensity* of light would mean the increase of the oscillating electric force in the wave. Acting on electrons near the surface of the metal (these are the electrons which carry electric current through metallic wires), this stronger electric force should throw them out with higher kinetic energy. But the experiment showed that, even if the intensity of light was increased one hundredfold, photoelectrons were coming out with exactly the same velocity.

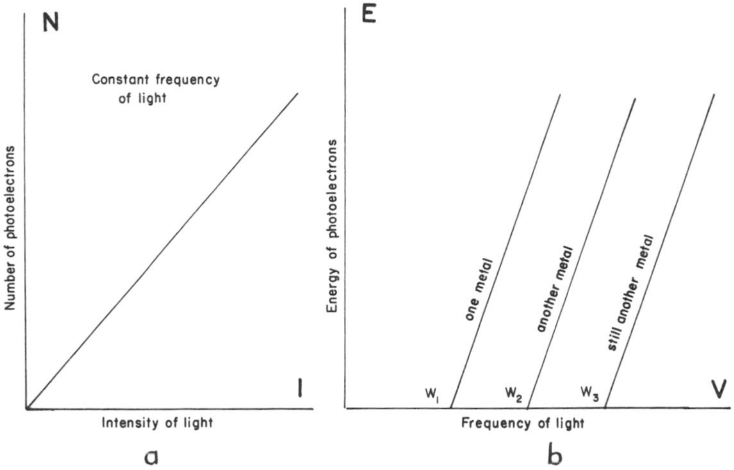


FIG. VII–11.

Experimentally found laws of photoelectric effect. (a) The dependence of the number of the photoelectrons on the intensity of light. (b) The dependence of the energy of the photoelectrons on the frequency of light.

On the other hand, the curve in Figure VII-11 shows a very definite relation between the velocity (or kinetic energy) of electrons and the *frequency* of the incident light, the relation for which there seems to be no reason in the classical electromagnetic theory of light.

Using the idea of light quanta carrying a definite amount of energy proportional to their frequency, one gets, however, the explanation of two empirical laws in quite a natural way. When an incident light quantum hits the surface of metal and interacts with one of the electrons, it must communicate to the electron all its energy, since there can be no energy less than one quantum. Larger intensity of the incident light means more light quanta of the same frequency, and therefore proportionally more electrons with the same kinetic energy. When the frequency of the incident light increases, the situation is different. Each light quantum has now more energy and communicating it to an electron throws it out of metal with higher velocity. Passing through the surface of metal, the electron loses a certain amount of energy received from the light quantum; this amount of energy depends on the nature of the metal and is known under the (very inappropriate) name "work function." Thus the energy of the photoelectron is given by a very simple formula:

$$E = h\nu - W$$

where $W$ stands for the work function of the metal in question. As long as $h\nu < W$ (or $E < 0$), electrons do not get enough energy from light quanta to cross the surface and nothing happens. As soon as $h\nu$ becomes larger than $W$, photoelectron emission begins, and the energy of electrons increases linearly with $\nu$. The slope of the curve in Figure VII-11b must be equal to quantum constant $h$, and indeed it is! Thus, with one stroke, Einstein explained the mysterious laws of photoelectric effect, and gave a vigorous support to Planck's original idea concerning the packages of radiant energy.

Another vigorous support of the light quantum hypothesis, which by that time was already deserving the name of a theory, was given by the work of an American physicist, Arthur Compton, Hawaiian-guitar player, tennis champion, and great investigator of the nature of cosmic rays. These later studies brought him the fame of being the strongest man in all Mexico. The circumstances, told to the author by Compton himself, were as follows. Studying the changes in the intensity of cosmic rays from the pole to the equator, Compton had to make a measurement of that intensity somewhere in the southern part of Mexico. The site for the measurements had to be somewhere away from cities in order to avoid the disturbances caused by power lines, traffic, etc., but on the other hand it should be a place with a sufficiently good supply of electric current. The solution was a Catholic monastery some distance south of Mexico City, a place which was quiet, had its own power station and storage batteries, and had an abbot who wished to further the progress of science. Compton arrived at the railroad station closest to the monastery with some twelve boxes loaded with scientific equipment; they were neat-looking wooden boxes, about the dimensions of a medium-sized suitcase, with metal handles to carry them. Two boxes contained four Kohlrausch electrometers—black metal spheres with little windows through which to observe a filament registering electric charge. The rest of the boxes were loaded with lead bricks used for radiation shielding.

It is well known to all visitors to Mexico that, upon arrival at the station, passengers are immediately surrounded by a crowd of barefoot men and boys shouting: *"Llevo su equipaje, señor?"* and pulling the traveling bag from your hand. In this case Compton picked up the two boxes containing Kohlrausch electrometers and nodded to the Mexicans to take the rest. And here was a procession: *Americano distinguido* walking lightly along the platform, swinging two instrument boxes in his hands, and a line of Mexicans, two to each box, bending under their weight. But that was not the end of the adventure. When the truck loaded with Compton and his instrument boxes arrived at the gates of the monastery, it was stopped by two Mexican soldiers who requested the inspection of the luggage. The point was that at that time the government of Mexico was engaged in a great quarrel with the Catholic Church, and guards were placed around all Catholic institutions. Opening the boxes, the soldiers found "four black bombs, and a lot of lead" which could presumably be used only to make bullets. Compton was arrested, and had to wait several hours in the local police station before long-distance calls to the American Embassy in Mexico City settled the matter. The intensity of cosmic rays in the monastery in question turned out to be exactly what was expected.

But to return to the Compton effect. Being a hard-boiled experimentalist, Compton liked to visualize collisions between light quanta and electrons as similar to those between ivory balls on a billiard table, except for the fact that, whereas all billiard balls are exactly alike (save for color), light quanta and electrons should be considered as balls of different masses. He argued that, in spite of the fact that the electrons forming the planetary system of an atom are bound to the central nucleus by attractive electric forces, these electrons would behave exactly as if they were completely free if the light quanta which hit them carried sufficiently large amounts of energy. Suppose that a black ball (electron) is resting on a billiard table and is bound by a string to a nail driven into the table's surface and that a player, who does not see the string, is trying to put it into the corner pocket by hitting it with a white ball (light quantum). If the player sends his ball with a comparatively small velocity, the string will hold during the impact and nothing good will come from this attempt. If the white ball moves somewhat faster the string may break, but in doing so it will cause enough disturbance to send the black ball in a completely wrong direction. If, however, the kinetic energy of the white ball exceeds, by a large factor, the strength of the string that holds the black ball, the presence of the string will make practically no difference, and the result of the collision between the two balls will be the same as if the black ball were completely unbound.

Compton knew that the binding energy of the outer electrons in an atom is comparable to the energy of the quanta of visible light. Thus, in order to make the impact overpoweringly strong, he selected for his experiments the energy-rich quanta of high frequency X-rays. The result of a collision between X-ray quanta and (practically) free electrons can indeed be treated very much in the same way as a collision between two billiard balls. In the case of an almost head-on collision, the resting ball (electron) will be thrown at high speed in the direction of the impact, while the incident ball (X-ray quantum) will lose a large fraction of its energy. In the case of a side hit, the incident ball will lose less energy and will suffer a smaller deflection from its original trajectory. In the case of a mere touch, the incident ball will proceed practically without deflection and will lose only a small fraction of its original energy. In the language of light quanta, this behavior means that, in the process of scattering, *the quanta of X-rays deflected by large angles will have a smaller amount of energy and, consequently, a larger wave length.* The experiments carried out by Compton confirmed, in every detail, the theoretical expectations and thus gave additional support to the hypothesis of the quantum nature of radiant energy.

## BOHR'S ATOM

In the year 1911 there arrived in Manchester a young (25-year-old) Danish physicist named Niels Bohr (Fig. VII-9), who, during his studies in the University of Copenhagen, applied his experience as a nationally known football player to the problem of the "straggling" of alpha particles through a crowd of atoms which try to tackle and stop it. At this time Rutherford was just going through his epoch-making experiments which led to the discovery of the atomic nucleus. Bohr liked Rutherford's ideas, and Rutherford said to a friend: "This young Dane is the most intelligent chap I have ever met." Thus, they became friends and remained fellows-in-arms forever after.

It is practically impossible to describe Niels Bohr to a person who has never worked with him. Probably his most characteristic property was the slowness of his thinking and comprehension. When, in the late twenties and early thirties, the author of this book was one of the "Bohr boys" working in his Institute in Copenhagen on a Carlsberg (the best beer in the world!) fellowship, he had many a chance to observe it. In the evening, when a handful of Bohr's students were "working" in the Paa Blegdamsvejen Institute, discussing the latest problems of the quantum theory, or playing Ping-pong on the library table with coffee cups placed on it to make the game more difficult, Bohr would appear, complaining that he was very tired, and would like to "do something." To "do something" inevitably meant to go to the movies, and the only movies Bohr liked were those called *The Gun Fight at the Lazy Gee Ranch* or *The Lone Ranger and a Sioux Girl*. But it was hard to go with Bohr to the movies. He could not follow the plot, and was constantly asking us, to the great annoyance of the rest of the audience, questions like this: "Is that the sister of that cowboy who shot the Indian who tried to steal a herd of cattle belonging to her brother-in-law?"

The same slowness of reaction was apparent at scientific meetings. Many a time, a visiting young physicist (most physicists visiting Copenhagen were young) would deliver a brilliant talk about his recent calculations on some intricate problem of the quantum theory. Everybody in the audience would understand the argument quite clearly, but Bohr wouldn't. So everybody would start to explain to Bohr the simple point he had missed, and in the resulting turmoil everybody would stop understanding anything. Finally, after a considerable period of time, Bohr would begin to understand, and it would turn out that what he understood about the problem presented by the visitor was quite different from what the visitor meant, and was correct, while the visitor's interpretation was wrong.

Bohr's addiction to Western movies resulted in a theory which is unknown to all but his movie companions of the period. Everybody knows that in all Western movies (Hollywood style at least) the scoundrel always draws first, but the hero is faster and always shoots down the scoundrel. Niels Bohr ascribed that phenomenon to the difference between willful and conditioned actions. The scoundrel has to decide when to grab for the gun, which slows his actions, while the hero acts faster because he acts without thinking when he sees the scoundrel reach for the gun. We all disagreed with that theory, and the next morning the author went to a toy shop to buy a pair of cowboy guns. We shot it out with Bohr, he playing the hero, and he killed us all.

Another example of Bohr's slowness of thinking was demonstrated by his inability to find a quick solution to crossword puzzles. One evening the author drove up to Bohr's country house in Tisvileleje (north Jutland) where Bohr had been working all day long with his assistant, Leon Rosenfeld (from Belgium), on an important paper concerning uncertainty relations (see later) for the electromagnetic field. Both Bohr and Rosenfeld were completely exhausted from the day's work, and, after dinner, Bohr suggested "for relaxation" to work on a crossword puzzle from some British magazine. It did not go too well and, about one hour later, Fru Bohr ("Fru" is Danish for "Mrs.") suggested that we all go to sleep. At some unknown hour of the night, Rosenfeld and I, who shared a guest room upstairs, were awakened by a knock on the door. We jumped up in the darkness, crying "What? What's happened?" There came a muffled voice through the door: "It is me, Bohr. I do not mean to disturb you, but I just want to say that the English industrial city with seven letters, ending in *ich* is Ipswich!"

"I do not mean to . . . but . . ." was Bohr's favorite expression, and many a time he would walk in with an open magazine in his hand, saying: "I do not mean to criticize, but I just want to understand how can a man write such a nonsense!"

One more factual story about Niels Bohr before we go to his theory of the atom. Once, late at night (about 11 P.M. Copenhagen time) the author was returning with Bohr, Fru Bohr, and a Dutch physicist, Cas Casimir, from a dinner given by one of the members of Bohr's institute. Cas was an expert "human fly" (in German, a "fasadenkläterer") and could often be seen in the Institute's library perched close to the ceiling with a book in his hand, his two legs stretched along the top bookshelves. We were walking along a deserted street and passed a bank building. The façade of the bank, formed of large cement blocks with what Alpinists call footholds between them, invited

Casimir's attention, and he went up about two floors high. When he came down, Bohr wanted to match his feat and went slowly up the bank's wall. Feeling a little worried, Fru Bohr, Casimir, and I were standing below, watching Bohr's slow progress up the wall. At this moment two Copenhagen policemen on their night beat approached quickly from behind, ready for action. They looked up at Bohr hanging halfway between the first and the second floor, and one of them said: "Oh, that's only Professor Bohr!" and, completely reassured, the two guardians of law and order walked quietly away.

After these preliminary remarks, let us discuss Bohr's theory of the atom, published in 1913 and based on Rutherford's discovery that atoms have massive positively charged nuclei, with swarms of electrons revolving around in the fashion of a tiny planetary system. The first difficulty Bohr encountered with that picture was that atoms could not exist longer than a negligible fraction of a second. Indeed, an electron rushing around an orbit is equivalent to an electric oscillator and is bound to emit electromagnetic waves, rapidly losing its energy. It was easy to calculate that, as a consequence, atomic electrons would move along spiral trajectories and fall on the nucleus in a matter of one hundred millionths of a second. And they certainly do not, since atoms are quite stable configurations. The situation was just as paradoxical as Jeans' ultraviolet catastrophe, and it was apparent to Bohr that the solution of the difficulty must be sought along the same lines. If radiant energy can exist only in certain minimum amounts or the multiples thereof, why not make the same assumption concerning the mechanical energy of electrons circling the nucleus? In this case, the motion of electrons in the normal state of an atom would correspond to these minimum amounts of energy, while the excited states would correspond to a larger number of these mechanical energy quanta. Thus, an atomic mechanism should behave, in a way, as an automobile transmission box; one can put it in low, in second, or in top gear, but not in between. If the motion of atomic electrons and the emitted light are both quantized, then the transition of an electron from a higher quantum level to a lower one in the atom must result in the emission of light quantum with $h\nu$ equal to the energy difference between the two levels. Inversely, if $h\nu$ of an incident light quantum is equal to the energy difference between the ground and excited state in a given atom, the light quantum will be absorbed and the electron will move from the lower level to the higher one.

These energy exchange processes between matter and radiation are represented schematically in Figure VII-12a,b, which leads to a very important conclusion. If a light quantum with the energy $h\nu_{32}$ can be emitted in the transition of an electron from the energy state $E_3$ to the energy state $E_2$, and if the transition from $E_2$ to $E_1$ results in the emission of light quantum with the energy $h\nu_{21}$, then we should be able to observe, in some cases at least, a light quantum with the energy $h\nu_{32} + h\nu_{21} = h(\nu_{32} + \nu_{21})$, corresponding to a direct transition from $E_3$ to $E_1$. Similarly, the emission of the light quantum with the energies $h\nu_{31}$ and $h\nu_{32}$ would lead us to expect the possibility of

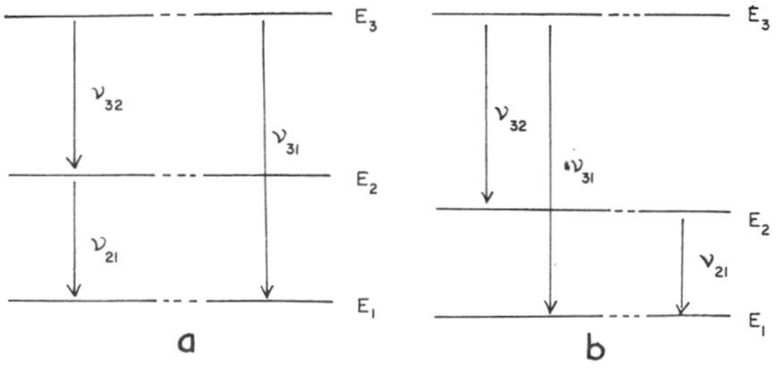


FIG. VII–12.

Illustration of Rydberg's principle. (a) If an electron can jump from the energy level $E_3$ to the level $E_2$ emitting frequency $\nu_{32}$, and then jump again from $E_2$ to $E_3$ emitting frequency $\nu_{21}$, there must also be possible a direct transition from $E_3$ to $E_1$ emitting the frequency $\nu_{31} = \nu_{32} + \nu_{21}$. (b) If an electron can jump from $E_3$ to either $E_2$ emitting frequency $\nu_{32}$ or to $E_1$ emitting frequency $\nu_{31}$ there must also be possible the transition from $E_2$ to $E_1$ with frequency $\nu_{21} = \nu_{31} - \nu_{32}$.

light emission with light quantum $h\nu_{31} - h\nu_{32} = h(\nu_{31} - \nu_{32})$. Striking out $h$, we may say that, if a certain two emission frequencies are observed in the spectrum of a given atom, their sums and differences may also be expected. But, this is exactly the so-called "Rydberg's combination principle," discovered empirically by the German spectroscopist, of that name, long before the quantum theory came into being.

All the above-described facts left no doubt that Bohr's fundamental concept of the quantization of mechanical energy was correct, and it remained only to find out what are the rules of that quantization. To do this, Bohr took the case of the simplest of all atoms, the atom of hydrogen, which, according to the earlier discussion, consists of a single electron revolving around the nucleus carrying a single positive charge, or a proton as we call it now. The visible spectrum of hydrogen consists of four lines—one red, one blue, and two violet—but study in the ultraviolet disclosed a large number of lines with shorter wave lengths. This spectrum is shown in Plate II, where spectral lines are arranged in order of increasing vibration frequency. Such sequences of lines, which are getting more and more crowded and are approaching a definite limit on the high frequency side, were known in spectroscopy as *series*, and the hydrogen series is the most typical and regular of them. In 1885 a German school teacher, J. J. Balmer, discovered that the lines in the hydrogen spectrum (now known as "Balmer series") can be expressed by the very simple formula:

$$\nu = R\left(\frac{1}{4} - \frac{1}{n^2}\right)$$

where $R$ is a numerical constant and $n$ takes the values 3, 4, 5, 6, etc. (apparently $n$ cannot be 1 or 2 since in that case $\nu$ would be negative or zero). Multiplying this formula by $h$, in order to get the energy of the emitted light quantum on the left, we obtain:

$$h\nu = Rh\left(\frac{1}{4} - \frac{1}{n^2}\right)$$

which was rewritten by Bohr as:

$$h\nu = Rh\left(\frac{1}{2^2} - \frac{1}{n^2}\right)$$

according to a well-known fact of arithmetic.

From the earlier discussion it follows that $-\frac{Rh}{n^2}$ should represent the energy levels of the electron in a hydrogen atom, between which the transitions leading to the emission of Balmer lines are taking place. We write "minus" in front of both quantities because the orbital energy of electrons in an atom is negative, which means simply that their kinetic energy is smaller than their potential energy in the electric field, so that they cannot get out. Which kind of motion around the nucleus would correspond to these energy values?

The simplest way to answer this question is to remember that the potential energy of Coulomb forces changes inversely with the dis-

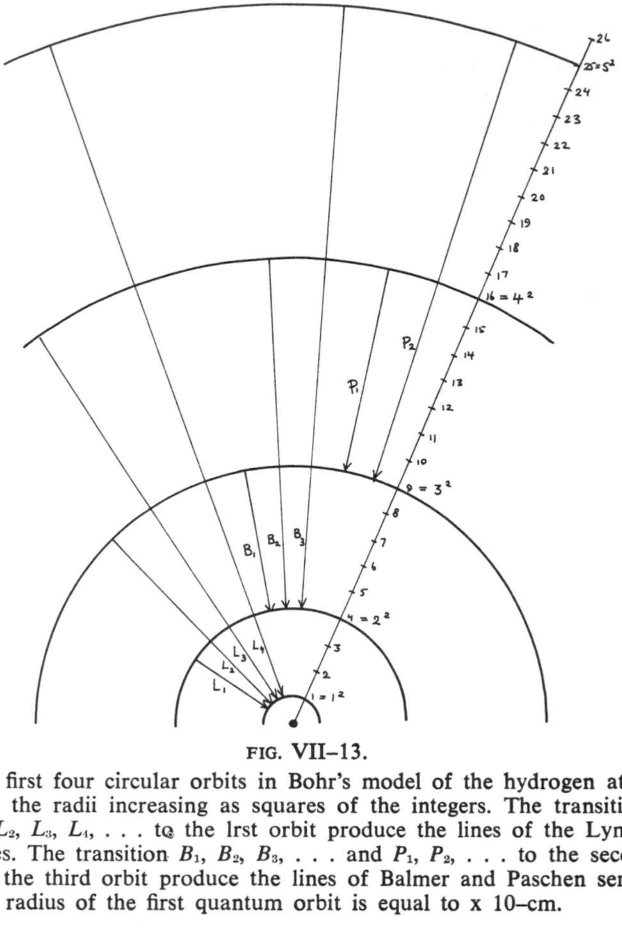


FIG. VII–13.

The first four circular orbits in Bohr's model of the hydrogen atom with the radii increasing as squares of the integers. The transitions $L_1$, $L_2$, $L_3$, $L_4$, . . . to the lrst orbit produce the lines of the Lyman series. The transition $B_1$, $B_2$, $B_3$, . . . and $P_1$, $P_2$, . . . to the second and the third orbit produce the lines of Balmer and Paschen series. The radius of the first quantum orbit is equal to x 10–cm.

tance from the center. Since the terms in the Balmer formula change as the inverse squares of the integer number *n*, we conclude that the radii of successive quantum orbits must increase as $n^2$. For the case of circular orbits, which Bohr discussed first, the relative sizes are shown in Figure VII-13. The transitions of the electron to the second orbit from the orbits located beyond it correspond to the lines of Balmer series, but what about other possibilities? The transitions from the orbits numbered 2, 3, 4, etc., to the first orbit must form a series of lines similar to Balmer series but located in the far ultraviolet part of the spectrum. On the other hand, the transitions from the higher orbits to orbit 3 must give a series in the far infrared. Both series were discovered by the spectroscopists, Theodore Lyman and Friedrich Paschen, and their existence gave strong support to Bohr's jumping electron theory.

Knowing that the radii of quantum orbits (under the assumption that they are circles) increase as the squares of integers, Bohr could find out which mechanical quantity is "quantized," i.e., increased by the same amount from one orbit to another. It turned out to be the product of the mechanical momentum of the electron by the length of its orbit, the quantity which in classical mechanics is known as "action." And the change of "action" from one quantum orbit to another turned out to be exactly equal to the quantum constant $h$ used by Planck in his theory of thermal radiation, and by Einstein in his explanation of photoelectric effect.

Soon it became clear that Bohr's original model with concentric circular quantum orbits must be generalized by adding some quantized ellipses. This generalization was carried out by the German physicist Arnold Sommerfeld. Figure VII-14 shows a complete set of possible quantum orbits of the electron in a hydrogen atom. The first circular orbit (*solid line*) remained intact. To the second circular orbit (*dashed line*) Sommerfeld added three elliptical orbits, moving along which the electron has the same energy as on the circular orbit. To the third circular orbit are added eight elliptical orbits (only three of them are shown in the picture), all of them corresponding to the same energy as that of the circular one. And beyond that were more and more elliptical orbits added to the circular orbits of higher order. The situation was becoming more and more complicated but, remarkably enough, it fitted better and better with the observed facts. The atom was no longer similar to the planetary system, in which Jupiter could suddenly jump over to the orbit of Venus, but was described by an abstract design only distantly related to the circles and ellipses of classical mechanics.

During the first decade of its development, Bohr's theory had tremendous success in explaining the properties of complex atoms, their optical spectra, chemical interactions, etc. But, through all its successes, the theory retained its original skeletal nature, and all

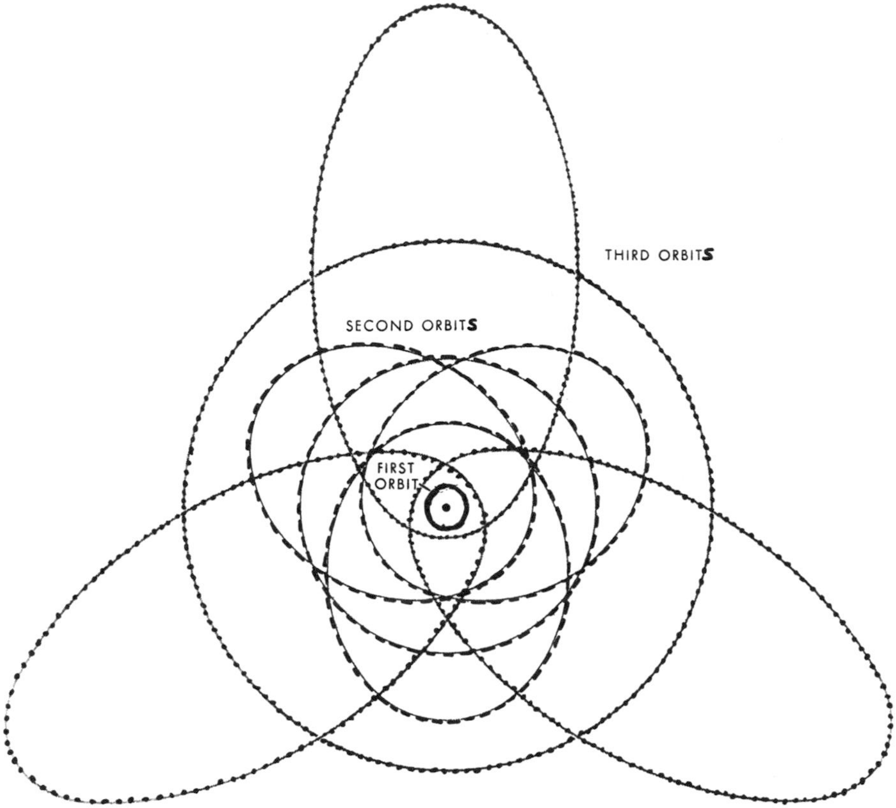


FIG. VII–14.

Circular and elliptical quantum orbits in the hydrogen atom. The first circular orbit (*solid line*) corresponds to the lowest energy of the electron. The next four orbits, one circular and three elliptical (*dashed lines*) correspond to the same energy, which is higher than on the first orbit. The next nine orbits (*dotted lines*), only four of which are shown in the figure, correspond to still higher energy (the same for all nine).

attempts to describe in the flesh the transitions of electrons from one energy state into another, and to calculate the intensities of spectral lines emitted as the result of these transitions, did not lead anywhere. The situation is very well characterized by verses written in the early twenties by a Russian theoretical physicist, Vladimir A. Fock, which are reproduced here in free English translation:

### HAIL TO NIELS BOHR*

Fill up the tankards and brighten the fire!
Drink to him—drink to him—toast him once more!
Plucking the strings of our latter-day lyre
We sing of our hero and idol—NIELS BOHR!
  Prosperous days to you,
  Honor and praise to you,
BOHR, the Colossus we fear and adore!

Endless your merits are; none can deny it.
What if your theory is madly obscure?
You have proclaimed it; we dassent defy it.
Your words, like God's words, are totally pure.
  NIELS, you Apollo, you,
  Humbly we follow you!
Laws you devise are devised to endure.

Mechanics—your servant—is meekly obedient.
None of its precepts, without your okay,
Can hope to stand fast. Should you find it expedient,
Even Energy's law might be wafted away.
  From the dais you sit upon,
  Leaning, you spit upon
Continuous Motion, a dragon to slay.

Causality's law, in a risky position,
Vies for your favor, but who can foretell
If one of these days, by your sudden volition,
It too will be turning and toasting in hell.
  Then there's the matter of
  Lesser men's chatter of
Ninety-two Bodies—and this you dispel.
  All of the tedious,
  Encyclopaedious
Talk of Divergence of Series you quell.

Quantum reveres you. At edicts you utter
Each tiny electron examines his plight;
Confined to his orbit, he flies in a flutter,
Rushes and radiates, estimates height,

* *Rendered into English verse by B.P.G. from the unpublished Russian verse composed by V. A. Fock in the early 1920's.*

Pre-judges the motion,
And dreams up the notion—
Mixing up Cause and Conclusion, in fright—
To leap from his mother orb
Straight to another orb,
Seeking, if vainly, a refuge in flight.

Hail to NIELS BOHR from the worshipful nations!
You are the Master by whom we are led.
Awed by your cryptic and proud affirmations,
Each of us, driven half out of his head,
Yet remains true to you,
Wouldn't say boo to you,
Swallows your theories from alpha to zed,
Even if—(Drink to him!
Tankards must clink to him!)—
None of us fathoms a word you have said!

## BOHR'S ATOMIC MODEL, AND THE PERIODIC SYSTEM OF ELEMENTS

Having discussed the motion of a single electron in a hydrogen atom, we should turn to the question of what happens in the case of the atoms which contain 2, 3, 4, and many more electrons. For the nuclei carrying a larger electric charge, the general pattern of quantum orbits remains the same as in the case of hydrogen atoms except that, due to the larger attractive force exerted by the nucleus, the diameters of all orbits shrink more and more as we go to the elements of higher and higher atomic numbers.

How is the larger number of electrons in the atoms of heavier elements accommodated on these shrinking quantum orbits? In terms of classical physics the answer to this question is almost trivial. The most stable of any mechanical system is the one in which the system cannot lose any more energy by dropping to a still lower energy level. Thus all the additional electrons in heavier atoms might be expected to drop into the first quantum orbit and play "ring-around-the-rosy," or more exactly, "ring around the nucleus." And because we know that the diameter of that ring becomes smaller in heavier elements, we might predict that it would also become more and more tightly packed with electrons. The fact is that this does not happen; regardless of the charge of the nucleus, the overall size of atoms remains approximately the same.

This problem drew the attention of the German physicist Wolfgang

Pauli (Fig. VII-15), whose corpulent and jovial figure was a familiar and welcome sight in Bohr's Institute of Theoretical Physics. Pauli was a first class theoretical physicist, and among his friends his name will always be associated with the mysterious phenomenon known as the Pauli effect. It is well known that all theoretical physicists are very clumsy in handling experimental equipment, and usually break expensive and elaborate apparatus merely by a touch. Pauli was such a good physicist that things broke down when he merely walked into a laboratory. The most persuasive case of Pauli

FIG. VII–15.
**Enrico Fermi (*left*) and Wolfgang Pauli.**

effect occurred one day when the equipment in the laboratory of Professor James Frank at the Physics Institute of the University of Göttingen unexpectedly blew up and went to pieces without any apparent reason. Subsequent investigation showed that this catastrophe took place exactly at the time when a train carrying Pauli from Zurich to Copenhagen stopped for five minutes at the Göttingen railroad station.

Thinking about the motion of electrons within an atom, Pauli formulated his now famous principle (which he himself called the "exclusion principle") according to which each quantum orbit can hold no more than two electrons. The principle demands that if both these vacancies are filled, the next electrons must be accommodated in other orbits. When all the orbits in a given shell are filled, the orbits in the next shell (corresponding to a higher energy level) begin to fill.

As we proceed along the natural sequence of elements toward the heavier and heavier atoms, the radii of quantum orbits shrink because of the increasing charge of the nucleus, but, on the other hand, more and more orbits become occupied by electrons. Thus the size of atoms remains on the average the same all the way from the lightest to the heaviest elements. There are, however, small variations of atomic size as we proceed from one completed shell to another (noble gas configurations). This causes small periodic changes of density of various elements which go parallel with the periodic changes in their chemical properties.

The electron shells of all the species of atoms in the periodic table are filled according to this fixed hierarchy of energy states. The first shell, which represents the lowest available energy state, is the first to fill. In the helium atom this shell is completely filled by the 2 electrons chasing each other around the first quantum orbit. The next element, lithium, has 3 electrons, one of which, according to the exclusion principle, must be added in the second shell, consisting of one circular and three elliptical orbits. Since these four orbits can hold a total of 8 electrons and the inner orbit holds 2, both the first and second shells will be filled in the neon atom, which has 10 electrons. The extra electrons in still heavier elements must be added in a third set of circular and elliptical orbits, and so on. Pauli's exclusion principle thus explains the internal structure of elements in terms of the way in which their consecutive electron shells are filled. The principle also underlies the external, or chemical, identity of an atom and the periodicity of chemical properties in the sequence of atomic species in the table of elements. These characteristics are determined by the number of electrons in the outer shells of atoms, which make contact when atoms collide with one another.

When the Pauli principle was originally formulated, electrons were believed to be nothing more than point charges of negative electricity. It was soon discovered, however, that electrons must also be considered as tiny magnets; they possess a magnetic moment because they spin rapidly as they orbit around a nucleus. Having learned to regard electrons as tiny magnets, we must take into account both the electrical forces which are mainly responsible for their orbital motion and the magnetic forces set up by their spin.

An electron tends to spin in one of two ways: either in the direction in which it travels along its orbit, or in the opposite direction. It was shown that 2 electrons following the same orbit must spin in opposite directions. This discovery requires us to formulate the Pauli principle in a somewhat different way. Because electrons spinning in opposite directions set up weak magnetic fields which slightly alter each other's orbits, we now say that the two electrons originally permitted to travel in the same orbit actually follow two different (though very similar) orbits. It is therefore more rational to regard the permitted orbits as close pairs split apart by weak magnetic interactions.

This view of atomic shell structure gives us a simple explanation of the nature of the chemical valence of different elements. We can show, on the basis of the quantum theory, that atoms which have an almost completed shell have a tendency to take in extra electrons in order to finish this shell and that atoms which have just the beginning of a new electron shell have a tendency to get rid of these extra electrons. For example, chlorine (atomic number 17) has 2 electrons in the first shell, 8 in the second, and 7 in the third, which makes the outer shell short 1 electron. On the other hand, a sodium atom (atomic number 11) has 2 electrons in the first shell, 8 in the second, and only 1 electron as the beginning of the third shell. Under these circumstances, when a chlorine atom encounters a sodium atom, it "adopts" the latter's lonely outer electron and becomes $Cl^-$, while the sodium atom becomes $Na^+$. The two ions are now held together by electrostatic forces and form a stable molecule of table salt. Similarly, an oxygen atom that has two electrons missing from its outer shell (atomic number $= 8 = 2 + 6$) tends to adopt two electrons from some other atom and can thus bind two monovalent atoms (H, Na, K, etc.) or one bivalent atom such as magnesium (atomic number $= 12 = 2 + 8 + 2$), which has 2 electrons to lend. An example of chemical binding of this kind is shown in Figure VII-16. It also becomes clear why the noble gases, which have all their shells completed and have no electrons to give or to take, are chemically inert.

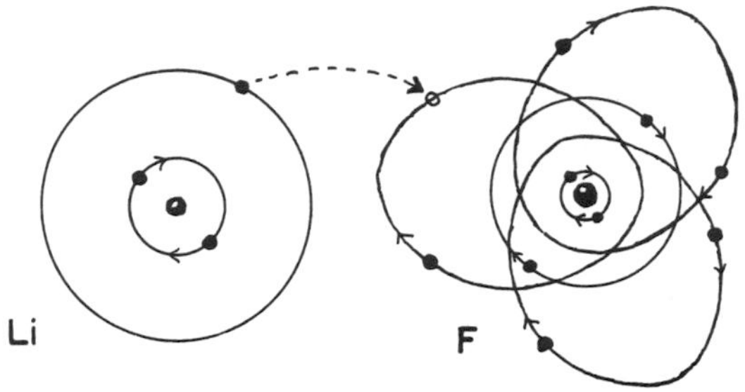


FIG. VII–16.

Chemical binding between a lithium (Li) atom and a fluorine (F) atom in the lithium fluoride (LiF) molecule. An extra electron from the lonely shell in the Li atom jumps over to a vacant place in the crowded shell of the F atom.

### THE WAVES OF MATTER

In the year 1924, a 32-year-old French aristocrat, Marquis Louis de Broglie (pronounced something like "broccoli") (Fig. VII-17), who started his scientific career as a student of medieval history and only later acquired an interest in theoretical physics, presented to the faculty of the University of Paris a doctoral thesis containing extraordinary ideas. De Broglie believed that the motion of material particles is accompanied and guided by certain pilot waves which propagate

FIG. VII–17.

P. A. M. Dirac (*left*) and Louis de Broglie.

through space along with the particles. If this is so, the selected quantum orbits in Bohr's atomic model could be interpreted as the orbits which satisfy the condition that *their length contains an integer number of these pilot waves;* one wave in the first quantum orbit, two in the second, etc. (Fig. VII-18). We have seen earlier that, for the simple case of circular motion, Bohr's quantum orbits satisfy the condition that their lengths multiplied by the momentum (mass times

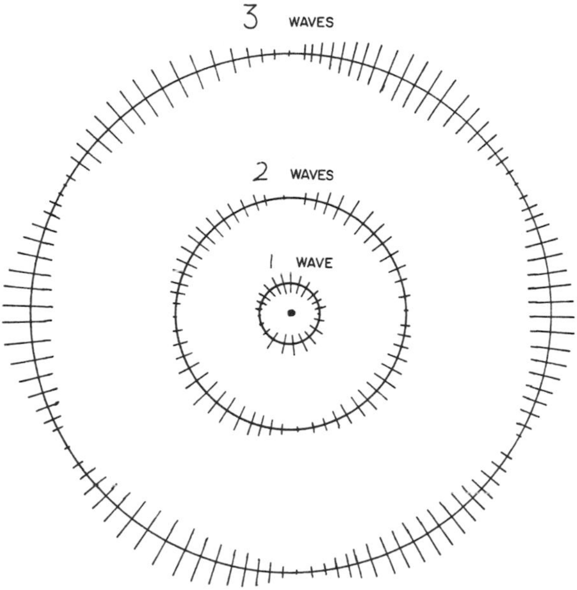


FIG. VII–18.
**De Broglie's waves running along Bohr's orbits.**

velocity) of the moving electron is equal to $h$ for the first orbit, $2h$ for the second, $3h$ for the third, etc. These two statements become identical if we assume that the length of the pilot wave is equal to $h$ divided by the momentum of the particle:

$$\frac{h}{mv}$$

and that is what Louis de Broglie did assume. For the orbits of intermediate radii, a pilot wave running around cannot "catch itself by its tail," and consequently, that type of motion cannot exist. Thus, with one bold stroke, de Broglie changed the skeletal quantum orbits of Niels Bohr to a fleshier notion of organ pipes, drum membranes, etc. Quantum mechanics of particles acquired properties similar to waves of sound or light.

This revolutionary proposal could be subjected to an experimental test. If electrons are guided by de Broglie waves in their motion within the atom, they must also show some wave properties while flying along straight lines through space. For electron beams of a few kilovolts used in laboratories, de Broglie wave length was expected to be about $10^{-8}$ cm, being comparable with that of X-rays, so that one could employ the X-ray diffraction technique to check whether there are any waves accompanying electrons or not.

An experiment in this direction was carried out in 1927 by Sir J. J. Thomson's son George (later Sir George) and the American physicists C. J. Davisson and L. H. Germer, who directed at a crystal a beam of electrons accelerated in an electric field. The result was a picture like the one in Plate IV, *lower,* showing beyond any doubt that one deals here with the wave phenomenon of diffraction. The wave length estimated from the diameters of diffraction rings coincided exactly with the wave length given by de Broglie's formula $h/mv$. It was also decreasing and increasing when the electrons in the beam were speeded up or slowed down. A few years later, a German physicist, Otto Stern, repeated the Davisson and Germer experiments, using, instead of electrons, a beam of sodium atoms, and found that the diffraction phenomenon described by de Broglie's formula also takes place in this case. Thus, it became quite certain that small particles, such as electrons, or atoms, are guided in their motion by "pilot waves," the nature of which was, however, completely obscure at that time.

De Broglie's views were generalized and put on a strictly mathematical basis in 1926 by an Austrian physicist, Erwin Schrödinger (Fig. VII-19), who incorporated them in the famous Schrödinger's equation, applicable to the motion of particles in any field of forces. The use of Schrödinger's equation in the case of hydrogen as well as in the case of more complex atoms reproduced all the results of Bohr's quantum orbital theory, and, in addition, treated the questions (such as intensities of spectral lines) which the old theory could not tackle. Instead of circular or elliptical quantum orbits, the interior of an atom was now described by the so-called $\psi$ functions corresponding to various types of de Broglie waves which can exist in the space surrounding the atomic nucleus.

Simultaneously with Schrödinger's first paper, which was published in a leading German magazine, *Annalen der Physick,* there appeared in a competing magazine, *Zeitschrift der Physick,* a paper on the quantum theory written by a young German physicist (24 years old at that time), Werner Heisenberg (Fig. VII-19). It is difficult to describe Heisenberg's theory with any degree of popularity.

FIG. VII–19.
Werner Heisenberg (*left*) and Erwin Schrödinger.

The main idea was that mechanical quantities, like position, velocity, force, etc., should not be represented by ordinary numbers like 5 or 7½ or 13⁵⁄₇, but by the abstract mathematical structures known as "matrices," each matrix being like a crossword-puzzle array of ordinary numbers followed by an infinite sequence of lines and columns. One can develop rules of addition, subtraction, multiplication, and division of these matrices, which are quite similar to those of ordinary algebra, but there is one great exception. In matrix algebra the product *A* times *B* is not necessarily equal to the product *B* times *A*, which is the result of larger complexity in matrix multiplication procedure. The nearest analogy is that of the human language, in which Douglas Malcolm is not the same as Malcolm

Douglas, and flat top is not the same as top flat. Well, Heisenberg showed that if one considers all the quantities in the equations of classical mechanics to be matrices, and also introduces an additional condition that: momentum $\times$ velocity $-$ velocity $\times$ momentum $= hi$ where $h$ is the quantum constant and $i = \sqrt{-1}$ is our old friend an imaginary unit, one obtains a theory which describes correctly all known quantum phenomena.

The simultaneous appearance of two papers which were arriving at exactly the same results by using two entirely different methods brought the world of physics into consternation, but it was soon found that the two theories are mathematically identical. In fact, Heisenberg's matrices represent the tabulated solutions of Schrödinger's equation, and in solving various problems of quantum theory, one can use the wave mechanics and matrix mechanics intermittently.

### UNCERTAINTY RELATIONS*

What is the physical meaning of de Broglie waves which guide the material particles in their motion? Are they real waves like the waves of light, or only mathematical fictions introduced just for convenience in describing physical phenomena in microcosm? This question was answered a couple of years after the formulation of wave mechanics by W. Heisenberg, who asked himself how quantum laws introducing minimum amounts of radiant and mechanical energy affect the basic notions of classical mechanics.

Heisenberg went to the root of the trouble: the attempt to apply ordinary rules and methods of observation to phenomena on the atomic scale. In the world of everyday experience we can observe any phenomenon and measure its properties without influencing the phenomenon in question to any significant extent. To be sure, if we try to measure the temperature of a demitasse with a bathtub thermometer, the instrument will absorb so much heat from the coffee that it will change the coffee's temperature substantially. But with a small chemical thermometer we may get a sufficiently accurate reading. We can measure the temperature of an object as small as a living cell with a miniature thermocouple, which has almost negligible heat capacity. But in the atomic world we can never overlook the disturbance caused by the introduction of the measuring apparatus.

* This section follows closely the author's article "The Uncertainty Principle," published in the Jan. 1958 issue of *Scientific American*.

The energies on this scale are so small that even the most gently performed measurement may result in substantial disturbances of the phenomenon under observation, and we cannot guarantee that the results of measurements actually describe what would have happened in the absence of the measuring devices. The observer and his instruments become an integral part of the phenomenon under investigation. Even in principle there is no such thing as a physical phenomenon per se. In all cases there is an absolutely unavoidable interaction between the observer and the phenomenon.

Heisenberg illustrated this by a detailed consideration of the problem of trying to track the motion of a material particle. In the gross world we can follow the flight of a Ping-pong ball without affecting its trajectory one iota. We know that light exerts pressure on the ball, but we do not have to play Ping-pong in a dark room (assuming it were possible), because the pressure of light is much too small to make any difference in the ball's flight. But substitute an electron for the Ping-pong ball, and the situation becomes quite different. Heisenberg examined the situation with a "thought experiment," a method of reasoning used by Einstein in his discussion of the theory of relativity.

In such a mental exercise the experimenter is allowed an "ideal workshop" in which he can make any kind of instrument or gadget —provided that its design and functioning do not contradict basic laws of physics. For example, he can have a rocket that moves with almost the speed of light, but not more than the speed of light; or he may use a light source which emits just a single photon, but not half a photon. Heisenberg equipped himself with an ideal setup for observing the flight of an electron (Fig. VII-20). He imagined an electron gun which could shoot a single electron horizontally in a completely evacuated chamber—barren of even a single air molecule! His light came from an ideal source which could emit photons of any desired wave length and in any desired number. And he could watch the movement of the electron in the chamber through an ideal microscope which could be tuned at will over the whole range of the spectrum, from the longest radio waves to the shortest gamma rays.

What will happen when an electron is fired in the chamber? According to our classical textbooks on mechanics, the particle should follow a trajectory known as a parabola. But actually, the moment a photon strikes it the electron will recoil and change its velocity.

Observing the particle at successive points in its motion, we shall find it taking a zigzag course because of the photon impacts. Let us, then, since we have an ideally flexible instrument, minimize the impacts by reducing the photons' energy, which we can do by using light of lower frequency. In fact, by going to the limit of infinitely low frequency (which is possible in our apparatus), we can make the disturbance of the electron's motion as small as we wish. But here comes a new difficulty. The longer the wave length of the light, the less able we are to define the object, because of the diffraction effect. So we can no longer find the exact position of the electron at

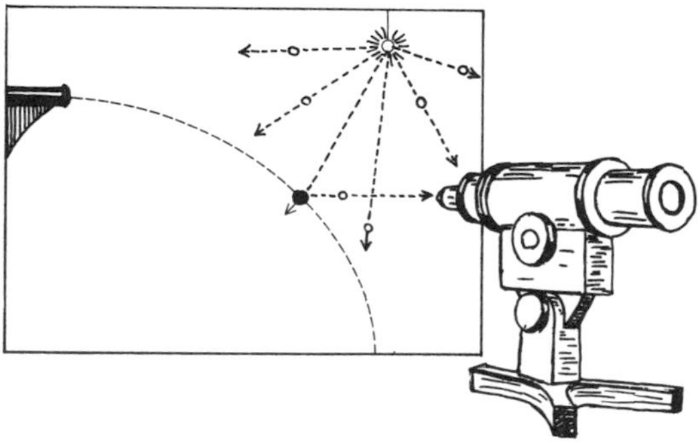

FIG. VII-20.
Heisenberg's idealized experiment for observing the trajectory of a particle.

any given instant. Heisenberg showed that the product of the uncertainties in position and in velocity can never be smaller than Planck's constant divided by the mass of the particle.

$$\Delta v \Delta x \geqslant \frac{h}{m}$$

So with very short waves we can define the positions of a moving particle sharply but we will interfere greatly with its velocity, while with very long waves we can determine its undisturbed velocity but we will be very uncertain about its positions. Now we can choose a middle ground between these uncertainties. If we use some optional intermediate wave length of light, we will disturb the particle's trajectory only moderately and still be able to define its path to a fairly close approximation (Fig. VII-21). The observed path, expressed in classical terms, will not be a sharp line, but rather a band with smeared out boundaries. Describing the trajectory of an electron in this way gives us no difficulty in a case such as a television picture tube, where the "thickness" of the electron's path to the screen is very much smaller than the diameter of the spot formed on the screen by the electron beam. Here we can represent the electron's trajectory satisfactorily by a line. But we cannot describe the orbit of an electron inside an atom in the same terms. The band of uncertainty is about as wide as the distance of the orbit from the nucleus!

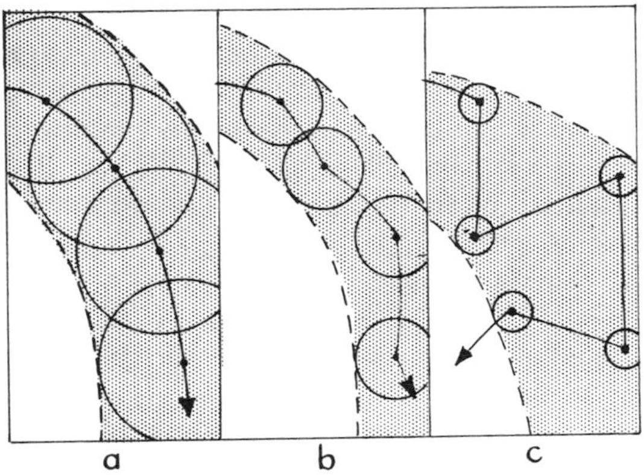


FIG. VII–21.
The path of a particle in Heisenberg's imaginary experiment. (a) The wave length of light is too long, and each position measurement is very inexact. (b) Optimum conditions. (c) The frequency is too high, and the particle is kicked about too much.

Suppose we give up the attempt to track a moving particle with light and try the cloud-chamber method instead. In our hypothetical workshop we build an ideal "cloud chamber" which is completely evacuated of material particles but is filled with very tiny imaginary "indicators" that become "activated" whenever an electron passes close by. The activated indicators would show the track of the moving particle just as water droplets do in a real cloud chamber.*

Classical mechanics would say that in principle the indicators could be made small enough and delicately responsive enough so that they

* The real cloud chamber, used by nuclear physicists, will be described in the next chapter.

would subtract no significant amount of energy from the moving particle and we could observe its trajectory with any desired precision. But quantum mechanics finds a fundamental objection to this procedure. One of its rules is: The smaller the mechanical system, the larger are its quanta (minimum amounts) of energy. Thus as the size of the "indicators" was reduced (for more precise measurement of the electron's position), they would take more energy from the passing particle. The situation is quite analogous to the fatal difficulty in trying to track a particle by means of light, and we again arrive at the same relation for the uncertainties in the position and velocity.

Where does all this leave us? Heisenberg concluded that *at the atomic level we must give up the notion of the trajectory of an object as a mathematical (i.e., infinitely thin) line.* This concept is accurate enough when we deal with phenomena in the realm of ordinary experience, where we can think of a moving object as held in its path by a kind of railroad track. But in the small world of electrons in an atom, individual motions and events are not so firmly predetermined. Small material particles such as electrons and protons move over a range under the guidance of waves, which should be considered just as the broadened-up line trajectories of classical mechanics. The important point is that the guidance is performed in a *stochastic* rather than a strictly *deterministic* way. We can calculate only the *probability* that an electron will strike a given point on a screen, or that any other material particle will be found in a given place at a given instrument, but we cannot tell for sure which way it will go in a given field of forces.

It must be made clear that the word "probability" here is used in a rather different sense from the way it is usually understood in classical physics and everyday life. When we say in a game of poker that there is a certain probability of drawing a royal flush, we mean only that we have to estimate the chances because we do not know the arrangement of the cards in the pack. If we knew exactly how the cards were stacked, we could predict definitely whether we would get a royal flush or not. Classical physics assumed that the same was true of a problem such as the behavior of gas molecules: their behavior had to be described on the basis of statistical probability only because of incomplete knowledge—if we were given the positions and velocities of all the particles, we could predict events within the gas in full detail. The uncertainty principle cuts the ground from under that idea. We cannot predict the motions of individual particles because we can never know the initial conditions exactly in the first place. It is impossible *in principle* to obtain an exact measurement of both the position and the velocity of a particle on the atomic scale.

Is the wave function $\psi$ (or rather, the square of it), guiding the path of a material particle, a definite "physical entity" which *exists* in the same sense as atoms of sodium or intercontinental ballistic missiles exist? The answer depends on what one means by the word "existence." Wave functions "exist" in the same sense as the trajectories of material bodies. The orbits of the earth around the sun, or of the moon around the earth, *do exist* in the mathematical sense representing the continuum of points occupied consecutively by a moving material body. But they *do not exist* in the same sense as the railroad tracks which guide the motion of a train across the country. In particular, wave function has no mass, being nothing more than a smeared-out trajectory.

Probably the nearest analogy in the field of classical physics is given by the notion of entropy. Entropy is a mathematical function, invented by theoretical physicists and connected with the mathematical probability of any given pattern of molecular motion which determines the direction in which thermodynamical processes habitually proceed: from the smaller values of entropy to the larger ones. But the entropy is not a "physical entity" in the same sense as mass or energy, and whereas we can speak about 1 gm of matter, or 1 gm of energy (since Einstein) it does not make any sense to speak about 1 gm of entropy. It makes just as little sense to speak about 1 gm of de Broglie waves, or 1 gm of the Schrödinger function!

A look at Heisenberg's formula shows why we can disregard the principle of uncertainty and safely trust the good old principle of determinism when we deal with matter on the macroscopic scale. The product of the uncertainty in position by the uncertainty in velocity is equal to Planck's constant $h$ divided by the mass of the particle. Planck's constant is an extremely small quantity: its numerical value amounts to only about $10^{-27}$ in centimeter-gram-second units. When we consider a particle weighing as much as 1 mg, we can in principle simultaneously determine its position within a trillionth of a centimeter and its velocity within a trillionth of a centimeter per second—or $30\mu$ per century!

Heisenberg's principle was developed by Bohr into a new philosophy of physics. It called for a profound change in our ideas about the material world—ideas that we acquire in ordinary experience from early childhood. But it allowed many puzzles of atomic physics to make sense.

Many physicists readily accepted the new view. Others did not like it at all. To the latter group belonged Albert Einstein. His philosophical convictions about determinism did not permit him to elevate uncertainty to a principle. And just as skeptics were trying to find contradictions in his theory of relativity, Einstein tried to discover contradictions in the uncertainty principle of quantum physics. However, his efforts led only to strengthening of the principle of uncertainty. This is interestingly illustrated by an incident that took place at the Sixth International Solvay Congress on Physics, in Brussels in 1930.

In a discussion at which Bohr was present, Einstein performed a "thought experiment." Arguing that time was a fourth coordinate of space-time and that energy was a fourth component of momentum (mass × velocity), he said that Heisenberg's uncertainty equation implied that the uncertainty in time was related to the uncertainty in energy, the product of the two being at least equal to Planck's constant $h$. Einstein set out to prove that this was not the case—that the time and the energy could be determined without any uncertainty. Consider, he said, an ideal box, lined with perfect mirrors, which could hold radiant energy indefinitely. Weigh the box. Then at a chosen instant some time later a clockworks, preset like a time bomb, will open an ideal shutter to release some light. Now weigh the box again. The change of mass tells the energy of the emitted light. In this manner, said Einstein, one could measure the energy emitted and the time it was released with any desired precision, in contradiction to the uncertainty principle.

The next morning, after an almost sleepless night, Bohr delivered a mortal blow to Einstein's disproof. He offered a counter thought experiment with an ideal apparatus of his own (which, as Bohr's student, the author later actually built in wood and metal, for Bohr's use in lectures on the subject, Fig. VII-22). Bohr attacked the question of weighing Einstein's box. A spring scale equipped with a pointer recording the weight on a vertical column placed alongside is, he said, as good as any. Now since the box must move vertically with a change in its weight, there will be an uncertainty in its vertical velocity and therefore an uncertainty in its height above the table, Bohr pointed out. Furthermore, the uncertainty about its elevation above the earth's surface will result in an uncertainty in the rate of the clock, for according to the theory of relativity the rate depends on the clock's position in the gravitational field. Bohr proceeded to show that the uncertainties of time and of the change in the mass of

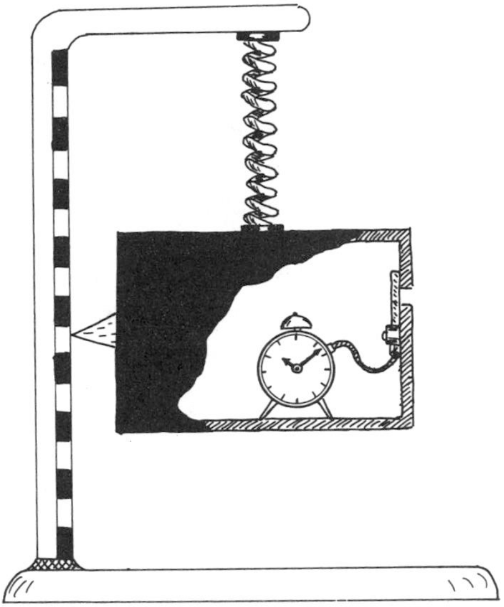

FIG. VII–22.
Einstein-Bohr scales for measuring the weight of light.

the box would indeed have the relation which Einstein had tried to disprove.

Einstein, bitten by his own argument, had to agree that the Bohr-Heisenberg concept was free of internal contradictions, but to the very end of his life he refused to accept the uncertainty principle and remained hopeful that physics would some day return to the deterministic point of view.

## HOLES IN NOTHING

Paul Adrien Maurice Dirac (Fig. VII-17) received his degree in electrical engineering in the early twenties, and right away found himself unemployed. Unable to find a job, he applied for a graduate fellowship at Cambridge University and was accepted. Less than ten years later he received a Nobel Prize in physics for his important contributions in quantum mechanics. Dirac was, and still is, an "ivory tower" type of scientist and, while he was always glad to chat with his fellow men about his trip to the Orient or any other ordinary topic, preferred to pursue his studies all by himself. But his remarks during scientific meetings were always sharp and to the point. Once, during a theoretical physics conference in Copenhagen, he listened to a Japanese physicist, Y. Nishina, who covered the blackboard with calculations and finally arrived at an important formula pertaining to the scattering of short-wave radiation by free electrons. Dirac called Nishina's attention to the fact that in the formula which he finally derived on the blackboard the third term in brackets had a negative sign whereas in the original manuscript the term was positive. "Well," retorted Nishina, "it is correct as it stands in the manuscript. Deriving this formula here on the blackboard I must have made a mistake in sign in some place." "In the *odd number* of places," corrected Dirac. Indeed, three, five, seven, etc. mistakes in the signs would have brought the same result.

Once, during the question period after Dirac's lecture in the University of Toronto, a Canadian professor from the audience raised his hand. "Dr. Dirac," he said, "I do not understand how you derived this formula at the upper left corner of the blackboard." "This is a statement and not a question," said Dirac. "Next question, please."

His ability for fast mental gymnastics is demonstrated by an unusual solution for a puzzling problem which was occupying the minds of mathematicians and physicists at the University of Göttingen during one of Dirac's visits there. The problem was to write all numbers from 1 to 100 by using all available algebraic notations +, −, power, radical, etc., but no figures except four 2's. Thus, for example, 1 can be written as $\frac{2 \times 2}{2 \times 2}$. For 2 we have $\frac{2}{2} + \frac{2}{2}$, for 3 and 5, $2^2 - \frac{2}{2}$, $2^2 + \frac{2}{2}$, and for 7 $\frac{2}{.2 \times 2} + 2$. . . .

Being given this problem, Dirac very quickly found a general solution for writing *any* number by using *only three 2's*. The solution is:

$$N = -\log_2 \log_2 \sqrt{\sqrt{\cdots\sqrt{\sqrt{2}}}}$$

where the number of radicals is equal to the given number $N$. For those who know some algebra the proof of the above is self-evident.

But among all his important mathematical discoveries, Dirac was especially proud of one which, however, did not contribute any to his fame. Chatting with the wife of a faculty member, he was watching her knit a scarf or something. Back in his study, he tried to reproduce in his mind the rapid motion of needles in the hands of the lady, and came to the conclusion that there is another possible way to handle the needles. He hurried back to tell about his discovery, and was disappointed to find that both methods, "knitting" and "purling," had been known to women for centuries.

But having missed an important discovery in the field of topology, Dirac contributed a lot to the field of relativistic quantum theory. Wave mechanics, which was at that time only a few years old, was originally formulated by Schrödinger for the case of nonrelativistic motion, i.e., for the motion of particles with velocities small as compared with the velocity of light, and theoretical physicists were breaking their heads in the attempt to unite the two great theories: relativity and quantum. In addition Schrödinger's wave equation considered the electron as a point, and all the attempts to apply it to a spinning electron which possesses the properties of a little magnet were not leading to any satisfactory result.

In his famous paper published in 1930, Dirac formulated a new equation, now carrying his name, which permits two birds to be killed with one stone. It satisfies all the relativistic requirements, being applicable to an electron no matter how fast it is moving, and at the same time it leads automatically to the conclusion that the electron *must behave* as a little magnetized spinning top. Dirac's relativistic wave equation is too complicated to be discussed here, but the reader may rest assured that it is perfectly all right.

But, good as Dirac's equation was, it immediately led to a very serious complication, just because it united so successfully relativity and quantum. The trouble arose from the fact (not discussed in Chap. VI) that relativistic mechanics leads to the mathematical possibility of two different worlds: one a "positive" world in which we live, and another a strange "negative" world which can only challenge our imagination. In this "negative" world all objects have *negative mass,* which means that, being pushed one way, they will start moving in the opposite direction. By obvious analogy, one could call the electrons with negative mass "donkey electrons." Strange things would happen in this world of negative mass. In order to make an object move forward, we have to pull it back, and in order to have it stop we should push it forward. Consider two electrons at rest located close to each other. Due to their electric charges, there are repulsive forces between them. If both electrons are "ordinary" ones, these forces will give them accelerations in the opposite directions, and the electrons will fly apart at high speed. If, however, one of these two electrons is a "donkey," the repulsive force will cause it to move *toward* the other electron, while this other electron is flying away. Since both accelerations are numerically equal, the two electrons will speed away with ever-increasing velocity, the donkey electron chasing the normal one. No, there is no contradiction with the law of the conservation of energy. Kinetic energy of the ordinary electron is $\frac{1}{2}mv^2$ while for the donkey we have:

$$-\frac{1}{2}mv^2$$

Thus the total energy of the system is $\frac{1}{2}mv^2 - \frac{1}{2}mv^2 = 0$, i.e., just the same as when they were at rest.

Nobody ever observed donkey electrons, donkey stones, or donkey planets; this is just a fictitious extra solution of Einstein's equations of mechanics. And before Dirac united relativity and quantum, there was no cause for worry. In fact, a normal electron at rest has the energy $m_0c^2$, and, when it moves with the velocity $v$, its kinetic energy must be added to it. A donkey electron, on the other hand, has the rest energy $-m_0c^2$, and its motion will result in additional *negative* kinetic energy. Thus, the energy diagram for the two kinds of electrons looks as it is shown in Figure VII-23. It breaks up into two parts, the upper for an ordinary electron and the lower for donkeys, the two parts separated by a gap between $+m_0c^2$ and $-m_0c^2$, which does not correspond to any kind of possible motion at all. Thus, *if the motion of particles is continuous,* there is no way that it can change from the upper part of the diagram to the lower one, and one could dismiss the difficulty by simply saying: "Our electrons are well-behaved particles with positive mass, and we don't give a darn about the other mathematical possibility!"

We cannot get out of the difficulty so easily, however, if we unite relativity and quantum theories. In fact, according to the quantum theory, electrons just love *to jump* from one energy level to another, even though there is no continuous transition between the two states of motion. If electrons can jump from one Bohr orbit to another, emitting their energy in the form of light quanta, why can't they jump from the upper energy levels to the lower ones in Figure VII-23? But if this were possible, every single normal electron would jump down into the donkey stables of negative energy, and, as a result of losing more and more energy by radiation, move faster and faster, gaining

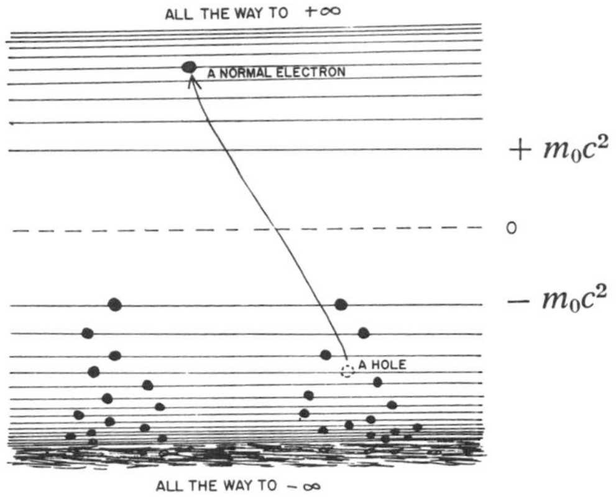


FIG. VII–23.
Dirac's ocean of "donkey electrons" showing the formation of an electron pair (one positive and one negative).

negative kinetic energy. . . . Of course this does not happen, but why not?

The only way Dirac could handle that difficulty was to assume that all the states of negative energy are completely filled up by donkey electrons, and that the electrons from the positive energy states are prohibited from coming down by the Pauli exclusion principle. Of course, this meant that vacuum is not vacuum any more, but is thickly filled by donkey electrons moving in all possible directions with all possible velocities! In fact, each unit volume of vacuum must contain an *infinite number* of these self-contradictory particles! Why do we never notice them? The explanation is rather enigmatic. Imagine a deep-water fish which never comes to the surface of the ocean and therefore does not know that the water ends somewhere above it. If this fish is intelligent enough to speculate on its surroundings, it would not even think about water as a "medium," but would consider it as "free space." Similarly, it can be argued that physicists do not perceive the presence of this infinitely dense herd of donkey electrons, because they are distributed quite uniformly through space. Of course, this idea smelled of the old-fashioned world ether, but it was worth investigating. Returning to our intelligent deep-water fish we can imagine that it formed the notion of gravity by observing empty beer bottles, other refuse, and even whole ships coming down to the ocean bottom. But then one day some air trapped in a sunken ship's cabin was released, and our intelligent deep-water fish observed a school of glittering silvery bubbles rising up toward the surface of the ocean. The fish would of course be very much surprised and, after due reflection, would come to the conclusion that these silvery spheres must have a negative mass. Indeed, how else can they move up when gravity pulls everything down?

Well, Dirac had similar ideas about his ocean filled to capacity by electrons in the negative energy state. Suppose there is a bubble in Dirac's ocean, that is to say, one of the donkey electrons is missing. How would a physicist perceive that? Since the absence of a negative charge is equivalent to the presence of a positive charge, he would see it as a positively charged particle. Also, according to the bubble analogy, the sign of the mass will be reversed, and the lack of negative mass will be perceived as the presence of positive mass. Could it be that such a bubble in Dirac's ocean is nothing else but an ordinary proton? It was a brilliant idea, but it did not work. Dirac tried to explain the much larger mass of the "would be proton" bubbles by high viscosity resulting from the interactions between the donkey electrons, but failed to do it. The mass of the positively charged bubble particle was stubbornly turning out to be exactly equal to the mass of an ordinary electron. The difficulties were augmented by the calculation of Pauli, who had shown that, if the proton really were a bubble in Dirac's ocean, the hydrogen atom could not exist but for a negligibly small part of a second. Indeed, if the hydrogen atom were "a droplet rotating around a bubble," the droplet would fall in and fill the cavity of the bubble, and the hydrogen atom would be annihilated in a flash. In this connection, Pauli proposed what is known as the "second Pauli principle," according to which any new idea formulated by a theoretical physicist becomes immediately applicable to all atoms forming his body. According to that principle, Dirac's body would be annihilated within a small fraction of one microsecond after he conceived that idea, and other theoretical physicists would be saved from hearing about it. . . .

In the year 1931, an American physicist, Carl Anderson, was studying the tracks produced in a cloud chamber by high energy electrons in cosmic ray showers. In order to measure the velocity of these electrons, he placed the cloud chamber in a strong magnetic field, and, to his great surprise, the photographs showed that one half of the electrons were deflected one way, while another half were deflected the opposite way. Thus, there was a mixture of 50% positively charged and 50% negatively charged electrons, both having the same mass. Those were the holes in Dirac's ocean which failed to graduate as protons, but emerged as particles in their own right. The experiments with positive electrons, or *positrons* as they are often called, quickly confirmed all the predictions based on Dirac's hole theory. The pair of one positive and one negative electron could be produced by the impact of high-energy light quanta (gamma rays or cosmic rays) against the atomic nuclei, and the probability of these events coincided exactly with the values calculated on the basis of Dirac's theory. Flying through ordinary matter, positrons were observed to be annihilated in collisions with ordinary electrons, the energy equivalent to their mass being liberated in the form of high-energy photons. Every single detail, *in fact,* was exactly as it was predicted.

But what about the fantastic theory which considers positive electrons to be the holes in an infinitely dense distribution of electrons with negative mass? Well, the theory is a theory, and is justified by its agreement with experimental evidence, no matter whether it pleases us or not. It has been shown since the appearance of Dirac's original paper that it is, *in fact,* not necessary to assume that existence of infinitely dense oceans of electrons with negative mass, and that positrons can be considered for all practical purposes as the holes in an absolutely empty space.

### ANTIMATTER

After the discovery of positive electrons, physicists dreamed about the possibility of negative protons which would stand in the same relation to ordinary positive protons as positrons to electrons. But, since protons are almost two thousand times heavier than electrons, their production would require energies ranging up to several billion electron volts. This started a number of ambitious projects of constructing particle accelerators* which could supply that amount of energy to nuclear projectiles, and in the United States the cornerstones were laid for two such superaccelerators: a *bevatron* at the Radiation Laboratory of the University of California at Berkeley and a *cosmotron* at Brookhaven National Laboratory at Long Island, N. Y. The race was won by the West Coast physicists Emilio Segré, O. Chamberlain *et al.*, who announced in October, 1955, that they had observed negative protons being ejected from targets bombarded by 6.2 bev (billion electric volt) atomic projectiles.

The main difficulty in observing the negative protons formed in the bombarded target was that these protons were expected to be accompanied by tens of thousands of other particles (heavy mesons) also formed during the impact. Thus, the negative protons had to be filtered out and separated from all the other accompanying particles. This was achieved by means of a complicated "labyrinth" formed by magnetic fields, narrow slits, etc. through which only the particles possessing the expected properties of antiprotons could pass. When the swarm of particles coming from the target (located in the bombarding beam of the bevatron) was passed through this "labyrinth," only the negative protons were expected to come out through its opposite end. When the machine was set into operation, the four experimentalists were gratified to observe the fast particles coming out at a rate of about one every 6 minutes from its rear opening. As further tests have shown, the particles were genuine negative protons formed in the bombarded target by the high-energy bevatron beam. Their mass was found to have a value of 1,840 electron masses, which is known to be the mass of an ordinary positive proton.

Just as the artificially produced positive electrons get annihilated in passing through ordinary matter containing a multitude of ordinary negative electrons, negative protons are expected to get annihilated by encountering positive protons in the atomic nuclei with which they collide. Since the energy involved in the process of proton-antiproton annihilation exceeds, by a factor of almost two thousand, the energy involved in an electron-antielectron collision, the annihilation process proceeds much more violently, resulting in a "star" formed by many ejected particles.

* See Chapter VIII, the section entitled First Nuc-crackers.

The proof of the existence of negative protons represents an excellent example of an experimental verification of a theoretical prediction concerning properties of matter, even though at the time of its proposal the theory may have seemed quite unbelievable. It was followed in the fall of 1956 by the discovery of *antineutrons,* i.e., the particles that stand in the same relation to ordinary neutrons as negative protons do to positive ones. Since in this case the electric charge is absent, the difference between neutrons and antineutrons can be noticed only on the basis of their mutual annihilation ability.

As protons, neutrons, and electrons forming the atoms of ordinary matter can exist in the antistates, one can think about antimatter formed by these particles. All physical and chemical properties of antimatter should be the same as that of ordinary matter, and the only way to tell that two stones are anti in respect to each other is to bring them together. If nothing happens, they are the same kind of matter; if they are annihilated in a tremendous explosion they are "anti's."

The possible existence of antimatter poses tremendous problems for astronomy and cosmology. Is all the matter in the universe of the same kind, or are there patches of our kind of matter and antimatter scattered irregularly through infinite space? There are strong arguments that within our stellar system of the Milky Way all matter is of the same kind. In fact, if it were not so, the annihilation processes between the stars and the diffused interstellar material would produce a strong observable radiation. But is our nearest neighbor in space, the Great Andromeda Nebula, and are hundreds of millions of other stellar galaxies scattered through space within the range of the 200 in. telescope of Palomar Mountain Observatory made of the same kind of matter or do we have here a fifty-fifty mixture? If all matter in the universe is of the same kind, why is this so? And, if it is partially ordinary matter and partially antimatter, how did these two mutually exclusive fractions separate from one another? We do not have the answers to any of these questions, and it can only be hoped that future generations of physicists and astronomers will be able to solve the mystery.

## QUANTUM STATISTICS

The quantum theory of motion had a severe impact on the kinetic theory of heat discussed in Chapter IV of this book. Indeed, if the electrons moving within the atom may have only certain discrete values of kinetic energy, the same must apply to the gas molecules moving within a closed container. Thus, considering the distribution of energy between the molecules of gas, one could no longer assume that gas molecules may possess any kinetic energy (Fig. VII-24a), as was assumed in the classical theories developed by Boltzmann, Maxwell, Gibbs, and others. On the contrary, there should be definite

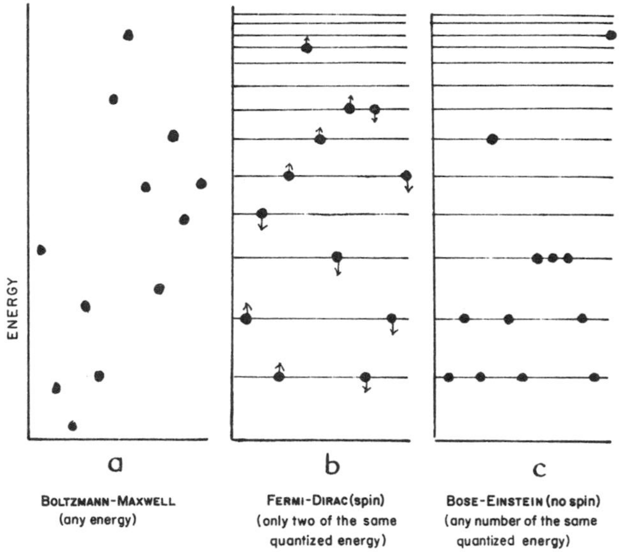


FIG. VII–24.

Three kinds of statistical treatment of the energy of twelve gas particles.

quantum levels, determined by the size of the container, and no energies in between these levels could be allowed. The situation was complicated by the fact that some particles (like electrons) obey the Pauli principle, which prohibits more than two of them to occupy the same quantum level, while other particles (like air molecules) are not subject to this restriction. This fact led to two different kinds of statistics: the so-called Fermi-Dirac statistics applicable to the particles obeying the Pauli principle, and the Bose-Einstein statistics applicable to the particles which do not obey it. Figure VII-24b, c, is intended to clarify the difference between these two types of statistics. The entire development of quantum statistics is very exciting, but extremely difficult to explain without using "technical" terminology.

Thus, all that can be stated here is that both new statistics practically do not differ from the good old classical statistics in all the everyday cases such as atmospheric air. Deviations are expected, and observed, only in such cases as electron gas in metals and in the so-called "white dwarf" stars where the situation is governed by Fermi-Dirac rules, and in ordinary gases at a temperature very close to absolute zero, where Bose-Einstein rules prevail. It may be hoped that the readers of this book who are sufficiently enchanted by the subject will proceed in the study of modern physics in a more advanced way. In that case the problems of quantum statistics will become as clear as crystal to them after just a half dozen years of study.

# CHAPTER VIII *The Atomic Nucleus and Elementary Particles*

## DISCOVERY OF RADIOACTIVITY

EARLY in 1896 the French physicist Henri Becquerel, having heard about Roentgen's recent discovery of X-rays, decided to see if something similar to X-rays is also emitted from the fluorescent materials which were known to glow under the action of incident light rays. For these studies, he selected crystals of a mineral known as uranyle (double sulfate of uranium and potassium) which he was studying before because of its strongly expressed fluorescent properties. Since Becquerel believed that the radiation is the result of external illumination, he placed a crystal of uranyle on a photographic plate wrapped in black paper and put it on the window sill. When he developed the plate after a few hours' exposure to sunlight, it clearly showed a darkish spot under the place at which the uranyle crystal had been placed. He repeated the experiment several times, and the dark spot was always there, even though he put more black paper around the photographic plate.

On February 26 and 27 (1896) the Paris sky was covered with heavy clouds, rain was falling intermittently, and life on the boulevards was hiding under the awnings of the cafés and restaurants. Unhappy Professor Becquerel put the freshly wrapped photographic plate with the uranyle crystal on it in the drawer of his desk to wait for better weather. The sun did not appear again until March 1st, and even then was often obscured by the passing clouds. Nevertheless, Becquerel again exposed his contraption to the sun's rays, and went in to the dark room to survey the results. It was something unbelievable! Instead of darkish impressions which were obtained earlier during the full day's exposure to the brillant sunshine, there was a charcoal black spot under the place where the uranyle crystal had been placed! Apparently the darkening of the plate had nothing to do with the exposure of the uranyle crystal to the sun's rays, and the darkening of the photographic plate was going on uninterruptedly all the time it was sitting with a piece of uranyle on it in the closed drawer of Becquerel's desk.

It was a penetrating radiation similar to X-rays, but it was emanating all by itself and without any external excitation from the atoms, presumably those of uranium in Becquerel's crystal. Becquerel tried to heat up the crystal, chill it down, grind it to a powder, dissolve it in acids, and do to it everything that he could think of, but the intensity of the mysterious radiation persistently remained the same. It became clear that this new property of matter, which received the name of *radioactivity,* has nothing to do with the physical or chemical way in which the atoms are put together, but is a property hidden deep inside of the atom itself.

## RADIOACTIVE ELEMENTS

During the first years after the discovery of radioactivity, a large number of chemists and physicists were busy studying the new phenomenon. Madame Marie Sklodowska Curie, Polish born, educated in chemistry, and the wife of the French physicist Pierre Curie, carried out an extensive test of all chemical elements and their compounds for radioactivity, and found that thorium emits radiation similar to that of uranium. Comparing radioactivity of uranium ores with that of metallic uranium, she noticed that ores are about five times more radioactive than would be expected from their uranium content. This indicated that the ores must contain small amounts of some other radioactive substances much more active than uranium itself, but, to separate them, very large amounts of expensive uranium ores were needed. Madame Curie succeeded in obtaining from the Austrian government a ton of worthless residues (at that time) from the state uranium producing plant in Joachimschal (Bohemia) which, being deprived of uranium, still retained most of its radioactivity. Being led by Theseus' thread of penetrating radiation, Madame Curie managed to separate a substance having chemical properties similar to those of bismuth, which she called *polonium* in honor of her native country. Still more work, and another substance chemically similar to barium was separated and received the name of *radium;* it was two million times more radioactive than uranium.

The pioneers, both in new and unknown countries and new fields of science, often fall victim of the hidden perils they meet on their way. Madame Curie's death, at the age of 67, was due to leukemia, a disease which is now known to be caused by exposure to the penetrating radiation. When physicists learned better how to be careful with radiation, photographic films were placed between the sheets of Madame Curie's laboratory books. The developed films have shown numerous fingerprints caused by radioactive deposits on the sheets touched by Madame Curie's fingers.

The discovery of polonium and radium was followed by the discoveries of more and more radioactive substances. Among them was *actinium,* a close relative of the fissionable uranium, which was separated by Debierne and Giesel, and *radiothorium* and *mesothorium,* separated by Otto Hahn who, some forty years later, discovered the phenomenon of uranium fission.

### RADIOACTIVE FAMILIES

On the physical side of the picture, the work was progressing in the study of the properties of penetrating radiation. In 1899 the 28-year-old Ernest Rutherford found that there are three different kinds of rays :

1) Alpha ($\alpha$) rays, which could be stopped by a sheet of paper and were proved to be the ions of helium. (They were actually the nuclei of helium atoms, but Rutherford did not know it until he carried out the scattering experiments [see p. 223] twelve years later.)

2) Beta ($\beta$) rays, which could pass through aluminum foils a few millimeters thick, and which turned out to be the streams of very fast-moving electrons.

3) Gamma ($\gamma$) rays, which could penetrate lead shields many centimeters thick, and are similar to X-rays, having, however, much shorter wave length.

It is customary to reproduce in physics textbooks (including those written earlier by the author) a diagram like that in Figure VIII-1, which shows the deflection of $\alpha$, $\beta$, and $\gamma$ rays passing through a

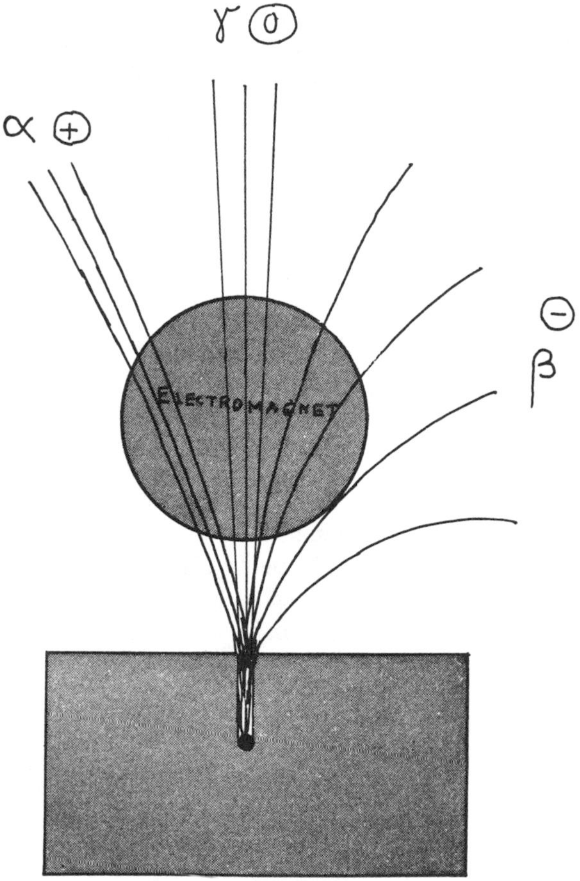


FIG. VIII–1.
Alpha, beta, and gamma rays.

magnetic (or electric) field. The alpha beam bends to the left (positive charge), beta beam to the right (negative charge), and gamma beam remains undeflected (electromagnetic waves). It is doubtful, however, that such an experiment was ever conducted during the early studies of radioactivity (a noticeable deflection of alpha particles requires exceptionally strong electromagnets which were constructed only much later), and the difference between alpha and beta was established by much more intricate indirect methods.

Early in the game, Rutherford and his collaborator, Frederick Soddy, came to the conclusion that the phenomenon of radioactivity is the result of spontaneous transformation of one chemical element into another. The emission of an alpha particle, with charge $+2$ and mass 4, results in the formation of an element which is located two steps to the left in Mendeleev's system, and whose atomic weight is four units less. The emission of a beta particle (a negative electron) brings the element one step to the right in Mendeleev's system, and does not change its atomic weight at all. The emission of gamma rays is simply the result of a disturbance of the atom caused by the ejection of a positively or a negatively charged particle.

The series of successive alpha and beta decays degrades the heavy unstable atoms of radioactive elements, reducing their atomic number and weight until they finally reach a stable state which is an atom of lead. Since an alpha decay changes atomic weight by 4 units while beta decay does not change atomic weight at all, there could exist four families of radioactive elements:

1) Those with atomic weight being a multiple of four: $4n$
2) those with atomic weight $4n + 1$
3) those with atomic weight $4n + 2$
4) those with atomic weight $4n + 3$

The atomic weight of uranium is 238, i.e., $4 \times 59 + 2$. Thus, uranium and all members of its family derived by alpha and beta decay belong to the third of the above-mentioned categories. Atomic weight of thorium is 232, i.e., $4 \times 58$, so that the thorium family belongs to the first category. Protactinium, which decays into actinium, and the other members of the actinium family, has atomic weight 231, i.e., $4 \times 57 + 3$, thus belonging to the third category. A radioactive family with atomic weight $4n + 1$ (second category) does not exist in nature, but can be produced artificially in atomic piles.

Through the hard work of the early students of radioactivity, the heraldic trees of the existing radioactive families were constructed. Page 277 shows the decay scheme of the uranium family, which starts with old man uranium-238 and, after eight alpha and six beta transformations, finishes with the stable lead 206.

The two figures above the name of each radioactive element give its atomic number and atomic weight, while the figure below is its half-lifetime in years, days, hours, minutes or seconds. Similar decay

| | | | | | | | | |
|---|---|---|---|---|---|---|---|---|
| | | | | | | | 90-234<br>UX 1<br>245d | 92-238<br>U 1<br>$4.5 \times 10^9$y |
| | | | | | | | 91-234<br>UX 2<br>11m | |
| | | 82-214<br>RaB<br>27m | 84-218<br>RaA<br>3m | 86-222<br>Rn<br>3.8d | 88-226<br>Ra<br>1,590y | 90-230<br>Io<br>$8.3 \times 10^4$y | 92-234<br>U II<br>$2.7 \times 10^5$y | |
| | 81-210<br>RaC″<br>13m | 83-214<br>RaC<br>20m | | | | | | |
| | 82-210<br>RaD<br>22y | 84-214<br>RaC′<br>$10^{-4}$s | | | | | | |
| | 83-210<br>RaE<br>50d | | | | | | | |
| 82-206<br>Pb<br>stable | 84-210<br>Po<br>137d | | | | | | | |

ALPHA DECAY

BETA DECAY

schemes can be constructed for thorium, protactinium, and the artificially-produced (nameless) fourth family.

### THE SURVIVAL LAW

If one follows the life history of a large group of babies or puppies or ducklings or any other kinds of animals born on the same day, one finds that they do not also die on the same day. Some live a bit longer, some live a somewhat shorter time, and, if one plots the percentage of individuals which are still alive at a certain date, one obtains a typical survival curve shown in Figure VIII-2a. It indicates that there exists

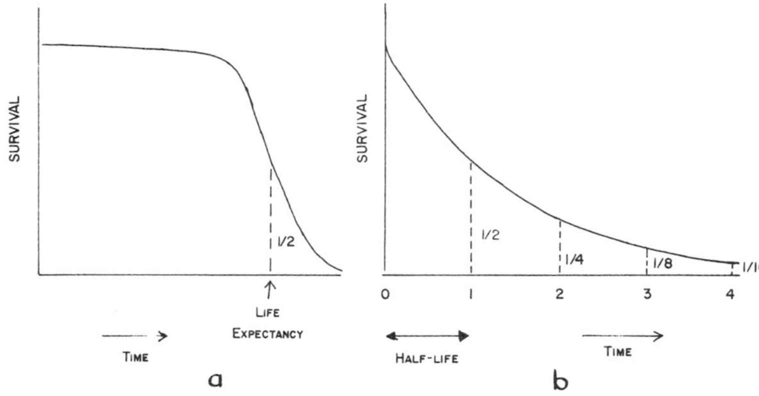


FIG. VIII–2.
Survival curves for animals (a) and nuclei (b).

a certain "life expectancy," as shown in the diagram, which is about 75 years for a man, 15 years for a dog, and only a few years for a duck. The curve indicates that there are comparatively small chances of their dying much before a certain age, and equally small chances of their living very long after reaching it.

In the case of the radioactive atoms the situation is entirely different, and the chances for a member of the radioactive family freshly formed by the "reincarnation" of its predecessor (either through alpha or beta transformation) to be "reincarnated" into the next member of the family are independent of the period of time which passed since it was formed. The situation is similar to the case of soldiers involved in a continuous battle with the enemy, where a certain percentage are killed every day, and there is no way of telling whose number will be up tomorrow. In this case we cannot speak any more about "life expectancy," but must introduce a rather different notion of "half life," i.e., a period of time during which one half of the soldiers will be killed or one half of the unstable radioactive atoms will decay. The curve representing such a process is shown in Figure VIII-2b, and is, as the mathematicians call it, an "exponential curve." Different radioactive elements have widely varying half lives. Uranium itself decays at the rate of 50% in 4.5 billion years, radium in 1,590 years, whereas half of the atoms of RaC′ break up in only one ten-thousandth of a second. The existence of the three natural radioactive families is due to the longevity of their forefathers, uranium I (or $_{92}U^{238}$), thorium ($1.3 \times 10^{10}$ years), and protactinium's "granddaddy" ($5 \times 10^8$ years),* which is comparable with the age of the universe.

No family of the type $4n + 1$ exists in nature because, as was found when these nuclei were produced artificially in atomic piles, the head of the family has a considerably shorter life, and all the family must have completely decayed a long time ago.

### THE LEAKY BARRIERS

The explanation of the slowness of alpha transformations was given independently by the author of this book, who at that time was working in Germany, and by the team of Ronald Gurney (Australia) and Edward Condon (United States), and is based on wave mechanics. It was known that atomic nuclei are surrounded by high barriers of electric force which were first investigated by Rutherford's experiments on scattering of alpha particles. As an alpha particle approaches the nucleus, it is subject to the repulsion which is proportional to the product of nuclear charge ($Ze$) by the charge of alpha particle ($2e$) divided by the square of the distance between them. When the particle comes in contact with the nucleus, cohesive forces between it

* Protactinium itself has a half-life period of only 12,000 years, but the precursor of the actinium family, from which protactinium is derived by one $\alpha$ and one $\beta$ decay, has a life of ½ billion years. It does not have a name referring to the actinium family genealogy, but, being an isotope of uranium, is simply called U-235. Emitting an $\alpha$ particle, $_{92}U^{235}$ transforms into $_{90}UY^{231}$ which, through the following $\beta$ decay, results in $_{91}Pa^{231}$. U-235, carrying the name of the uranium family but actually belonging to the actinium family (type $4n+3$), is the famous "fissionable" uranium which made possible the development of "atomic" bombs and nuclear reactors.

and the particle forming the nucleus pull it in and hold it tightly inside. The potential curve corresponding to these two kinds of forces is shown in Figure VIII-3 and looks like a rampart or a barrier with a steep wall inside and gentler slopes on the outside. In order to get into the nucleus, incident alpha particles have to climb to the top of that barrier, and then drop down into the nuclear interior. Similarly, any particle leaving the nucleus must climb the internal wall of the barrier, and then roll down along its outer slope. Studying scattering of an alpha particle in uranium, Rutherford found that the barrier

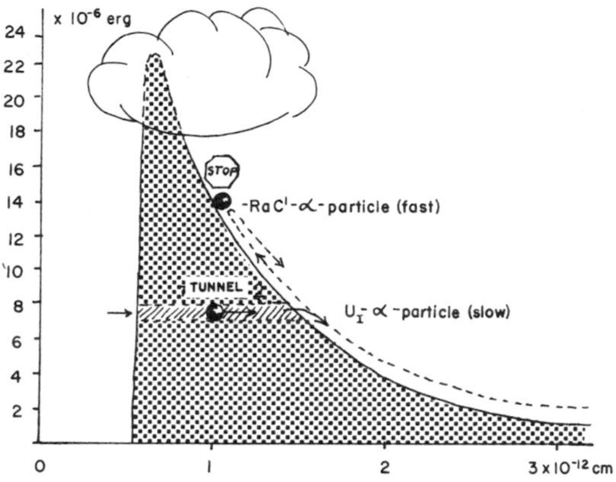


FIG. VIII–3.

Potential barrier around the uranium nucleus, as obtained from Rutherford's scattering experiments.

surrounding the nuclei of that element must be at least $14 \times 10^{-6}$ erg high, since fast alpha particles emitted from RaC′ and having that energy did not show any sign of reaching the top. On the other hand, alpha particles emitted by uranium itself have the energy of only: $8 \times 10^{-6}$ erg. How could the outgoing particles with such small energy roll over a barrier which is several times higher? According to classical mechanics this is, of course, quite impossible. If one builds a wooden barrier on the table and rolls a ball toward it whose energy is only half as large as necessary to reach the top, the ball will always climb half way up the slope and then roll back. But wave mechanics comes to a different conclusion, and to understand it we must involve the analogy between the de Broglie waves and the waves of light. In geometrical optics one is familiar with the notion of "total internal reflection." If a ray of light traveling through the glass (Fig. VIII-4a) falls on the interface *AB* between the glass and air, under a comparatively small angle of incidence, it will be refracted entering into the air, and its new direction will be closer to the interface *AB*. If, however, the angle of incidence is larger than a certain critical value,

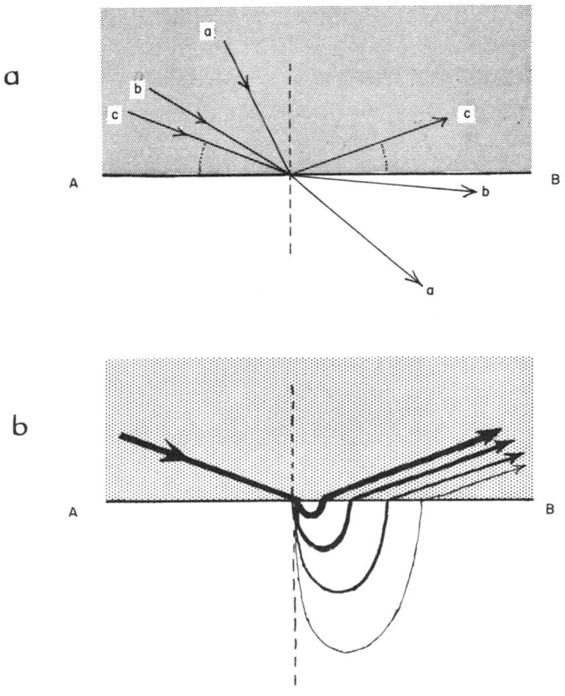


FIG. VIII–4.
Total internal reflection of light according to geometrical optics (a) and wave optics (b).

no light will enter into the air and the ray will be totally reflected from the interface.

Considering this phenomenon from the point of view of the wave nature of light, one comes, however, to a different conclusion. It turns out that some of the light *does* enter into the air beyond the interface *AB*, but it does not penetrate too far and is thrown back from an air layer only a few wave lengths thick. Figure VIII-4b, in which the lines are no longer rays of light but represent the flow line of radiant energy, shows what happens. If we bring another piece of glass, close the interface *AB*, some of the flowlines passing through the air will enter into the second piece of glass. This phenomenon can be observed experimentally if the distance between interfaces is equal to just a few wave lengths of light (i.e., a few microns).

Just as wave optics permits the penetration which is completely prohibited by geometrical optics, wave mechanics helps the material particles to perform deeds which were completely impossible if classical mechanics were a hundred per cent correct.

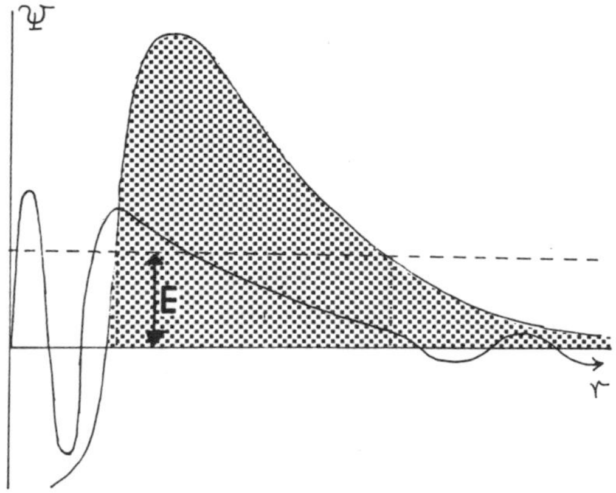


FIG. VIII–5.
Wave mechanical penetration of an alpha particle through the nuclear potential barrier.

Alpha particles, sitting inside the nucleus, are in a state of very fast motion, constantly colliding with the walls of the potential barrier surrounding them. The de Broglie wave, guiding the motion of these particles, slowly leaks through the walls of the barrier, making it possible for alpha particles to get through even if they cannot get over the top (Fig. VIII-5). The penetrability of nuclear potential barriers is extremely low, and in the case of the uranium nucleus only one attempt out of $10^{38}$ is successful. Since, moving within the space of only $10^{-12}$ cm across with the velocity of $10^9 \frac{\text{cm}}{\text{sec}}$, an alpha particle hits the internal wall of the barrier $10^{21}$ times per second, it takes $\frac{10^{38}}{10^{21}} = 10^{17}$ sec, or several billion years, until it succeeds in getting out. In the case of RaC′ nucleus the penetrability of the barrier is higher, and one out of only $10^{17}$ attempts is successful. Thus the corresponding life becomes $\frac{10^{17}}{10^{21}} = 10^{-4}$ sec as observed. The calculations of the half-life periods of different radioactive elements on the basis of that theory led to a perfect agreement with the observed figures.

It goes without saying that wave mechanical phenomena of that kind are of importance only in the world of atoms and nuclei. In the case of the above-described experiment in which the ball is rolled up the wooden slope with not enough velocity to get over the top, there is still a chance that it will go through the potential barrier as an old-fashioned ghost goes through the wall of a castle, but these chances can be calculated to be about $10^{-10^{27}}$, i.e., a number which has $10^{27}$ zeros after the decimal point. If one tries to write that number down, the first significant figure will appear in the neighborhood of the most distant galaxies seen through the 200-in. telescope. Thus, do not try to roll the ball up the slope.

## NUCLEAR CONSTITUTION AND NEUTRONS

The interpretation of the phenomenon of radioactivity as a spontaneous decay of atomic nuclei left no doubt that the nuclei are complex mechanical systems composed of many constituent particles. The fact that atomic weights of the isotopes of all elements are represented very closely by integer numbers indicated that protons must play the role of one of the nuclear components. But the protons alone did not suffice. In fact, the nucleus of carbon, for example, having atomic weight 12, must contain 12 protons. But since the charge of the carbon nucleus is only 6, there must be 6 negative charges present, and it was assumed that those negative charges were supplied by 6 electrons which join the 12 protons to form a carbon nucleus. However, the assumption of the presence of electrons within atomic nuclei was leading to very serious difficulties from the point of view of the quantum theory. Indeed, since the energies of the quantum states of an electron are rapidly increasing with the decreasing dimensions of the region to which the electron is confined, one had to expect that electrons moving within atomic nuclei should have the energies of billions of electron volts. This straightforward conclusion of the quantum theory looked very odd, since, whereas the energies of that order were observed in the case of cosmic rays, the energies involved in nuclear phenomena were millions (and not billions!) of electron volts. When Niels Bohr told Ernest Rutherford about that "fact of life," they decided that the only way to save the situation was to assume the existence of *chargeless protons* which were tentatively called "neutrons." With this assumption it was not at all necessary to have electrons inside of the atomic nucleus, and the composition of the carbon nucleus, for example, could be written as: ${}_6C^{12} = 6$ protons $+ 6$ neutrons.

In the mid-twenties, a vigorous program was started in Cavendish Laboratory with the purpose of kicking these hypothetical "neutrons" out of the nuclei of some light elements, thus giving a direct proof of their existence. But the results were negative, the work in this direction was suspended, and the discovery of neutrons was delayed by quite a number of years. And it was only in 1932 that Rutherford's student, J. Chadwick, studying the mysterious, highly penetrating radiation (first observed by W. Bothe in the case of alpha bombardment of berillium target) proved that it is formed by a stream of neutral particles with the mass closely equal to that of a proton. Thus, after early miscarriages, the neutron was finally born within the walls of Cavendish Laboratory.

### BETA DECAY AND NEUTRINOS

While the emission of alpha particles represents a real nuclear decay resulting in a product with smaller atomic weight, the emission of beta rays is nothing but electric adjustment of the atomic nucleus resulting from the emission of one or more alpha particles. In the discussion in the previous section, we have seen that atomic nuclei consist of protons and neutrons, and that, in heavier elements, the number of neutrons exceeds the number of protons. For example, in ${}_{88}Ra^{226}$ the number of neutrons is $226 - 88 = 138$ with the number of protons only 88, the ratio being $\frac{138}{88} = 1.568$. The nucleus of ${}_{86}Rn^{222}$ formed by alpha-decay of radium has only 136 neutrons and 86 protons with the ratio $\frac{136}{86} = 1.581$. Thus, in the process of alpha-decay, the ratio of neutrons to protons increases, and after several alpha-steps it may become larger than is convenient for the peaceful co-existence of the two kinds of particles. In this case a neutron transforms itself into a proton by emission of a negative electron or a beta particle. It may be noticed from the table on p. 277 that beta transformations always occur in pairs. It is because neutrons and protons in the nucleus are subject to the same Pauli principle as the atomic electrons, and each quantum level is occupied by two of them (with opposite spin). Thus, when the level becomes unstable, two particles undergo beta transformation, one after another.

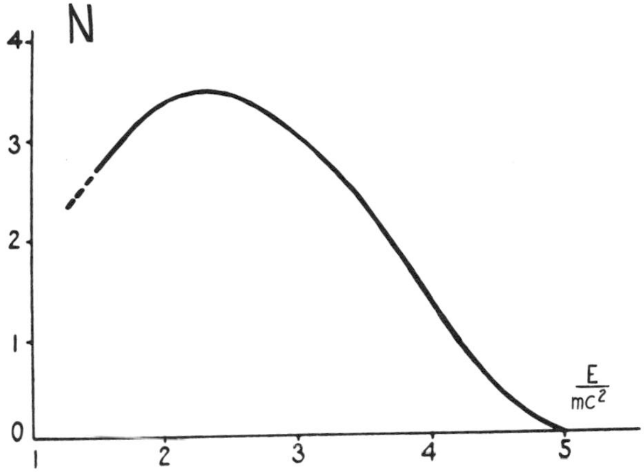


FIG. VIII–6.
The beta-ray spectrum of $In^{114}$. The number ($N$) of electrons is plotted against their energy ($E$) expressed in $mc^2$ units.

In the year 1914 a young British physicist, James Chadwick, was working in the University of Berlin under the guidance of the noted German physicist, Fritz Geiger (the inventor of the Geiger counter). His job was to study the spectrum of beta rays emitted by various radioactive substances which seem to differ radically from alpha and gamma rays by showing a continuous distribution of their kinetic energies ranging from almost zero to rather high values (Fig. VIII-6). As Chadwick's work was completed and sent out for publication in the fall of that year World War I broke out, and Chadwick was promptly arrested as an enemy alien and sent to a prison camp for the duration. The first year in the camp was dull, since the young and talented physicist could not make any friends among his fellow prisoners, consisting mostly of businessmen, traveling salesmen, etc. Then, after a big battle somewhere in France, a new prisoner appeared in the camp. He was C. D. Ellis, a brilliant officer of His Majesty's Highland Regiment, who had been captured on the field of battle. The two Britishers became friends and, to kill time, Chadwick started teaching Ellis the facts of nuclear physics. When the war was over they both returned to England and Ellis enrolled as a graduate student of Cambridge University where Chadwick was a lecturer. A few years later, Ellis published a paper which represented an important extension of Chadwick's work.

One of the possible explanations of the continuous energy spectrum of beta rays could be the wide-ranging energy losses suffered by beta particles while escaping from radioactive substances in which they originate. Ellis devised a very clever experiment in which all beta rays emitted by a radioactive substance were absorbed in a piece of lead, and the heat produced carefully measured. The result of this experiment showed that the total energy liberation per particle was exactly equal to the *average* energy of electrons in the continuous spectrum, proving that no losses in the material were taking place. Thus, the physicists faced a paradoxical situation. While in the series of radioactive transformations the emitted alpha particles always have sharply defined energies which are equal to the differences between the internal energy contents of the mother and daughter nuclei, the energies of beta particles vary in wide limits. What happens to the energy difference between the two nuclei of the same radioactive element, one of which emits a fast beta particle, and another a slow one? Niels Bohr, who was very much excited about that paradoxical situation, went so far as to suggest that the law of conservation of energy may not hold in the case of radioactive beta transformation and, while in the case of the emission of a slow beta particle a certain amount of energy may disappear into the thin air, in the case of the emission of a very fast beta particle an additional amount of energy can be created from nothing. According to this hypothesis, the law of conservation of energy in elementary nuclear processes would hold only *on the average,* thus making it impossible to construct the perpetual motion machine of the first kind (see Chap. IV) based on the processes of radioactive decay.

Wolfgang Pauli, who was more conservative in his views on that subject, suggested an alternative which would balance the energy bookkeeping of nuclear processes. He considered it a possibility that the emission of a beta particle is always accompanied by an emission of another "mystery particle" which escapes the observations and carries away the balance of energy. If one assumes that these nuclear "thieves of Baghdad" have no electric charge and have a mass as small as or smaller than that of an electron, they would indeed easily escape, with their share of energy, the most careful roadblocks set by physicists. Pauli gave to these hypothetical "thieves" the name of "neutrons." (This was before the particles now called neutrons were discovered by Chadwick in 1932.) But all these discussions remained within the realm of conversations and private correspondence, and the name never got "copyrighted," by being printed in a scientific magazine. When after the discovery of (Chadwick's) neutrons, Enrico Fermi (Fig. VII-15), at that time a professor at the University of Rome, was reporting Chadwick's paper at a seminar, somebody from the audience asked him whether Chadwick's neutrons are the same particles Pauli used to speak about. "*No,*" answered Fermi, "*le neutrone di Chadwick sonno grande. Le neutrone di Pauli erano piccole; egli devono star chiamato* neutrini." (In Italian, "neutrino" is a diminutive of "neutrone.")

Since it has become customary in this book to quote little stories about great physicists, and since Enrico Fermi was one of the greatest physicists of our time, here is a story about him, based on his own words. For his early discoveries in the field of physics, he was elected a member of the Royal Italian Academy of Sciences and was given the title *Eccellenza* by Benito Mussolini. Once he drove his little Fiat to an Academy meeting which was to be addressed by Mussolini himself, so that the main gates leading to the courtyard were guarded by two *carabinieri*. They crossed their carbines in front of Fermi's little car and asked him who he was. "They will not believe me if I say that I am an *Eccellenza,*" Fermi thought, "because all the *Eccellenzas* look much more dignified, and travel in big chauffeur-driven limousines." So he smiled at the *carabinieri* and said that he was the driver of *Eccellenza* Fermi. This worked, and they let him drive in and wait until his master should come out from the meeting.

Coming back to neutrinos, one should say that this particle was really very elusive and for a long time the nuclear physicist hunting it could only see the damage it did, but could not catch the particle itself. Only, in 1955, Fred Reines and Cloyd Cowan from Los Alamos Scientific Laboratory managed to make a catch. The most intensive source of neutrinos is presented by atomic piles where legions of neutrinos are emitted as a result of the beta decay of fission products formed in the chain reaction. While even the most penetrating gamma rays and fastest neutrons are effectively stopped by thick concrete shielding surrounding the pile, the neutrinos fly through that shielding as easily as a swarm of mosquitoes through a chicken fence. To detect them, Reines and Cowan placed outside of the shielding a huge container filled with hydrogen and surrounded by batteries of particle counters of different kinds. It was expected that a fast neutrino, colliding with a proton, would kick out a positive electron, turning proton into neutron: $P + \nu \rightarrow n + \overset{+}{e}$; but the theoretically estimated probability of such a process was extremely small. To detect that process they used neutron and positron counters connected in such a way that they would give a signal only when both are hit simultaneously by a neutron and a positron. Since the probability of a chance coincidence was extremely low, a simultaneous hit of the two counters could result only from the reaction written above. Conducting the experiment with the pile in full operation, they were getting several signals per minute, but the signals dropped quickly when the pile was shut off. From their observations they found that the effective cross section of the process in which a neutrino turns a proton into a neutron is only $10^{-43}$ cm$^2$, which means that, in order to cut the intensity of a neutrino beam by a factor two, one should use a water-shield hundreds of light years thick.

The theory of neutron-proton transformation with the emission of an electron and a neutrino developed by Fermi stands in an excellent agreement with all experimental data concerning beta decay. It also serves as a prototype for all decay theories developed later in connection with various transformation processes among the elementary particles.

## FIRST NUC-CRACKERS

Ever since Rutherford realized that the phenomenon of radioactivity represents a spontaneous transformation of one chemical element into another, he was besieged by the desire to get a crack at the atomic nucleus of some stable element and to turn it into another element, thus realizing an ancient dream of alchemists. When World War I broke out in 1914, the British Admiralty asked Rutherford to transform Cavendish Laboratory, of which he had just recently become a director, into a war research institution to develop antisubmarine warfare methods against German U-boats. Rutherford refused on the basis that he had a much more important task of breaking the nucleus of the atom. It is true that this work of Rutherford paved the way for the development of the most powerful war weapons, the atomic and hydrogen bombs, but it is not true that Rutherford foresaw these developments. In fact, shortly before his death in 1937, Rutherford had a heated debate with a Hungarian physicist, Leo Szillard, about the possibility of large-scale liberation of nuclear energy, and insisted that this could never be done. To prove his point, Szillard went to a patent office and took a patent on large-scale nuclear reactions. Three years later the fission of the uranium nucleus was discovered, and in another six years the first A-bomb exploded over Hiroshima and brought World War II to a close. Rutherford was undoubtedly watching these developments, sitting on an ethereal cloud and listening to the music of harps, but it is more than likely that the old man thought: "So what? Now these . . . chaps are using my discoveries for killing one another!"

But, returning to the year 1919, we must see what Rutherford was doing about cracking the nucleus. Since the rampart of Coulomb repulsion surrounding the atomic nucleus becomes higher and higher as one moves along Mendeleev's system of elements, the best chance would be to bombard the lighter nuclei. Also the high energy alpha particles from the fast-decaying radioactive elements would do a better job than the slower ones. Thus, in his first attempt Rutherford decided to shoot alpha particles from RaC′ at the nuclei of nitrogen gas, and noticed to his great satisfaction that, apart from the numerous alpha particles scattered by nitrogen nuclei, there were also a few fast-moving particles of another kind which Rutherford identified as protons. Rutherford's first observations were made by scintillation method, but soon the study of nuclear transformations was considerably facilitated by the use of a brilliant invention, the Wilson or cloud chamber, of C. T. R. Wilson, whose studies were mentioned in a previous chapter in connection with J. J. Thomson's experiments. It is based on the fact that whenever an electrically charged fast-moving particle passes through the air (or any other gas), it produces ionization along its track. If the air through which these particles pass is saturated with water vapor, the ions so produced serve as the centers of condensation for tiny water droplets, and we see long thin tracks of fog stretching along the particles' trajectories. The scheme of a cloud chamber is shown in Figure VIII-7. It consists of a metal cylinder, *C,* with a transparent glass top, *G,* and a piston, *P,* the upper surface of which is painted black. The air between the

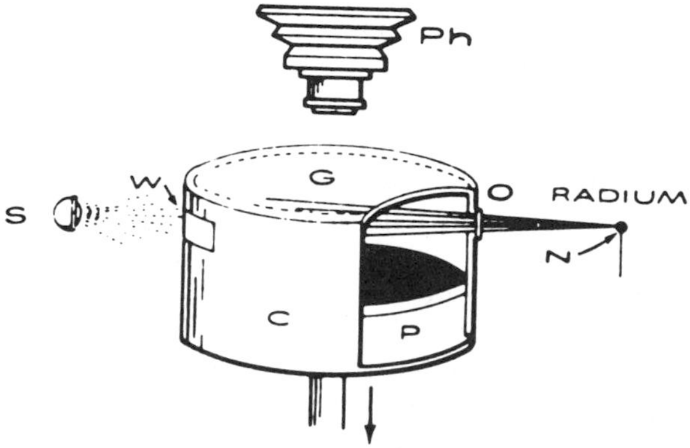


FIG. VIII–7.
The scheme of C. T. R. Wilson's cloud chamber.

piston and the glass top is initially almost saturated with water vapor and is brightly illuminated by a light source, *S,* through a side window, *W*. Suppose now that we have a small amount of radioactive material on the end of a needle, *N,* which is placed near the opening, *O*.

The particles that are ejected by the radioactive atoms will fly through the chamber ionizing the air along their paths. However, since the air is not quite saturated with water vapor, no condensation occurs, and the positive and negative ions produced by the passing particles recombine rapidly into neutral molecules. Suppose, however, that the piston is pulled rapidly down for a certain distance. The expansion of the air enclosed between the piston and the glass top will lower the air's temperature and will cause the condensation of water vapor in the very same way that clouds are formed as a result of rising streams of humid air in the terrestrial atmosphere. But, since the condensation of water vapor is considerably helped by the presence of ions produced by the charged particles passing through the chamber at this moment, the fog formation will take place along the particles' tracks, and thin, long streaks of fog will stand out clearly against the black background in the beam of illuminating light. This picture can be viewed directly by looking through the glass top or photographed by a camera, *Ph.*

In Plate V, *upper,* we see the first photograph of artificial nuclear disintegration taken in 1925 by Rutherford's student, P. M. S. Blackett. Numerous tracks diverging from a point beyond the edge of the picture are caused by alpha particles from radioactive material placed at that position. This material represented a mixture of RaC and RaC′ which was formed from RaC by the process of alpha transformation. Alpha particles from RaC move comparatively slowly and are stopped by air in the middle of the photograph. The RaC′ alpha particles, about the fastest particles emitted by radioactive elements, can penetrate thicker layers of air and their tracks finish at the top of the picture. In the upper middle part of the photograph we see a fork, due to the transformation of nitrogen nucleus under the impact of an alpha particle. The thin, long track going to the left belongs to a proton kicked out of the nucleus, while the thick track going up was shown to be caused by a fast-moving nucleus of oxygen. The alchemical transformation taking place here can be represented by a formula: ${}_7N^{14} + {}_2He^4 \rightarrow {}_8O^{17} + {}_1H^1$ where, according to a conversion, lower indices represent atomic number, and the upper, atomic weight. The atom ${}_8O^{17}$ represents the heavier isotope of ordinary oxygen ${}_8O^{16}$ and is present in small amounts in atmospheric air. Measuring the energies of ${}_1H^1$ and ${}_8O^{17}$ produced in this reaction, which can be done on the basis of the lengths of their tracks, one found that it is less than the initial energy of the alpha particle by the amount of 1.26 mev. For the combined masses on both sides of the above reaction equation, we find

| | |
|---|---|
| $He^4 = 4.00388$ | $H^1 = 1.00813$ |
| $N^{14} = 14.00755$ | $O^{17} = 17.00453$ |
| 18.01143 | 18.01266 |

Thus the energy balance, in this case negative, is $-0.00125$ units which is equivalent to $-1.16$ mev. This figure agrees within experimental errors with the above-quoted figure for the energy loss in the reaction. Measurements of that kind represented the first direct experimental proof of the Einstein law of the equivalence of energy and mass. Thus, in this reaction, nuclear energy is not liberated but lost. In other cases, however, such as alpha bombardment of aluminum, considerable amounts of nuclear energy are gained.

Since alpha particles are the only heavy projectiles emitted by natural radioactive elements, the early work on artificial nuclear transformations was limited to this type of reaction only. In 1939, the author of this book, working with Lord Rutherford in Cambridge, calculated, on the theory of the potential barrier, that protons would be much better projectiles, both because of their smaller electric charge and because of their smaller mass. Calculations have shown, in fact, that protons, accelerated by an electric potential of one million volts and moving with the energy several times smaller than alpha particles from RaC′, should produce noticeable disintegration of light elements. Rutherford asked his students, J. Cockcroft (now Sir John) and E. T. S. Walton, to construct a high tension machine which would produce beams of protons of that energy, and the first "atom smasher" went into operation in 1931. Directing the beam of protons at a lithium target, Cockcroft and Walton proved that at each successful impact two newly produced alpha particles were flying in opposite directions from the collision point. The reaction obviously was: ${}_3Li^7 + {}_1H^1 \rightarrow 2\,{}_2He^4$. Substituting lithium for boron they observed triple prongs (Plate V, *lower*), which indicated that, being hit by a proton, the boron nucleus breaks up into three equal fragments: ${}_5B^{11} + {}_1H^1 \rightarrow 3\,{}_2He^4$.

The pioneering work of Cockcroft and Walton was followed by development of larger and larger particle accelerators, based on a variety of ingenious principles. One species of atom smasher is called Van de Graaff after the inventor and is based on a simple principle of electrostatics, according to which an electric charge brought through an opening into the inside of a hollow metallic sphere will become distributed along its outer surface. Indeed the mutual repulsion between the electrons brought in will drive them as far as possible from each other. Van de Graaff's machine consists of large insulated metallic spheres, and a running belt which is being continuously charged on the outside, and discharged after entering into the sphere. Although the electric tensions which can be produced by Van de Graaff's machine are limited to a few million volts, they have been developed into compact, sturdy gadgets very convenient for many types of laboratory work.

Another, much more ingenious, device for accelerating nuclear particles was developed by Ernest Orlando Lawrence, after whom the Radiation Laboratory of the University of California is now

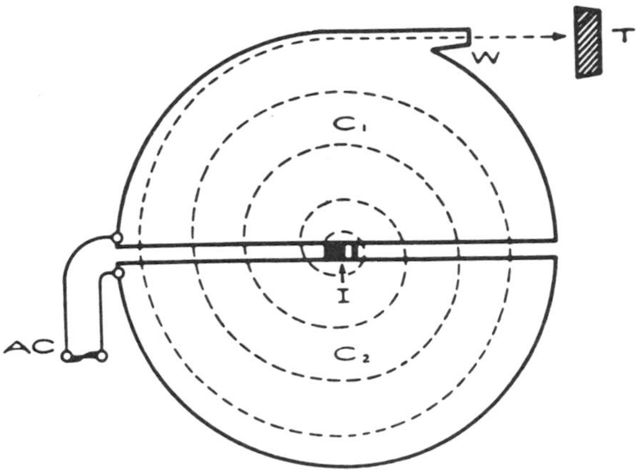


FIG. VIII–8.
The principle of the cyclotron.

named. It is based on an entirely different principle and utilizes the multiple acceleration of charged particles moving along a circle in a magnetic field. The principle of the cyclotron is shown in Figure VIII-8. It consists essentially of a circular metal chamber cut into halves, $C_1$ and $C_2$, and placed between the poles of a very strong electromagnet. The half chambers, $C_1$ and $C_2$, are connected with a source of alternating high electric potential, AC, so that the electric field along the slit separating them periodically changes its direction. The ions of the element to be used as atomic projectiles are injected in the center of the box, *I*, at a comparatively low velocity, and their trajectories are bent into small circles by the field of the magnet. The gimmick of the cyclotron is that, for a given magnetic field, the period of revolution of an electrically charged particle along its circular trajectory is independent of the velocity with which that particle is moving. Since the increase in the radius of the path and the length of the circular trajectory is exactly proportional to the increase in velocity, the time necessary for one revolution remains the same.

If things are arranged in such a way that the period of revolution of the ions injected into the field of the magnet is equal to the period of alternating tension produced by the AC source, the particles arriving at the boundary between the two half chambers, $C_1$ and $C_2$, will be subject each time to an electric force acting in the same direction that the particles are moving. Thus, each time the ion passes through that boundary it will be given additional acceleration and its velocity will gradually increase. Gathering speed, the ions will move along an unwinding spiral trajectory and will finally be ejected through the window, $W$, in the direction of the target, $T$.

The upper photograph of Plate VI shows a cyclotron under development at the University of Colorado which is expected to produce proton beams with the energy of about 30 mev. The upper part of the large electromagnet and the supporting beam can be clearly seen. The University of California's bevatron (lower part of Plate VI) and Long Island's cosmotron represent further developments of the original cyclotron's principle.

Describing the results of their experiments on nuclear transformations caused by the bombardment of different materials by fast-moving nuclear projectiles, physicists always talk about "effective cross sections" or simply "cross sections." To understand this notion, let us consider the case of an antiaircraft battery trying to shoot down an approaching hostile air vehicle. If the enemy was so stupid as to send a blimp (Fig. VIII-9a), any hit at its body will be fatal, and the "effective cross section" would be equal to the geometrical cross section of the vehicle. In the case of an airplane, however (Fig. VIII-9b), the shell fragments could go through many parts of its frame without resulting in a "kill." There are only a few areas, such as the head and body of the pilot, essential parts of the engine and the steering, which must be hit to bring the plane down. The combined profile area of these spots is known as "effective cross section," and may be considerably smaller than the object itself. Thus, for example, the "kill cross section" of Achilles was limited to just a few square inches on the heel of his left foot.

In considering the kill probability, be it of the enemy aircraft or of the atomic nucleus, one is interested only in the fraction of the total profile area which should be hit, and not necessarily in the exact location of the sensitive spots. The situation is similar to that of two duelers, one of whom was very thin while the other was very fat. The fat man objected that the situation was unfair, since he presented a much larger target area for the pistol bullet than his opponent. "All right," said the thin man, "ask your second to draw with

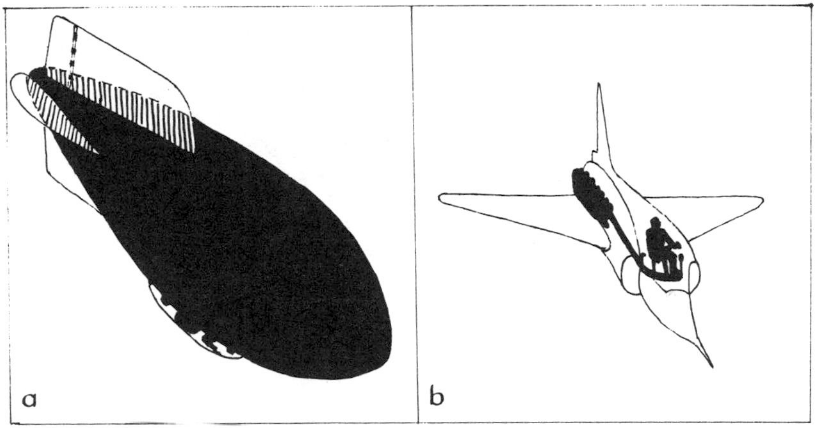


FIG. VIII–9.
The "kill cross section" (*black area*) of a blimp (a) and a plane (b) (assuming self-sealing gasoline tanks).

chalk my silhouette on your coat, and the hits beyond that line will not count."

The radius of an atomic nucleus is of the order of $10^{-12}$ cm, so that its geometrical cross section is about $10^{-24}$ cm$^2$. The cross section of exactly $10^{-24}$ cm$^2$ is known as a "barn," because it is so big, and if a nucleus is cracked each time it is hit, the effective cross section is about one barn. If, however, because of one reason or another there is only one kill, say per one hundred hits, we say that the effective cross section is 0.01 barns or $10^{-26}$ cm$^2$. In further discussion, the reader will find examples of even smaller cross sections in the nuclear bombardment processes.

NUCLEAR STRUCTURE AND STABILITY

While the atomic electrons fly freely through space, maintaining distances which are several thousand times larger than their diameters, protons and neutrons forming atomic nuclei are packed as tightly as herrings in a barrel (Fig. VIII-10). Thus, whereas in the case of an atom one may speak about electronic atmosphere which possesses many properties of ordinary gas, the material of the nucleus should be more closely compared with a droplet of liquid in which the molecules are held together by the forces of cohesion. This "droplet model" of the nucleus, which was proposed by the author

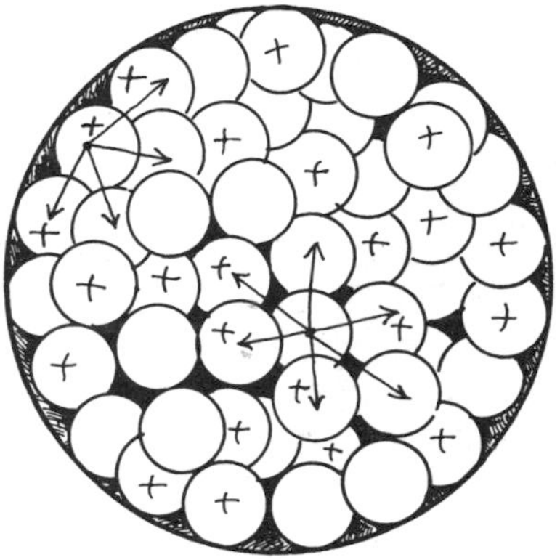

FIG. VIII–10.
Atomic nucleus made of protons and neutrons. Those in the interior experience no force, while those on the surface are being pulled inward.

of this book some thirty years ago, permits us to understand many nuclear properties. First of all, while gases are easily compressible because of a lot of empty space between the particles, liquids change their volume very little, no matter how high the pressure to which they are subjected. And, indeed, we have seen before that, as we go along Mendeleev's system, the volume of atoms remains essentially unchanged with more and more electrons being packed on the quantum orbits with smaller and smaller diameters. On the other hand, the measurements show that the radius of the atomic nucleus increases as the cubic root of its mass, so that the volume increases as the mass and the density remain constant. The density of this nuclear fluid, the droplets of which form atomic nuclei, exceeds the density of water by a factor of $10^{14}$ and a jigger filled with it would weigh five billion tons! Like any other liquid, nuclear fluid manifests the phenomenon of surface tension, since the nucleons located on the surface are pulled in by the cohesive forces of other nucleons, thus tending to reduce the surface area to the minimum. But, just as in the case of density, surface tension of nuclear fluid is enormously larger than that of ordinary fluids. If we make a soap film on a frame formed by a U-shaped wire and a piece of straight wire put across it, the force of surface tension acting on the movable wire would be able to support the weight of about 70 mg/cm of its length. If we could do the same with nuclear fluid, the force would be ten billion tons. Because of surface tension, atomic nuclei have a very nearly spherical shape, as do rain droplets, and the vibrations and rotations of these tiny droplets must be responsible for the emission of gamma rays by the excited nuclei.

It was shown, however, by the Princeton physicist, John Wheeler, that nuclear fluid may not necessarily exist in the form of little spheres but can, in principle, assume different shapes. The point is in that case there exist, apart from the forces of nuclear cohesion, also the Coulomb repulsive forces between the positive charges of protons. Wheeler has shown in an unpublished paper that the existence of these repulsive forces permits atomic fluid to assume the shape of a doughnut. In fact, in this case, surface tension forces which tend to contract the doughnut into a sphere will be opposed by electric repulsion between the opposite sides of it, and the entire configuration will be perfectly stable. Such doughnut nuclei, which would be considerably larger than those of uranium, having an atomic weight of many thousands, will be surrounded by electrons which would move close to their surface along the trajectories similar to the wiring of a circular electromagnet.* Such doughnut nuclei do not exist in nature, and it is hard to believe that they could be manufactured in the future by the most skillful nuclear physicist. But if they could be made, points out Wheeler, one could use them as links to make long chains. The thread made of such nuclear chains would be extremely strong, and, being as thin as cobweb, would hold the weight of a battleship. But it would also be very

* Compare Fig. V-11.

heavy, and one yard length of it would weigh about 1,000 tons.

It does not seem that Wheeler's doughnut nuclei will ever find practical application, but the simpler nuclear shapes governed by the same two kinds of forces opened for us the age of atomic energy. Let us consider the balance between the surface tension and electric energy of an atomic nucleus. The total surface energy is, of course, proportional to its surface and increases as the nuclei become large. Since the density of nuclear fluid remains constant, its volume is proportional to its mass (atomic weight) and its radius to the cubic root of the mass. Thus the total surface energy, being proportional to the surface, increases as the square of the cubic root of the mass, or, in other words, as the mass in two-thirds power. To calculate Coulomb energy, we have to use a law of electrostatics which states that the energy of a charged spherical body is directly proportional to the square of its charge, and inversely proportional to its radius. Nuclear electric charge is given by its atomic number which is approximately proportional to atomic weight. Remembering that radius varies as the cubic root of atomic weight, we find that Coulomb energy increases approximately as atomic weight in the power $1\frac{2}{3}$. This is a much faster increase than that of the surface tension energy, and we conclude that, while in light nuclei electric repulsion forces may play a minor role, they would become quite important in heavier nuclei. Since surface tension forces tend to hold liquid droplets in one piece, and to fuse two droplets which come into contact into a single bigger one, we should expect that in the case of light elements energy will be liberated in the processes of nuclear fusion. On the other hand, in the case of heavy nuclei, the disruptive Coulomb forces would have the upper hand, and nuclear fission would be an energy-liberating process. Calculations show that the "fusion region" extends about one-third of the way up Mendeleev's system, with expected energy liberation being smaller and smaller as we approach the limit. The "fission region" which begins at that point corresponds first to rather low energy liberation, which increases rapidly, reaching the highest values for the heaviest elements. Thus, every chemical element represents a potential source of nuclear energy and the question is only how to start the nuclear reactions and to keep them going.

The liquid droplet model of atomic nuclei represents a very good approximation to reality, but one should not forget that protons and neutrons inside the nucleus are subject to the same quantum laws as electrons in the atom, which must cause some deviation from the simplified picture given above. And, indeed, such deviations were found in more detailed study of nuclear properties. Figure VIII-11 shows the change of binding energy per nucleon for the entire range from the lightest to the heaviest nuclei. One notices a regular decrease of binding energy in the first part of the sequence and a slow increase later on; this corresponds to the fusion and fission regions.

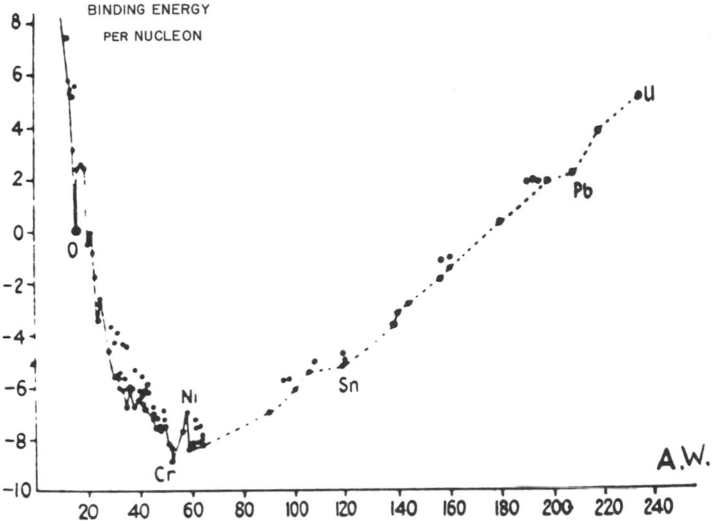


FIG. VIII–11.
Binding energy per nucleon as the function of atomic weight.

But one also notices that the curve is not quite smooth and there are a number of kinks indicating abnormally strong binding between the nucleons. These places correspond to the completed nucleon shells within the nucleus and are quite analogous to the completed electron shells in atoms. In the case of atoms, the elements with completed electron shells (the noble gases) are chemically inert since they are "completely" satisfied with their sets of electrons. A similar effect in the case of nuclei is shown in Figure VIII-12, which represents the relative probabilities for an incident neutron to be captured by the nuclei of different elements. For certain numbers of neutrons already present in the nucleus (50, 82, 126), the probability of capturing another neutron sharply drops down, indicating that these nuclei contain the completed neutron shells. The study of these and many other irregularities in nuclear properties leads to the conclusion that strongly bound internal shells are formed in the nucleus whenever the number of either neutrons or protons is equal to one of the following numbers: 2, 8, *14,* 20, 28, *50, 82,* and *126*. It must be noticed, however, that whereas

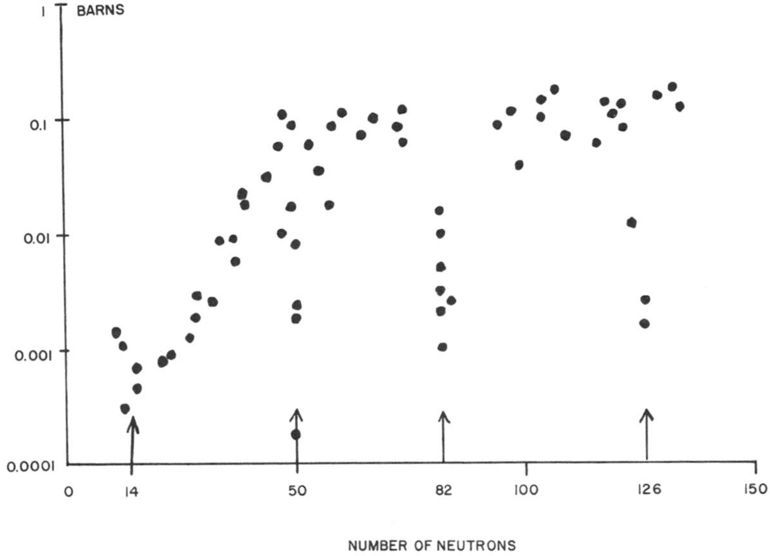


FIG. VIII–12.

"Capture cross section" of neutrons as the function of the number of neutrons in the nucleus.

in atoms each new electron shell is located essentially outside of the previous one, thus leading to an onionlike structure, neutron and proton shells in the nuclei interpenetrate each other, each of them occupying the entire nuclear volume. This lack of geometrical distinction between nucleonic shells makes their effect less pronounced and more difficult to study and to explain. This difficulty was, however, overcome simultaneously and independently by Maria Goepert Meyer in Chicago and Hans Jensen in Heidelberg, who were able to construct a complete system of nuclear shells standing in complete agreement with the observed facts. When they met to compare their results, they found that they both were born on the same day of the same year, and so they became very good friends.

### FISSION CHAIN REACTIONS

On January 27th, 1939, a small conference on theoretical physics, organized jointly by the George Washington University (where the author was teaching at that time) and the Carnegie Institution of Washington, was dragging on in Washington, D.C. On that day, Niels Bohr, who was one of the visiting dignitaries, received a letter from a German lady physicist, Lise Meitner, who at that time (because of Hitler) was working in Stockholm. She said she had received a letter from her former colleague, Otto Hahn, in Berlin, informing her that he and his assistant, Fritz Strassman, bombarding uranium with neutrons, found the presence of barium, an element about half way down Mendeleev's system. Meitner and her nephew, Otto Firsch (notice two Ottos in that business!), who went with her to Stockholm, thought that it may have been the result of fission, i.e., breaking into two, of a uranium nucleus struck by a neutron. As soon as Bohr read the telegram to the participants, the discussion immediately shifted from the comparatively unexciting topic of the conference to a heated argument as to whether the fission of the uranium nucleus can possibly lead to large-scale liberation of nuclear energy. Enrico Fermi, who was also taking part in the conference, went to the blackboard and was writing some formulas pertaining to the fission process. A correspondent from a Washington newspaper, who was happily dozing during the previous parts of the meeting, woke up and started taking notes, but Merle Tuve, a Carnegie Institution nuclear physicist, quickly showed him out the door, saying that the discussion was too technical for him. This was the first step in security regulations which were quickly imposed on the "atomic energy" development. But what the reporter heard before being shown out got into the newspapers, and the next morning the author was wakened by a long-distance call from Robert Oppenheimer in California, who wanted to know what it was all about. And so things started.

The article on the theory of nuclear fission by Niels Bohr and John Wheeler, which was published in the September, 1939, issue of *The Physical Review* and was the first and the last article on that topic published before the security curtain came down, was based on the droplet model of the nucleus discussed above. When the nucleus, hit by the incident neutron, begins to vibrate, passing through a sequence of elongated shapes, the equilibrium between the surface tension and electric forces becomes distorted; the former attempts to return the nucleus to its original spherical shape, while the latter attempts to increase the elongation. If the ratio of the large to the small axis of the ellipsoid exceeds a certain limit, a

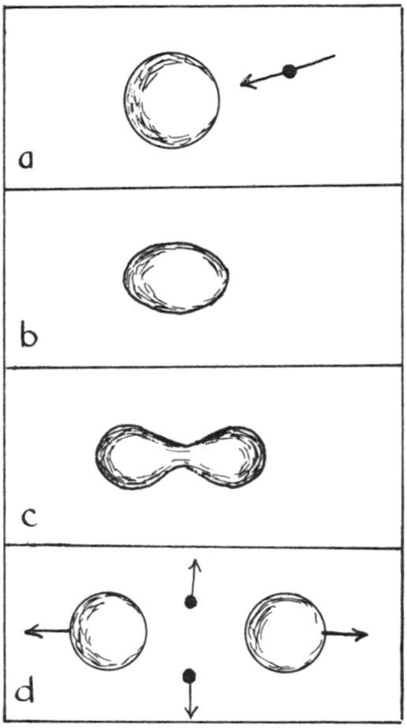


FIG. VIII–13.
Fission of a heavy nucleus, resulting from a neutron impact.

cleavage develops along the equatorial plane and the nucleus neatly breaks into two halves (Fig. VIII-13). It was soon found that the fission of the uranium nucleus is accompanied by the ejection of a couple (more exactly, 2.5) of neutrons which, in their turn, may hit two of the bystanding nuclei and break them up. This would produce four new neutrons which could crack another four nuclei. . . . Thus a chain reaction may develop which rapidly engulfs the entire piece of uranium with the liberation of tremendous amounts of nuclear energy.

It is difficult to write on the subject of what is commonly known as "atomic energy." In the early days when most of the facts and data were hidden behind a heavy security curtain, there was not much one could write about. Now, however, when oodles and oodles of information can be found in numerous books, and in magazine and newspaper articles, the subject becomes somewhat dull and trivial. Besides, although the fission of the uranium nucleus can be considered as a very interesting paragraph (but only a paragraph) in the story of physics, the developments of atomic bombs, reactors, and fertilizers belong rather to the field of technology. Thus, apart from the reproduction (in Plate VII) of two very beautiful (especially if they could be reproduced in color) photographs of one atomic bomb and one reactor (swimming pool type), only the most essential steps will be discussed in this section.

First, there was a disappointing fact, which emerged as early as the above-described Washington conference, that it is not the main isotope of uranium which shows the phenomenon of fission, but the very rare isotope $U^{235}$ which is present in the amount of only 0.7%. Besides, the main isotope $U^{238}$, which forms the remaining 99.3% of natural uranium, is not just a harmless admixture but has a very strong appetite for neutrons and captures them at such a rate that it quenches any chain reaction which may start in $U^{235}$. There were only two ways to cope with the situation: either separate $U^{235}$ from the harmful $U^{238}$, or by some odds and ends, attempt to run the reaction in natural uranium, trying to keep the greedy $U^{238}$ away from its prey.

Both methods were tried. In a top secret plant in Oak Ridge, Tennessee, various methods of separation of uranium isotopes were investigated, and finally the production was centered on the diffusion method, based on the fact that uranium compounds containing light isotope diffuse somewhat faster through the porous membranes than those containing the heavy isotope.

The odds and ends needed for running the reaction in natural uranium were mostly devised by Enrico Fermi, and were based on the principle of moderation. It was found that the heavy uranium isotope has a large appetite for comparatively fast-moving neutrons, while the light isotope prefers very slow ones. Since the neutrons emitted in the fission of the uranium nucleus have very high velocities, one had to slow them down to the level of $U^{235}$ appetite sufficiently fast so that they would not be swallowed by $U^{238}$. This could be achieved by mixing natural uranium with a large amount of the so-called "moderator," i.e., an element whose atoms, having no appetite at all for neutrons, take away a part of the neutron's kinetic energy in the collision process. The two best moderators turned out to be the atoms of deuterium (heavy hydrogen isotope) and the atoms of carbon, which determined the two types of piles (carbon, and heavy water) now in use. The first atomic pile, using carbon moderator (graphite bricks) and built under the supervision of Fermi under the grandstand of the University of Chicago Stadium, started operating on December 2, 1941. Of course, the nuclear chain reactions in moderated piles go very slowly, and the energy produced cannot be used either for military or for peaceful purposes. But there is a trick! While a fission chain reaction is going among $U^{235}$ nuclei, some of the neutrons are swallowed by the hungry $U^{238}$ nuclei which, by virtue of the moderator, are deprived of the Lucullan feast. What happens when $U^{238}$ nuclei swallow a neutron is given by the following alchemical equation:

$${}_{92}U^{238} + {}_{0}n^{1} \rightarrow {}_{92}U^{239} + \gamma$$

$${}_{92}U^{239} \rightarrow {}_{93}Np^{239} + \bar{e}$$

$${}_{93}Np^{239} \rightarrow {}_{94}Pu^{239} + \bar{e}$$

*Np* and *Pu* stand for *neptunium* and *plutonium,* the two "transuranium elements" produced in the atomic pile. Whereas neptunium is just a transitory stage in the process, plutonium is really something! It possesses the same properties as $U^{235}$, only very much more so. It breaks up more easily, being hit by a neutron, and its fission is accompanied by a larger number of secondary neutrons. And, what is of course most important, plutonium, having different chemical properties than uranium, can easily (so they say) be separated from the remaining uranium when the cooking process in the pile is over.

Today the production of fissionable material in the United States amounts to *x* tons per year, as compared with *y* tons per year produced in the Soviet Union.

## FISSION BOMBS AND REACTORS

The most important notion in all the discussions concerning fission chain reactions is that of the *critical size.* When a single fission process occurs inside of a given sample of pure $U^{235}$ or $Pu^{239}$, several fission neutrons (on the average 2.5 for uranium and 2.9 for plutonium) are ejected from the point where the nuclear breakup took place. The average distance that a fission neutron must travel through the material in order to run into another nucleus is about 10 cm, so that, if the size of the sample in question is less than that, most of the fission neutrons will cross the surface of the sample and fly away before they have a chance to cause another fission and produce more neutrons. Thus, no progressive chain reaction can develop if the size of the sample is too small. Going to larger and larger samples, we find that more and more fission neutrons produced in the interior have a chance to produce another fission by colliding with a nucleus before they escape through the surface, and for samples of a proper size the number of fission neutrons which produce another fission within the sample becomes large enough to cause the reaction rate to increase rapidly in time. The size of a sample of a given fissionable material for which the percentage of the neutrons giving rise to subsequent fission processes is high enough to secure a progressive chain reaction is known as *critical size* for that particular material. Since the number of neutrons per fission is larger in the case of plutonium than in the case of $U^{235}$, the critical size of plutonium samples is smaller than that of $U^{235}$ samples, because the former can afford larger losses of neutrons through the surface.

In order to produce a nuclear explosion, one should build up a highly supercritical sample of fissionable material within a time period which is short enough not to allow the chain reaction to develop to any considerable strength. One can do it, for example, by shooting one subcritical mass into another subcritical mass, with sufficiently high speed so that the chain reaction will not develop to any appreciable degree before the complete "assembly" is achieved. There are also more ingenious (but classified) methods of achieving the same result.

If one wants to run a fission chain reaction under controlled conditions in order to use it for power-producing purposes, the sample must be kept all the time as close as possible to the critical size. It must be kept in mind that a nuclear chain reaction is by its nature an explosive reaction, and that any attempt to run it at a steady rate is comparable with keeping a furnace burning, using TNT as a fuel. But, in fact, it can be done with a very small chance of mishap.

It can be achieved by the use of "control rods" containing neutron-absorbing materials (such as boron) which are automatically pushed in or pulled out from the narrow channels drilled through the reacting fissionable material as soon as the rate of neutron production drops below or exceeds the desired level.

Today nuclear reactors are being successfully used as power plants in the countries which have a shortage of coal and oil, such as Great Britain, and for the propulsion of ships such as "atomic" submarines in the United States and "atomic" icebreakers in the U.S.S.R.

### THERMONUCLEAR REACTIONS

For centuries the astronomers and the physicists have been wondering what makes the sun (and all other stars) shine. It was clear that an ordinary "burning" does not suffice since, even if the material of the sun were the best aviation gasoline, it could not last from the time of the Egyptian pyramids until today. About a hundred years ago, Herman von Helmholtz in Germany (of course) and Lord Kelvin in England (of course) suggested that the sun could maintain its radiation of light and heat by the result of a slow contraction of its body. Calculations have shown that the contraction of the sun from a very large original size to its present diameter would liberate enough energy to maintain its radiation for a few hundred million years. But the more recent estimates of the age of the solar system made it clear that even this large figure is not large enough, and that the sun must have been shining for at least several billion years. The only way to account for that longevity of the sun was to assume that it derives its energy from some kind of nuclear transformation, and in the year 1929 a British astronomer, Robert Atkinson, and an Austrian physicist, Fritz Houtermans, put their heads together to see if this could be true. Their idea was that thermal collisions between the atoms in the hot interior of the sun can induce some nuclear reactions running fast enough to supply the necessary amount of energy. The studies of a British astronomer, Sir Arthur Eddington, have shown that the temperature in the interior of the sun must be as high as 20 million degrees, which corresponds to the energy of thermal motion of about $4 \times 10^{-9}$ ergs per particle. This energy is several hundred times smaller than the energy of atomic projectiles used in conventional experiments on artificial transformation of elements, but one must take into account that, whereas artificially accelerated nuclear projectiles rapidly lose their initial energy and have only a small chance to hit the target nucleus before coming out of the game, thermal motion continues indefinitely and the particles involved in it collide with one another for unlimited periods of time. Using the theory of wave-mechanical penetration through the nuclear potential barriers, which was developed only one year earlier, Houtermans and Atkinson were able to show that, at the temperatures and densities obtaining in the solar interior, the *thermonuclear reactions* between the hydrogen nuclei (protons) and the nuclei of other light elements could liberate a sufficient amount of energy to account for the observed radiation of the sun. This theory was proposed before Cockcroft and Walton's experiments on artificial transformation of elements caused by proton bombardments, and very little information was available on what happens when various light nuclei are hit by protons. Houtermans and Atkinson proposed at that time that there must be some light nucleus which has the ability to capture protons and hold them in for a considerable period of time. After the fourth proton is captured, an $\alpha$ particle would be formed inside of the "proton trap" nucleus, and its ejection would liberate a large amount of nuclear energy. They tentatively called their article, which was published in 1929 in a German magazine, *Zeitschrift fur Physick*: "How Can One Cook Helium Nuclei in a Potential Pot?"* but the title was changed to a more conventional one by the magazine's editor, who had no sense of humor.

About ten years later, when sufficient information concerning the transformation of light nuclei struck by a proton was accumulated, the "proton-trapping nucleus" of Atkinson and Houtermans was identified as that of carbon. Hans Bethe in the United States and Carl von Weizsacker in Germany proposed independently the so-called *carbon cycle* shown in Figure VIII-14. In these series of nuclear reactions, four protons are captured consecutively by the nucleus of the carbon atom and, after the transformation of two of them into neutrons, are emitted as an $\alpha$ particle. The total period of the cycle is 6 million years, and the energy liberated in it amounts to $4 \times 10^{-5}$ erg. Since, according to the present data concerning the chemical composition of the sun, each gram of solar material contains about 0.0001 gm of carbon ($5 \times 10^{18}$ carbon atoms), the total rate of energy liberation by the carbon cycle amounts to 1 erg per gm per

* "Wie Kan Man Ein Helium Kern in ein Potencial Topf Kochen?"

sec which accounts for only 1% of the rate at which the energy must be produced in the solar interior.

Another process was proposed at the same time by Charles Critchfield, who was at that time a graduate student at George Washington University. His idea was that, if in a collision between two protons one of the protons is transformed into a neutron by the emission of a positive electron, a nucleus of deuterium (heavy hydrogen isotope) can be formed. Through the subsequent reactions, deu-

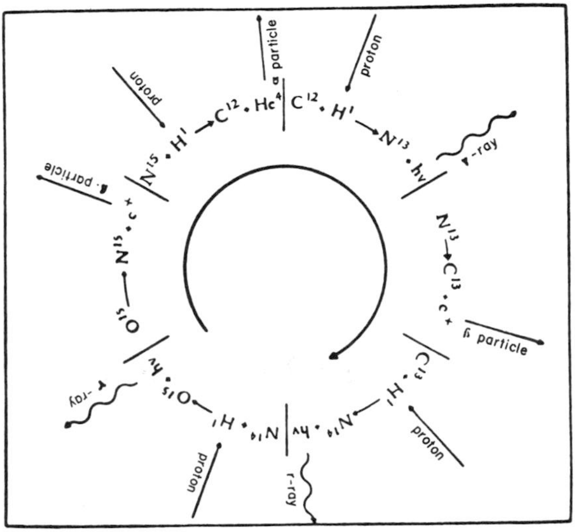


FIG. VIII–14.

Carbon cycle responsible for thermonuclear energy production in stars.

terium would be built into helium, thus achieving the same end as the carbon cycle but much faster. The reactions involved in this, the so-called H-H process, are:

$$ {}_1H^1 + {}_1H^1 \rightarrow {}_1D^2 + \overset{+}{e} + \nu $$
$$ {}_1D^2 + {}_1H^1 \rightarrow {}_2He^3 + \gamma $$
$$ {}_2He^3 + {}_2He^3 \rightarrow {}_2He^4 + 2\,{}_1H^1 $$

At the temperature of 20 million degrees, this reaction takes $3 \times 10^9$ years and liberates $4 \times 10^{-5}$ erg per proton. Since hydrogen constitutes about half of the solar material ($2 \times 10^{23}$ atoms per gm), the total rate of energy-liberation is about 100 erg per gm per sec, in good agreement with the observed value.

The predominance of H-H reaction over C-cycle in the sun is, however, not a general rule and is reversed in many stars. The point is that these two sets of thermonuclear reactions possess different sensitivity to temperature, and, while the rate of C-cycle is proportional to $T^{17}$, the H-H reaction goes only as $T^4$. Thus, in case of stars brighter than the sun, such as Sirius, which possess higher central temperatures, C-cycle takes the upper hand over H-H reaction. On the other hand, in stars fainter than the sun—and the majority of stars belong to that class—energy production is due entirely to H-H reaction.

The reader would be very much surprised if he (or she) should try to compare the rate of energy production inside of the sun with the rate that heat is produced in ordinary electric appliances, such as an electric coffeepot. One hundred ergs per gm per second is equivalent to about $2 \times 10^{-5}$ calories per gm per sec, and it would take $5 \times 10^7$ sec, or a year and a half, to bring 1 gm of cold water to the boiling point at that rate of heat supply! Thus, using an electric coffeepot in which the heating unit operates with the same efficiency as thermonuclear reactions inside of the sun, we would have to wait for years until water would be boiling, provided of course that the pot is perfectly insulated and no heat losses are taking place. The reason why the sun is so hot in spite of such a miserably low rate of heat supply is that it is so large. In fact, since the total heat production is proportional to the volume (i.e., to $R^3$), whereas the heat losses are proportional to the surface (i.e., to $R^2$) very large bodies get very hot, even if the rate of heat production per unit volume in their interior is very low.

It is clear from the above discussion that neither C-cycle nor H-H reaction, which provide the energy for stars illuminating our universe, are adequate for the ambitious *Homo sapiens* (Latin for a smart man) who wants to use nuclear energy for his own purposes. The key to the solution of that problem is provided by the heavy hydrogen isotopes, deuterium $D^2$, discovered by the American chemist Harold Urey and a still heavier isotope, tritium ${}_1T^3$. Deuterium is present in nature, even though in rather small amounts, and one out of three thousand molecules of water contains one atom of deuterium. Due to the development of methods for separation of isotopes, the cost of deuterium fell from that of expensive French perfume to that of cheap whisky, and there is a lot of water in the oceans. Tritium, being an unstable isotope, does not exist in nature (except for negligibly small amounts produced in the atmosphere by cosmic rays), and must be manufactured at a high cost in atomic piles. It is too expensive to be used as the principal fuel, but is helpful as the "nuclear kindling" for starting thermonuclear reaction in deuterium.

The possible reactions between the heavy hydrogen isotopes are:

$$ {}_1D^2 + {}_1D^2 \rightarrow {}_2He^3 + {}_0n^1 + 3.25 \text{ mev} $$
$$ {}_1D^2 + {}_1D^2 \rightarrow {}_1T^3 + {}_1H^1 + 4 \text{ mev} $$
$$ {}_1D^2 + {}_1T^3 \rightarrow {}_2He^4 + {}_0n^1 + 17.6 \text{ mev} $$

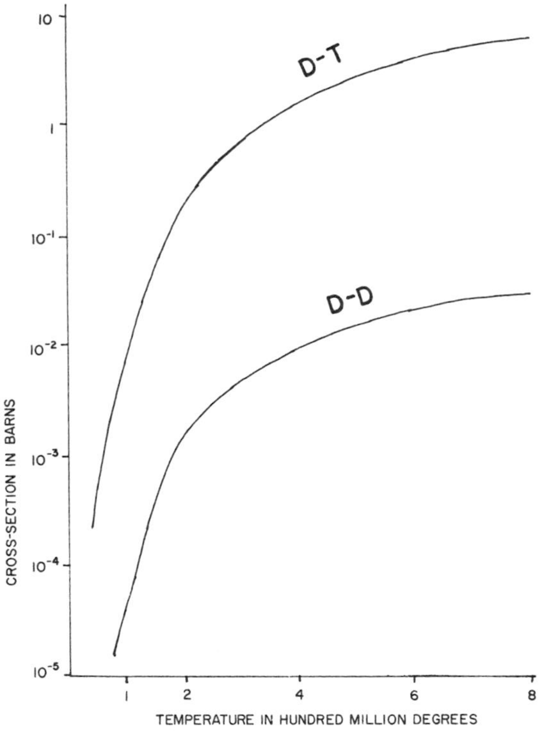


FIG. VIII–15.

Cross sections for D-D and D-T reactions as function of their thermal energy expressed in terms of absolute temperature.

and their effective cross sections, calculated on the basis of tunnel effect theory, are shown graphically in Figure VIII-15. Thus, all one has to do in order to produce thermonuclear reactions between the heavy hydrogen isotopes is to heat them to a temperature of a few hundred million degrees. This was achieved by Los Alamos scientists on November 1, 1952, when they exploded the first thermonuclear bomb at Elugelab, a coral island in the Pacific Ocean, and turned the island into a pool of water 1 mile wide and about 200 ft deep. This result was achieved by squeezing and heating an appropriate amount of heavy hydrogen by a powerful fission bomb explosion.

The situation becomes, however, much more complicated if one wants to run thermonuclear reactions under controlled conditions, and to use the liberated energy for constructive rather than for destructive purposes. It is clear that in this case the physical conditions under which thermonuclear reactions are taking place should be drastically changed. First of all, the reaction should run at extremely low densities to avoid the unbearably high gas pressure which would develop at the temperature of several hundred million degrees. In fact, at that temperature and the density of atmospheric air, deuterium gas will develop the pressure of about 100 million lb per sq. in., and no container would be able to keep it in place. The graph in Figure VIII-16 shows the rate of thermonuclear energy production for pure deuterium and for deuterium-tritium mixture at various gas densities. We see that, in order to achieve the rate of energy production of about 100 watts per cc, which is comparable with that in today's fission reactors, the density of deuterium can be as low as one ten thousandth of the density of atmospheric air which corresponds to about the best vacuum which can be produced in our laboratories. The second problem is to keep this hot rarefied gas away from the wall of the container, since otherwise the process of heat conduction into the walls will rapidly reduce the temperature of deuterium gas below the minimum value required for thermonuclear reaction.

This can be achieved in several different ways, all of which are based essentially on the use of strong magnetic fields. At the very high temperatures required in this case, the deuterium gas in the tube will be completely ionized and will consist entirely of negatively charged electrons and positively charged deuterons. (This state of

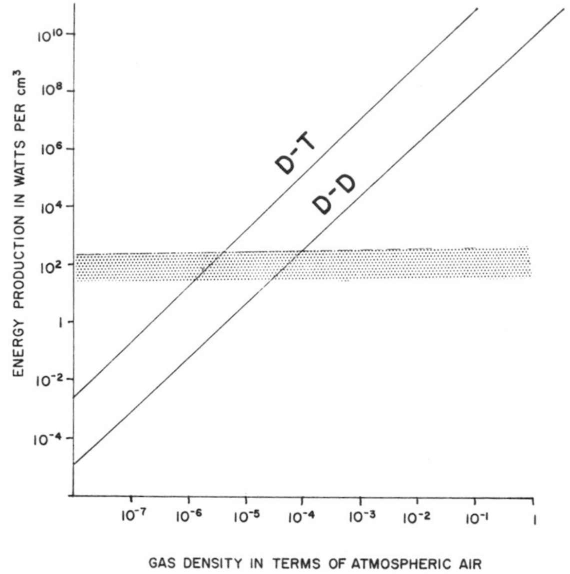


FIG. VIII–16.

The rate of nuclear energy liberation for varying gas density at the temperature of $7 \times 10^{9}$° K (about twice the internal temperature of the sun). The shaded band represents the energy production in the existing uranium and plutonium reactors.

matter is described nowadays by the term "plasma.") We know that when an electrically charged particle moves through a magnetic field, it experiences a force perpendicular to the direction of its motion and to that of the field. This force compels the particles to spiral along the direction of the magnetic lines, as is shown in Figure VIII-17a. Thus, by forming a strong axial magnetic field in a tube, we can effectively prevent free deuterons and tritons from coming close to the walls. If this can be achieved, the collisions between the particles spiraling along the tube are expected to result in D-D or D-T reactions with the release of nuclear energy and of large amounts of neutrons. Of course, in order to start such a process the gas in the tube must first be heated to a very high temperature by some outside agent.

The second possibility consists in using magnetic forces caused by short but strong electric discharges through the tube. It is known that two parallel electric currents flowing in the same direction are a

the pinch

b

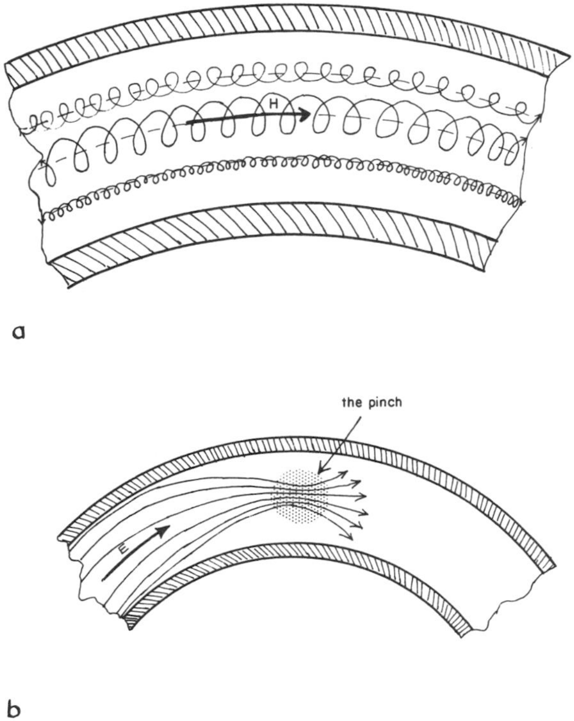

FIG. VIII–17.

Two principal methods which are now being developed for controlled thermonuclear reactions. "Stellarator" at Princeton, and "Perhapsotron" in Los Alamos.

magnetically attracted toward each other so that, in the case of a sufficiently strong current, the gas (or rather the plasma) inside the tube will have a tendency to detach itself from the walls and to be squeezed into a narrow jet along the axis. How this so-called "pinch effect" operates can be understood by inspecting Figure VIII-17b. In contrast to the previously described method, the pinch-effect device operates in jerks, as an automobile engine does, but it has the advantage that the gas in the tube is automatically heated by the electric discharge, and no outside heating is needed. It has been estimated that a current of several hundred thousand amperes lasting for a few microseconds would produce a "pinch" strong enough to cause a thermonuclear reaction in deuterium. The work in the above described directions is being carried out now in many laboratories of the world, and it is entirely possible that the problem of controlled thermonuclear reactions will be solved in short order.

### THE MESONS AND HYPERONS

In the early thirties, physicists were happy with just a few particles from which matter was made. Protons and neutrons for the atomic nuclei, electrons for their envelopes, and—oh, well—the neutrino, this problem child of that period. But in 1932 there appeared an article by a Japanese physicist, Hidekei Yukawa, which gave a headache to everybody concerned with the nature of nuclear cohesive forces. Yukawa suggested that these forces are due to a new particle which is continuously exchanged between protons and neutrons. It is very difficult, if at all possible, to describe in a simple way the complicated notion of the "exchange force." Probably the best one can do is to imagine two hungry dogs who come into possession of a juicy bone and are grabbing it from each other to take a bite. This tasty bone is continuously passing from the jaws of one of them into the jaws of the other, and in the resulting struggle the two dogs become inseparably locked. Yukawa's idea was that attractive forces between the nucleons are due to a similar struggle for the possession of that new tasty particle. That new particle could be electrically neutral, or could carry a positive or negative electric charge, and the exchange process would look in these cases as is shown in Figure VIII-18.

Yukawa has shown that, in order to account for the observed properties of nuclear forces, one should assume that this new particle has a mass intermediate between the mass of a proton and the mass of an electron, being about 10 times lighter than the former and about 200 times heavier than the latter. Nobody believed in the existence of these particles, which were tentatively called "yukons," until a couple of years later a Cal. Tech. physicist, Carl Anderson, discovered the presence of positively and negatively charged particles of that mass in the cosmic rays showering down to the ground from the upper reaches of the atmosphere.

Since its discovery, the name of the new particle has undergone several evolutionary stages. It was sometimes called "heavy electron," sometimes "light proton," and then somebody suggested the name

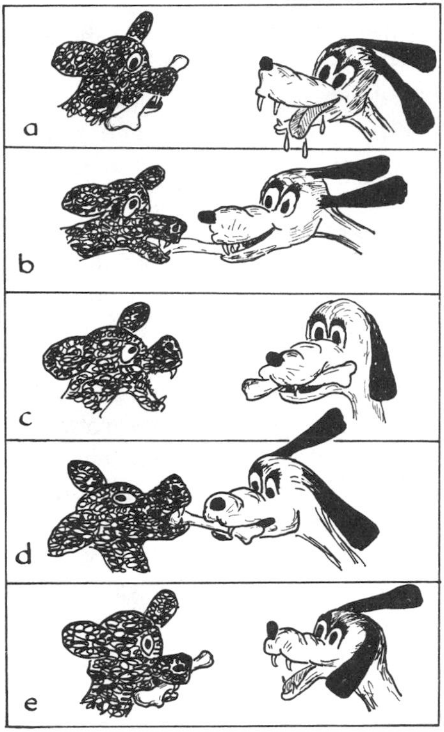


FIG. VIII–18.
The exchange of a meson (bone) between two nucleons.

"mesotron," derived from the Greek word *mesos* (μεσος), which means "in between." But Werner Heisenberg's father, who was a professor of classical languages, objected that the letters "tr" have no place in that name. Indeed, while the name "electron" was derived from the Greek *electra* (for "amber"), the Greek word *mesos* has no "tr" in it. Thus, against the protest of the French physicist, who did not want the name of the new particle to be mixed up with *maison* (French for "home"), the name of Yukawa's particle was finally settled as "meson."

From the very beginning, the mesons gave a lot of headaches to physicists, since there seemed to be something wrong with their absorption in the atmospheric air. For the particles of such extremely high energy (many billions of electric volts) absorption in different materials is expected to depend only on the total amount (mass) of the material they traverse. Indeed, since for these energies all atomic electrons with which these fast particles collide can be considered as being free (see the discussion about the Compton effect), it is only the number of electrons which counts and not the way they are attached to different atomic nuclei. Thus, if we measure the intensity of the cosmic ray beam on the top of a high mountain and at the base of it, the decrease of intensity must be determined only by the *weight* of the air column extending vertically from the lower location to the upper one. If the difference of barometric pressure between two locations is, let us say, 100 mm Hg, it follows that the weight of the column of air is equal to the weight of a column of mercury 100 mm high, and therefore the absorption of cosmic rays in a layer of mercury 100 mm thick must be the same as the absorption in the air between the top and the bottom of the mountain. This rule operated quite well in the case of cosmic ray electrons, but did not seem to work for the newly discovered particles.

An important experiment along these lines was performed in 1940 by Bruno Rossi *et al.* at Echo Lake (elevation 3,240 meters) near the top of Mt. Evans near Denver (elevation 1,616 meters) (Fig. VIII-19), the difference of barometric pressure between the two locations being 14.5 mm Hg or, what is the same, 2 meters of water. He used two identical meson counters—one in Denver and another on the mountain, the latter being submerged 2 meters under the surface of the lake.* Since lake water should produce in that case the same absorption as the layer of the air between the mountain lake and the streets of Denver, both counters were expected to show the same results. However, the experiment did not confirm this expectation, and the counter in Denver was consistently showing a much smaller number of mesons. The only possible explanation was that there was another reason than the atmospheric absorption which reduced the number of the mesons coming to the ground, and it was suggested by Enrico Fermi that the effect may be due to the intrinsic instability of mesons. In fact, if mesons break up in flight, the fraction coming through will depend on the time they travel. Since mesons coming down to Denver must have traveled an extra 1,624 meters, and since

* In the actual experiment an iron slab with the thickness equivalent to 2 meters of water was used, but it would be more poetic to use the water of this beautiful lake.

they travel practically with the velocity of light, the time interval involved is $\frac{1.6 \times 10^5}{3 \times 10^{10}} = 5 \times 10^{-6}$ sec. From this figure, and the observed decrease of intensity at the ground level, half the lifetime of mesons could be calculated, and this turned out to depend on their velocity. For very fast mesons with the energy of 250 mev, the lifetime was about $2 \times 10^{-3}$ sec, whereas for the slower ones with only

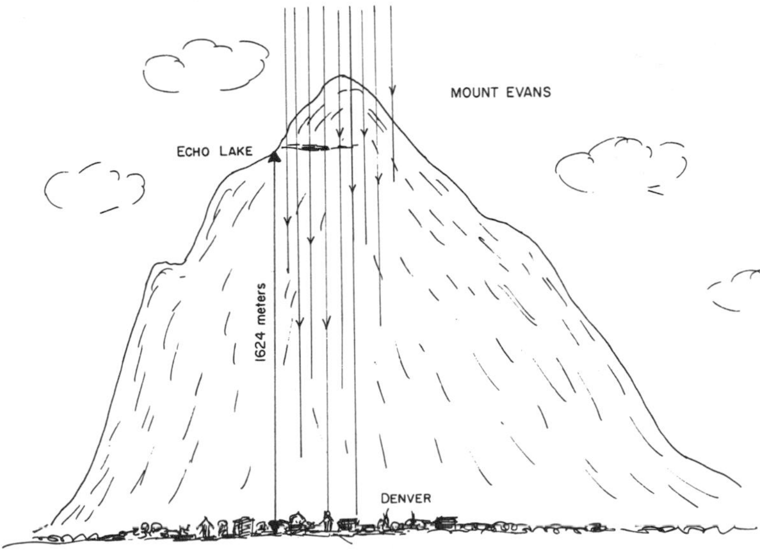


FIG. VIII–19.
Decay of mesons on the way from Echo Lake to Denver.

100 mev energy, a lifetime of only $5 \times 10^{-4}$ sec was observed. These observations gave the first experimental confirmation of Einstein's time dilation and the experimental results fitted very nicely with the formula $\Delta t' = \frac{\Delta t}{\sqrt{1 - \frac{v^2}{c^2}}}$. In the subsequent experiments one also could measure the lifetime of mesons after they have been brought to rest in a piece of absorbing material, and it was found to be as short as $2.5 \times 10^{-6}$ sec. If the fast-moving mesons in cosmic rays had that lifetime, they all would have decayed high in the atmosphere, and we

would never observe them on the ground!

What happens to a meson when it dies? This can only be answered by photographing the tracks of a meson and its decay products. In the case of particles of such high-penetrating power as cosmic ray mesons, it is not necessary to use the cloud chambers, which are in any case much too bulky to be sent up in balloons or rockets. The standard method in these studies is based on the use of photographic plates with thick emulsion layers. Many such plates are stacked into a pile, and, when a high-energy particle passes through such a pile, it affects the sensitive grains of the photographic emulsion lying along the way. Inspecting the developed plates under a microscope one observes long rows of darkened grains indicating the path of the particle. A composite thick emulsion photograph in Plate VIII, *upper,* shows a sequence of events out of which we will pick up at the moment only the last tracks in the picture. The one before last (running from bottom to upper left) belongs to a meson, a fact which can be established by counting the number of photographic grains affected per unit length of the track. The last track (running from the top down on the left) belongs to an ordinary electron produced at the point where the meson's track ends. The fact that the electron is thrown in the opposite direction proves that there must be one or more particles participating in the decay process and flying to the left. The fact that no other track is visible proves that these particles must be electrically neutral, and the detailed study of the directions and the energies involved leads to a conclusion that there are in fact two other particles, both being our old friends the neutrinos. Thus, the decay of a meson can be described by the equation:

$$\mu^{\pm} \rightarrow \overset{\pm}{e} + 2\nu$$

where + and − correspond to the positively and negatively charged mesons. Since the meson has 206 electron masses, the electron naturally has 1 electron mass, and the neutrino practically no mass at all, 205 electron masses remain unaccounted for. According to Einstein's principle of equivalence of mass and energy, this mass excess will be turned into about 100 mev of energy shared among the particles formed in the decay.

When mesons were first discovered, they were widely hailed as the particles which, according to Yukawa's exchange phenomenon theory, must be responsible for the cohesive forces between the nucleons. But, it soon turned out that the situation is not at all so simple. The difficulty arose in connection with a question of what mesons would do when slowed in their tracks within a thick block of absorbing material. In this case the fates of positive and negative mesons were expected to be quite different. Positive mesons, being repelled by the positively charged atomic nuclei of the material in which they were slowed down, will wander around as pariahs and, within a few microseconds, will decay into a fast positive electron and a pair of neutrinos. The high-energy positive electron will be shot out of the block, will cross one of the numerous counters with which meson-trappers fenced the block and will announce the death of a positive meson.

The slowed-down negative meson, on the other hand, will be caught on a quantum orbit of one of the nuclei and will become a temporary member of the atomic system. Calculations carried out by Enrico Fermi and Edward Teller have shown that such a capture will take place extremely fast, long before the slowed-down meson has a chance to break up. Since the radii of Bohr's quantum orbits are inversely proportional to the mass of the particle, this meson orbit will be about 200 times smaller than the innermost electron orbit, and the captured meson will move very close to the surface of the nucleus, looking somewhat like an earth satellite. Once on the orbit, the meson faces two possibilities: It can break up into a fast negative electron and two neutrinos, and the counters placed around the block will register the death of a negative meson. But, moving so close to the nucleus, the meson can also be swallowed by it. In fact, if the forces between protons and neutrons are due to a continuous exchange of mesons between them, there must be a reaction:

$$P^+ + \mu^- \rightarrow n^0 + \nu^0$$

From the strength of nuclear forces it can be estimated that this should be an extremely fast reaction taking only about $10^{-22}$ sec. Since the natural decay of the meson takes as long as about $10^{-6}$ sec, it follows that practically all mesons should be swallowed by the nuclei long before they die a natural death. At most, only one meson out of $10^{16}$ would have a chance to break up into an electron and neutrinos before being eaten up. Thus, no negative electrons should be shot out of the meson-slowing block. Experimental evidence sharply disagreed with that conclusion. Although the number of negative electrons shot out of the slowing-down block was smaller than that of positive electrons, for some material by a factor of 2, for others by a factor of 10, it was certainly not smaller by a factor of $10^{16}$! This meant that the nuclei's appetite for mesons is several million billion times smaller than is needed for a sufficiently strong exchange force in the Yukawa picture. So what could one do now? Mesons were predicted, mesons were discovered, but they were apparently the wrong kinds of mesons, and atomic nuclei had no more interest in them than lions in hay!

The help came from a thick emulsion photograph taken in 1947 by one of the balloons which a British physicist, C. F. Powell, was sending into the upper atmosphere. The photograph showed two tracks joined at their ends. One of them belonged to an ordinary meson with the mass 206, while the other must have been produced by a particle of the same charge but with the mass 273. The heavier particle was first called a "heavy meson" (like a "heavy middle-weight boxer") but was soon renamed a $\pi$ meson (or "pion"), while the earlier-discovered "light meson" was renamed a $\mu$ meson (or "muon").

Later studies have shown that a positive or negative pion decays into a (positive or negative) muon and one neutrino according to the equation:

$$\pi^{\pm} \rightarrow \mu^{\pm} + \nu^0$$

They are produced at the upper fringes of the atmosphere as a result of the impact of primary cosmic rays (which are essentially very high energy protons) against atomic nuclei, and have such a short half life ($2.6 \times 10^{-8}$ sec) that, even with the help of Einstein's time dilation, none of them ever comes to the earth's surface. The upper part of Plate VIII shows a production of a bunch of pions at the impact of an original cosmic ray proton against some nucleus in the photographic plate, and the track of one of them with successive transformations into a muon and into an electron. While there are only two types of muons: $\mu^+$ and $\mu^-$, there are three kinds of pions: $\pi^+$, $\pi^-$, and $\pi^0$, the last-mentioned breaking up into two high-energy radiation quanta:

$$\pi^0 \rightarrow 2\gamma$$

with a half life of only $10^{-16}$ sec.

During the following years more and more different kinds of particles were found showering on the heads of physicists. There appeared a K meson with the mass of 965 electrons, and several particles heavier than protons which received the name of "hyperons." Their names, decay modes, and half lives are shown in Table 1, and

Table 1. The Properties of the Elementary Particles of Matter

| Name and symbol | Mass (in electron masses) | Mean life-time (in seconds) | Decay scheme | Mass in terms of 137 $m_e$ |
|---|---|---|---|---|
| Xi; $\Xi^{\pm}$ | 2585 | $10^{-10}$ | $\Lambda_0+\pi^{\pm}$ | 18.88 |
| Sigma; $\Sigma^{\pm}$ | 2330 | $10^{-10}$ | $n+\pi^{\pm}$ | 17.02 |
| Lambda; $\Lambda^0$ | 2182 | $2.7\times 10^{-10}$ | $p^{+}+\pi^{-}$ or $n+p^{+}$ | 15.92 |
| Neutron; $n$ | 1838.6 | $10^3$ | $p^{+}+e^{-}+\nu$ | 13.40 (Neutron and Proton) |
| Proton; $p$ | 1836.1 | stable | | |
| Tauon; $\tau^{\pm}$ | 966.5 | $10^{-8}$ | $\pi^{\pm}+\pi^0+\pi^0$, etc. | 7.05 (Tauon and Theton) |
| Theton; $\theta^0$ | 965 | $10^{-10}$ | $\pi^0+\pi^0$ or $\pi^{+}+\pi^{-}$ | |
| Pion; $\pi^{\pm}$ | 273.2 | $2.6\times 10^{-8}$ | $\mu^{\pm}+\nu$ | 1.995 |
| Pion; $\pi^0$ | 264.2 | $10^{-16}$ | $2\gamma$ | 1.928 |
| Muon; $\mu^{\pm}$ | 206.7 | $2.2\times 10^{-6}$ | $e^{\pm}+2\nu$ | 1.511 |
| Electron; $e^{\pm}$ | 1 | stable | | |
| Neutrino; $\nu$ | 0 | stable | | |

there is no guarantee that more will not be discovered in the near future. The photographs of elementary events are becoming more and more intricate, as the one shown in Plate VIII, *lower*. This picture is taken by a new device known as a "bubble chamber" which is, in a way, a reversal of the cloud chamber. Instead of liquid droplets formed in a gas, one uses here the gas bubbles formed in a liquid medium such as liquid hydrogen. Although our factual knowledge about elementary particles is rapidly increasing, we hit a solid wall in any attempt to understand them, and all the theories developed in this direction are so far of purely phenomenological nature.

## THROUGH THE LOOKING GLASS

If one finds a left shoe, one is certain that the right shoe is somewhere under the bed or sofa; the same is true for gloves, and many other objects. But all men and women have their hearts on the left side, and their appendices on the right. A more fundamental fact of biology is that the protein molecules forming every living creature, be it an amoeba, a man, a herring, or a rosebush, have a left-handed symmetry, and that the right-handed plant and animal world does not exist on the surface of the earth. It is very odd since, whenever an organic chemist synthesizes proteins from the elements, he gets 50% left-handed, and 50% right-handed molecules. It may be that during the early stages of the development of life on our planet there existed two living worlds: a right-handed and a left-handed one. Being indigestible or even poisonous to each other, they may have fought a battle in which one side was completely destroyed.

But, in regular physics, the principle of mirror-symmetry (known as the "parity principle") was always satisfied, and to any physical process there could be found another process which looked exactly as a mirror image of the first one. In the year 1956 two young Chinese-American physicists, Chen Ning Yang and Tsung Dao Lee, suggested on the basis of theoretical consideration that this may not be true in the case of elementary particles.

As was mentioned several times before, elementary particles and neutrons in particular can be considered as little spinning tops rotating around their axes. This rotation can of course be clockwise or counter-clockwise, and the two states of motion can be transformed into each other simply by turning the top upside down. The electron emitted in the decay of a neutron flies preferentially along its rotation axis, and it was believed that electrons are emitted with equal probability in both directions (i.e., so to speak, through the north or the south pole). If this were true, the parity principle would be satisfied, and the mirror image of the decaying neutron would be identical with the original since all one would have to do to make them coincide would be to turn one of them upside down. If, however, the electron is always emitted in one direction (Fig. VIII-20a), the situation becomes entirely different. In fact, looking at the image of a decaying neutron in a mirror (Fig. VIII-20b), one would find that there is no way to turn it so that it would coincide with the original. If, in both cases, the electron is emitted upward, as in Figure VIII-20, the two neutrons rotate in opposite directions. If one turns (mentally) either the image or the original upside down, the two electrons would be emitted in opposite directions. The parity principle would be broken, and the behavior of elementary particles on the other side of the looking glass would not be identical with that in front of it.

To test Yang and Lee's hypothesis, a direct experiment was carried out in order to find out whether there is any correlation between the direction of the neutron's rotation and the direction of the electron's emission. A beta-decaying radioactive material was cooled down to a very low temperature and placed in a very strong magnetic field. Under these conditions, when the thermal agitation practically dies out, all atoms become oriented in a single direction along the magnetic lines of force. If electrons were emitted equally in both directions in respect to the neutron's rotation axis, equal numbers of them would be observed flying toward the north and south poles of the electromagnet. The experiment led, however, to an entirely opposite conclusion, and, as Yang and Lee predicted, all electrons were flying in

FIG. VIII–20.
A mirror image of a neutron decay.

the same direction. Soon thereafter the same result was obtained for the decay of $\mu$ meson.

This was the fall of the principle of parity; the world of elementary particles was proved to be lopsided. Where is the other half of it which corresponds to "through-the-looking-glass" physics? We do not know, and will not know until the basic nature of the elementary particles is understood.

## THE FUTURE OF PHYSICS

It is clear from what was said before that the future of physics lies in further studies and understanding of elementary particles and, whereas the experimental progress in this field is well under way, the theory is practically at a standstill. Twenty-five centuries ago Democritus postulated that matter consists of discrete minimal portions, and we are now becoming more and more persuaded of the correctness of that postulate. It was only about half a century ago that we learned that energy too has an "atomic" structure, and are now speaking about energy-quanta. In the course of the last sixty years physicists have learned how to quantize various kinds of energy. In the case of electromagnetic radiation, energy can take only the values of $nh\nu$ where $\nu$ is the vibration frequency and $n$ is an integer. In a simple hydrogen atom, the energy of different quantum states varies as $1/n^2$ where $n$ is an integer. In other, more complicated cases the correct answers are given by Schrödinger's and Dirac's equations. But, in the case of material particles, we are still in a state of complete ignorance. We do not know why an electric charge has always the same value: $4.77 \times 10^{-10}$ esu. We haven't the slightest idea why the masses of particles are quantized, having the relative values shown in Table 1. And we do not have any better idea than Democritus had why matter should consist of indivisible particles instead of being truly continuous.

The answers to the above questions will constitute the physics of the future, but for the last few decades not a single successful step has been made in obtaining these answers, and nobody can predict when a breakthrough can be expected. But, although one does not know a correct answer, one should not be blamed for speculating on this kind of problem. Let us take the elementary charge $e$, for example. It is known that $e^2$ divided by the product of the velocity of light $c$ and quantum constant $h$ is a pure number or a *dimensionless constant,* which means that no matter whether we express $e$, $c$, and $h$ in centimeter-gram-second units, in inch-pound-hour units, or in any other system of units (provided they are used consistently), this ratio remains the same. This ratio is known as the "fine structure constant" because it emerges in the description of the splitting of the Balmer series lines into several very close components, and its numerical value is given by 1 divided by 137. Why 137, and not 75 or 533? In physical formulas numerical coefficients always have some mathematical meaning. For example, if one studies the relation between the period $T$ of a pendulum, its length $l$, and the acceleration of gravity $g$, no matter which units one uses one always comes to the formula:

$$T = 6.283\sqrt{\frac{l}{g}}$$

What is this number 6.283? Well, if one tries to correlate it with various numbers known in mathematics, one finds that it is actually $2\pi$. And indeed, using the equations of theoretical mechanics for the derivation of that formula, we find that the coefficient *must be* $2\pi$. Similarly, deriving an expression for elementary charge by using the equations of the relativistic quantum theory which contain the constants $c$ and $h$, one should be able to come to a conclusion that the ratio $hc/e^2$ (inverse fine structure constant) is given by a certain mathematical expression which is numerically equal to 137. But, nobody knows at present *how* to develop such a theory, and, whereas it is not difficult to guess that 6.283 is $2 \times 3.141 \ldots$, it is much more difficult to guess what kind of animal the number 137 is!

Sir Arthur Eddington, who made invaluable contributions to the theory of the internal structure of stars, made quite a number of years ago a brave attempt to explain 137. His arguments run roughly as follows: We live in a four-dimensional world ($x$, $y$, $z$, $ict$), and $4 \times 4 = 16$. Thus, let us build a *matrix,* i.e., a square table, with 16 lines and 16 columns. Let us further assume that this matrix is "symmetrical" in respect to its diagonal, i.e., that the contents of a square in $n$th line and $m$th column are identical with that in a square in $m$th line and $n$th column. How many independent squares will we have? Well, it is not difficult to calculate. The matrix has altogether $16 \times 16 = 256$ squares. Out of these, 16 belong to the diagonal, which leaves 240. Thus, each triangular area on both sides of the diagonal will have 120 squares. Since squares on both sides of the diagonal are identical, this leaves us 120 independent squares which, with the 16 squares on the diagonal, makes 136. At the time that Eddington first arrived at this relation, the empirical value was believed to be 136. Only a few years later better measurements brought it up to 137, which forced Eddington to work up an "amendment theory" requiring one additional unit.

This idea of Eddington was ridiculed in a short paper by G. Beck, H. Bethe, and W. Riezler, published in the Jan. 9, 1931, issue of a German magazine, *Naturwissenschaften.* This paper, attempting to show how dangerous it is to juggle with numbers, read as follows:

"Some Remarks on the
Quantum Theory of Zeropoint Temperature"

Let us consider a hexagonal crystallattice. Its absolute zeropoint is characterized by the freezing of all its degrees of freedom, excluding, of course, the motion of electrons along Bohr's orbits. According to Eddington each electron possesses $\frac{1}{\alpha} = 137$ degrees of freedom. Besides the electrons, the crystallattice also contains an equal number of protons. In order to arrive at the zeropoint temperature, we must ascribe to each neutron (i.e. a proton plus an electron) $\frac{2}{\alpha} - 1$ degrees of freedom, since one degree of freedom is frozen during the electron's motion in its orbit. Thus, we obtain for the zeropoint temperature:

$$T_0 = -\left(\frac{2}{\alpha} - 1\right) \text{ degrees}$$

Assuming that $\frac{1}{\alpha} = 137$ we obtain for the zeropoint temperature:

$$T_0 = -273 \text{ degrees}$$

which agrees very well with the experimental value. We notice that our result is independent of a special choice of crystallattice.

Of course, the above numerical relation between 137 and 273 is a pure coincidence since, whereas 137 is a true pure number, the absolute zero temperature will be given by different numbers, depending on whether we use centigrade, Fahrenheit, or Réaumur temperature scales. After the paper was published the editor of the magazine, being informed by one of Berlin's physicists that it was a hoax, wrote a fiery letter to the authors, who were at that time working at Cambridge University. Back came a humble answer, which said the authors were very sorry for the misunderstanding, but that they were sure that the paper would be considered as a parody of the way *certain* physicists build their theories. Thus, the next issue of *Naturwissenschaften* carried a note from the editor to the effect that it was hoped that all readers understood that Beck, Bethe, and Riezler's article was just a parody. And then Sir Arthur Eddington exploded!

Here is a piece of poetry written at that time by the abovementioned Vladimir Alexandrovich Fock: *

* Rendered by B.P.G. into English verse, from the original Russified German.

*137–1840*

Though we may weigh it as we will,
Exhausted and delirious,
*One-hundred-thirty-seven* still
Remains for us mysterious.
But Eddington, *he* sees it clear,
Denouncing those who tend to jeer;
  It is the number of (says he)
  The world's dimensions. *Can it be?!*—
  The world enfolding you and me?
  The world that holds Sir Arthur E.?
  The very world we smell and see?—
Oh come, he can't be serious!

Well, here's a number of my own
(In tit for tat I revel):
*One-thousand-eight-four-oh.* I've shown
It's strictly on the level.
Sir Arthur, *keep* your puny sum,
It's yours from now to Kingdom Come!
  My *1* and *8* and *4* and *0*
  Will fit a world we've yet to know—
  So on and upward with the show!
  And on my cauldron down Below
  Let these four figures shine and glow,
Bewildering the Devil!

This all happened about thirty years ago, but today we still do not know why that figure is 137 and not something else, and whether Eddington's "explanation" was due to a pure coincidence or has an element of truth in it. Of course, one can class Eddington's effort as "numerology," which carries a bad connotation today, but there is a very close word: "number theory," which is a large and honored branch of pure mathematics. In their efforts to solve the riddles of nature, physicists often looked for the help of pure mathematics, and in many cases obtained it. When Einstein wanted to interpret gravity as the curvature of four-dimensional, space-time continuum, he found waiting for him Riemann's theory of curved multidimensional space. When Heisenberg looked for some unusual kind of mathematics to describe the motion of electrons inside of an atom, noncommutative algebra was ready for him. Only the number theory and topology (*analysis situs*) still remain purely mathematical disciplines without any application to physics. Could it be that they will be called to help in our further understanding of the riddles of nature?

But, returning to the problems of physics tomorrow, one would probably find more difficulties with the explanation of the masses of elementary particles than with their electric charges. In fact, any formula which would express a mass in terms of velocity ($c$), action ($h$), and a numerical constant, must also include a *length*. One can write:

$$\text{mass} = A \times \frac{\text{action}}{\text{velocity} \times \text{length}}$$

where $A$ is some reasonable number as 1, $\sqrt{2}$, $\frac{3}{5}\pi$, $\frac{1}{2}\pi^2$, etc. If we take $A$ to be about 1, the action equal to $h$ ($6.55 \times 10^{-27}$), and the velocity equal to $c$ ($3 \times 10^{10}$), and want to obtain for the average mass of a material particle, i.e., the mass of a meson ($2 \times 10^{-25}$), we must take the length to be equal to about $10^{-12}$ cm. Of course, if $A$ is not 1, but, let us say $2\pi(\cong 6)$ or $\pi^2(\cong 10)$, the lengths may be as small as $10^{-13}$ cm. The lengths of that order of magnitude are very common in physics of elementary particles. The "radius of an electron" calculated on the basis of classical electrodynamic theory is $2.8 \times 10^{-13}$ cm, whereas the distance at which nuclear forces begin to act between two particles is known to be $1.4 \times 10^{-13}$ cm. Thus, it seems that the length of several times $10^{-13}$ cm has a fundamental meaning in the problems of elementary particles.

For a few decades now, theoretical physicists have entertained the hope that the length of the order of $10^{-13}$ cm, which is usually called $\lambda$, plays the role of *elementary length* in the future development of the theory. Just as $c$ is the *highest possible velocity* in the theory of relativity, and $h$ is the *smallest possible action* in the quantum theory, $\lambda$ is destined to play the role of the *shortest possible distance* in the future theory of matter. It will be, so to speak, "the diameter of a mathematical point," and it will make no sense to speak about distances smaller than that. This possibility is a very interesting and exciting dream, which will probably come true, but nobody knows, at present, just when.

In working up toward a dramatic conclusion of this volume, which has become too long anyway, we can bring one more "numerological" relation observed in the field of elementary particles. We cannot understand what 137 means, so let us express the masses of all elementary particles in terms of 137 electron masses. The result is shown in Table 1 (page 321); and we see that all figures are extremely close to integer numbers, except for two, which are close to an integer-and-a-half. It may be a coincidence, but the chance of such a coincidence is one out of many billions! And, if it is not a coincidence, what meaning does it have? Can the sequence of the "sacred numbers":

$$19;\ 17;\ 16;\ 13\tfrac{1}{2};\ 7;\ 2;\ 1\tfrac{1}{2}$$

be explained on the basis of some reasonable theory? Can it be, for example, related to the number theory, having some connection with the sequence of prime numbers, or more intricate sequences of numbers? Or is it rather related to topology, having some connection with the number of vertices, edges, faces, and space boundaries of the four-dimensional polyhedra? We do not know. But let us hope that the work of the physicists of future generations will bring these problems to a victorious solution.

# *Sources*

The quotations from the writings of the great physicists of the past were taken *mostly* from:

*A Source Book in Greek Science.* By M. R. Cohen and I. E. Drabkin. McGraw-Hill, N.Y., 1948.

*Dialogue on the Great World Systems.* By Galileo Galilei. University of Chicago Press, 1953.

*Sir Isaac Newton's Mathematical Principles of Natural Philosophy and His System of the World.* University of California Press, 1934.

*Opticks or a Treatise of the Reflections, Refractions, Inflections and Colours of Light.* By Sir Isaac Newton. G. Bell & Sons, Ltd., London, 1931.

*Faraday's Diary.* Published for the Royal Institution of Great Britain, G. Bell & Sons, Ltd., London, 1932.

# *Index*

A CATALOG OF SELECTED

# DOVER BOOKS

IN ALL FIELDS OF INTEREST

# A CATALOG OF SELECTED DOVER BOOKS IN ALL FIELDS OF INTEREST

DRAWINGS OF REMBRANDT, edited by Seymour Slive. Updated Lippmann, Hofstede de Groot edition, with definitive scholarly apparatus. All portraits, biblical sketches, landscapes, nudes. Oriental figures, classical studies, together with selection of work by followers. 550 illustrations. Total of 630pp. 9⅛ × 12¼.
21485-0, 21486-9 Pa., Two-vol. set $29.90

GHOST AND HORROR STORIES OF AMBROSE BIERCE, Ambrose Bierce. 24 tales vividly imagined, strangely prophetic, and decades ahead of their time in technical skill: "The Damned Thing," "An Inhabitant of Carcosa," "The Eyes of the Panther," "Moxon's Master," and 20 more. 199pp. 5⅜ × 8½. 20767-6 Pa. $4.95

ETHICAL WRITINGS OF MAIMONIDES, Maimonides. Most significant ethical works of great medieval sage, newly translated for utmost precision, readability. Laws Concerning Character Traits, Eight Chapters, more. 192pp. 5⅜ × 8½.
24522-5 Pa. $4.50

THE EXPLORATION OF THE COLORADO RIVER AND ITS CANYONS, J. W. Powell. Full text of Powell's 1,000-mile expedition down the fabled Colorado in 1869. Superb account of terrain, geology, vegetation, Indians, famine, mutiny, treacherous rapids, mighty canyons, during exploration of last unknown part of continental U.S. 400pp. 5⅜ × 8½. 20094-9 Pa. $7.95

HISTORY OF PHILOSOPHY, Julián Marías. Clearest one-volume history on the market. Every major philosopher and dozens of others, to Existentialism and later. 505pp. 5⅜ × 8½. 21739-6 Pa. $9.95

ALL ABOUT LIGHTNING, Martin A. Uman. Highly readable non-technical survey of nature and causes of lightning, thunderstorms, ball lightning, St. Elmo's Fire, much more. Illustrated. 192pp. 5⅜ × 8½. 25237-X Pa. $5.95

SAILING ALONE AROUND THE WORLD, Captain Joshua Slocum. First man to sail around the world, alone, in small boat. One of great feats of seamanship told in delightful manner. 67 illustrations. 294pp. 5⅜ × 8½. 20326-3 Pa. $4.95

LETTERS AND NOTES ON THE MANNERS, CUSTOMS AND CONDITIONS OF THE NORTH AMERICAN INDIANS, George Catlin. Classic account of life among Plains Indians: ceremonies, hunt, warfare, etc. 312 plates. 572pp. of text. 6⅛ × 9¼. 22118-0, 22119-9, Pa. Two-vol. set $17.90

ALASKA: The Harriman Expedition, 1899, John Burroughs, John Muir, et al. Informative, engrossing accounts of two-month, 9,000-mile expedition. Native peoples, wildlife, forests, geography, salmon industry, glaciers, more. Profusely illustrated. 240 black-and-white line drawings. 124 black-and-white photographs. 3 maps. Index. 576pp. 5⅜ × 8½. 25109-8 Pa. $11.95

THE BOOK OF BEASTS: Being a Translation from a Latin Bestiary of the Twelfth Century, T. H. White. Wonderful catalog real and fanciful beasts: manticore, griffin, phoenix, amphivius, jaculus, many more. White's witty erudite commentary on scientific, historical aspects. Fascinating glimpse of medieval mind. Illustrated. 296pp. 5⅜ × 8¼. (Available in U.S. only) 24609-4 Pa. $6.95

FRANK LLOYD WRIGHT: ARCHITECTURE AND NATURE With 160 Illustrations, Donald Hoffmann. Profusely illustrated study of influence of nature—especially prairie—on Wright's designs for Fallingwater, Robie House, Guggenheim Museum, other masterpieces. 96pp. 9¼ × 10¾. 25098-9 Pa. $8.95

FRANK LLOYD WRIGHT'S FALLINGWATER, Donald Hoffmann. Wright's famous waterfall house: planning and construction of organic idea. History of site, owners, Wright's personal involvement. Photographs of various stages of building. Preface by Edgar Kaufmann, Jr. 100 illustrations. 112pp. 9¼ × 10.
23671-4 Pa. $8.95

YEARS WITH FRANK LLOYD WRIGHT: Apprentice to Genius, Edgar Tafel. Insightful memoir by a former apprentice presents a revealing portrait of Wright the man, the inspired teacher, the greatest American architect. 372 black-and-white illustrations. Preface. Index. vi + 228pp. 8¼ × 11. 24801-1 Pa. $10.95

THE STORY OF KING ARTHUR AND HIS KNIGHTS, Howard Pyle. Enchanting version of King Arthur fable has delighted generations with imaginative narratives of exciting adventures and unforgettable illustrations by the author. 41 illustrations. xviii + 313pp. 6⅛ × 9¼. 21445-1 Pa. $6.95

THE GODS OF THE EGYPTIANS, E. A. Wallis Budge. Thorough coverage of numerous gods of ancient Egypt by foremost Egyptologist. Information on evolution of cults, rites and gods; the cult of Osiris; the Book of the Dead and its rites; the sacred animals and birds; Heaven and Hell; and more. 956pp. 6⅛ × 9¼.
22055-9, 22056-7 Pa., Two-vol. set $21.90

A THEOLOGICO-POLITICAL TREATISE, Benedict Spinoza. Also contains unfinished *Political Treatise*. Great classic on religious liberty, theory of government on common consent. R. Elwes translation. Total of 421pp. 5⅜ × 8½.
20249-6 Pa. $7.95

INCIDENTS OF TRAVEL IN CENTRAL AMERICA, CHIAPAS, AND YUCATAN, John L. Stephens. Almost single-handed discovery of Maya culture; exploration of ruined cities, monuments, temples; customs of Indians. 115 drawings. 892pp. 5⅜ × 8½. 22404-X, 22405-8 Pa., Two-vol. set $15.90

LOS CAPRICHOS, Francisco Goya. 80 plates of wild, grotesque monsters and caricatures. Prado manuscript included. 183pp. 6⅝ × 9⅝. 22384-1 Pa. $5.95

AUTOBIOGRAPHY: The Story of My Experiments with Truth, Mohandas K. Gandhi. Not hagiography, but Gandhi in his own words. Boyhood, legal studies, purification, the growth of the Satyagraha (nonviolent protest) movement. Critical, inspiring work of the man who freed India. 480pp. 5⅜ × 8½. (Available in U.S. only)
24593-4 Pa. $6.95

ILLUSTRATED DICTIONARY OF HISTORIC ARCHITECTURE, edited by Cyril M. Harris. Extraordinary compendium of clear, concise definitions for over 5,000 important architectural terms complemented by over 2,000 line drawings. Covers full spectrum of architecture from ancient ruins to 20th-century Modernism. Preface. 592pp. 7½ × 9⅜. 24444-X Pa. $15.95

THE NIGHT BEFORE CHRISTMAS, Clement Moore. Full text, and woodcuts from original 1848 book. Also critical, historical material. 19 illustrations. 40pp. 4⅝ × 6. 22797-9 Pa. $2.50

THE LESSON OF JAPANESE ARCHITECTURE: 165 Photographs, Jiro Harada. Memorable gallery of 165 photographs taken in the 1930's of exquisite Japanese homes of the well-to-do and historic buildings. 13 line diagrams. 192pp. 8⅜ × 11¼. 24778-3 Pa. $10.95

THE AUTOBIOGRAPHY OF CHARLES DARWIN AND SELECTED LETTERS, edited by Francis Darwin. The fascinating life of eccentric genius composed of an intimate memoir by Darwin (intended for his children); commentary by his son, Francis; hundreds of fragments from notebooks, journals, papers; and letters to and from Lyell, Hooker, Huxley, Wallace and Henslow. xi + 365pp. 5⅜ × 8. 20479-0 Pa. $6.95

WONDERS OF THE SKY: Observing Rainbows, Comets, Eclipses, the Stars and Other Phenomena, Fred Schaaf. Charming, easy-to-read poetic guide to all manner of celestial events visible to the naked eye. Mock suns, glories, Belt of Venus, more. Illustrated. 299pp. 5¼ × 8¼. 24402-4 Pa. $7.95

BURNHAM'S CELESTIAL HANDBOOK, Robert Burnham, Jr. Thorough guide to the stars beyond our solar system. Exhaustive treatment. Alphabetical by constellation: Andromeda to Cetus in Vol. 1; Chamaeleon to Orion in Vol. 2; and Pavo to Vulpecula in Vol. 3. Hundreds of illustrations. Index in Vol. 3. 2,000pp. 6⅛ × 9¼. 23567-X, 23568-8, 23673-0 Pa., Three-vol. set $41.85

STAR NAMES: Their Lore and Meaning, Richard Hinckley Allen. Fascinating history of names various cultures have given to constellations and literary and folkloristic uses that have been made of stars. Indexes to subjects. Arabic and Greek names. Biblical references. Bibliography. 563pp. 5⅜ × 8½. 21079-0 Pa. $8.95

THIRTY YEARS THAT SHOOK PHYSICS: The Story of Quantum Theory, George Gamow. Lucid, accessible introduction to influential theory of energy and matter. Careful explanations of Dirac's anti-particles, Bohr's model of the atom, much more. 12 plates. Numerous drawings. 240pp. 5⅜ × 8½. 24895-X Pa. $5.95

CHINESE DOMESTIC FURNITURE IN PHOTOGRAPHS AND MEASURED DRAWINGS, Gustav Ecke. A rare volume, now affordably priced for antique collectors, furniture buffs and art historians. Detailed review of styles ranging from early Shang to late Ming. Unabridged republication. 161 black-and-white drawings, photos. Total of 224pp. 8⅜ × 11¼. (Available in U.S. only) 25171-3 Pa. $13.95

VINCENT VAN GOGH: A Biography, Julius Meier-Graefe. Dynamic, penetrating study of artist's life, relationship with brother, Theo, painting techniques, travels, more. Readable, engrossing. 160pp. 5⅜ × 8½. (Available in U.S. only) 25253-1 Pa. $4.95

ILLUSTRATED GUIDE TO SHAKER FURNITURE, Robert Meader. All furniture and appurtenances, with much on unknown local styles. 235 photos. 146pp. 9 × 12. 22819-3 Pa. $8.95

WHALE SHIPS AND WHALING: A Pictorial Survey, George Francis Dow. Over 200 vintage engravings, drawings, photographs of barks, brigs, cutters, other vessels. Also harpoons, lances, whaling guns, many other artifacts. Comprehensive text by foremost authority. 207 black-and-white illustrations. 288pp. 6 × 9. 24808-9 Pa. $9.95

THE BERTRAMS, Anthony Trollope. Powerful portrayal of blind self-will and thwarted ambition includes one of Trollope's most heartrending love stories. 497pp. 5⅜ × 8½. 25119-5 Pa. $9.95

ADVENTURES WITH A HAND LENS, Richard Headstrom. Clearly written guide to observing and studying flowers and grasses, fish scales, moth and insect wings, egg cases, buds, feathers, seeds, leaf scars, moss, molds, ferns, common crystals, etc.—all with an ordinary, inexpensive magnifying glass. 209 exact line drawings aid in your discoveries. 220pp. 5⅜ × 8½. 23330-8 Pa. $4.95

RODIN ON ART AND ARTISTS, Auguste Rodin. Great sculptor's candid, wide-ranging comments on meaning of art; great artists; relation of sculpture to poetry, painting, music; philosophy of life, more. 76 superb black-and-white illustrations of Rodin's sculpture, drawings and prints. 119pp. 8⅜ × 11¼. 24487-3 Pa. $7.95

FIFTY CLASSIC FRENCH FILMS, 1912–1982: A Pictorial Record, Anthony Slide. Memorable stills from Grand Illusion, Beauty and the Beast, Hiroshima, Mon Amour, many more. Credits, plot synopses, reviews, etc. 160pp. 8¼ × 11. 25256-6 Pa. $11.95

THE PRINCIPLES OF PSYCHOLOGY, William James. Famous long course complete, unabridged. Stream of thought, time perception, memory, experimental methods; great work decades ahead of its time. 94 figures. 1,391pp. 5⅜ × 8½. 20381-6, 20382-4 Pa., Two-vol. set $23.90

BODIES IN A BOOKSHOP, R. T. Campbell. Challenging mystery of blackmail and murder with ingenious plot and superbly drawn characters. In the best tradition of British suspense fiction. 192pp. 5⅜ × 8½. 24720-1 Pa. $4.95

CALLAS: PORTRAIT OF A PRIMA DONNA, George Jellinek. Renowned commentator on the musical scene chronicles incredible career and life of the most controversial, fascinating, influential operatic personality of our time. 64 black-and-white photographs. 416pp. 5⅜ × 8¼. 25047-4 Pa. $8.95

GEOMETRY, RELATIVITY AND THE FOURTH DIMENSION, Rudolph Rucker. Exposition of fourth dimension, concepts of relativity as Flatland characters continue adventures. Popular, easily followed yet accurate, profound. 141 illustrations. 133pp. 5⅜ × 8½. 23400-2 Pa. $4.95

HOUSEHOLD STORIES BY THE BROTHERS GRIMM, with pictures by Walter Crane. 53 classic stories—Rumpelstiltskin, Rapunzel, Hansel and Gretel, the Fisherman and his Wife, Snow White, Tom Thumb, Sleeping Beauty, Cinderella, and so much more—lavishly illustrated with original 19th century drawings. 114 illustrations. x + 269pp. 5⅜ × 8½. 21080-4 Pa. $4.95

PLANTS OF THE BIBLE, Harold N. Moldenke and Alma L. Moldenke. Standard reference to all 230 plants mentioned in Scriptures. Latin name, biblical reference, uses, modern identity, much more. Unsurpassed encyclopedic resource for scholars, botanists, nature lovers, students of Bible. Bibliography. Indexes. 123 black-and-white illustrations. 384pp. 6 × 9. 25069-5 Pa. $8.95

FAMOUS AMERICAN WOMEN: A Biographical Dictionary from Colonial Times to the Present, Robert McHenry, ed. From Pocahontas to Rosa Parks, 1,035 distinguished American women documented in separate biographical entries. Accurate, up-to-date data, numerous categories, spans 400 years. Indices. 493pp. 6½ × 9¼. 24523-3 Pa. $10.95

THE FABULOUS INTERIORS OF THE GREAT OCEAN LINERS IN HISTORIC PHOTOGRAPHS, William H. Miller, Jr. Some 200 superb photographs capture exquisite interiors of world's great "floating palaces"—1890's to 1980's: *Titanic, Ile de France, Queen Elizabeth, United States, Europa,* more. Approx. 200 black-and-white photographs. Captions. Text. Introduction. 160pp. 8⅞ × 11¾. 24756-2 Pa. $9.95

THE GREAT LUXURY LINERS, 1927-1954: A Photographic Record, William H. Miller, Jr. Nostalgic tribute to heyday of ocean liners. 186 photos of Ile de France, Normandie, Leviathan, Queen Elizabeth, United States, many others. Interior and exterior views. Introduction. Captions. 160pp. 9 × 12. 24056-8 Pa. $10.95

A NATURAL HISTORY OF THE DUCKS, John Charles Phillips. Great landmark of ornithology offers complete detailed coverage of nearly 200 species and subspecies of ducks: gadwall, sheldrake, merganser, pintail, many more. 74 full-color plates, 102 black-and-white. Bibliography. Total of 1,920pp. 8⅝ × 11¼. 25141-1, 25142-X Cloth. Two-vol. set $100.00

THE SEAWEED HANDBOOK: An Illustrated Guide to Seaweeds from North Carolina to Canada, Thomas F. Lee. Concise reference covers 78 species. Scientific and common names, habitat, distribution, more. Finding keys for easy identification. 224pp. 5⅜ × 8½. 25215-9 Pa. $6.95

THE TEN BOOKS OF ARCHITECTURE: The 1755 Leoni Edition, Leon Battista Alberti. Rare classic helped introduce the glories of ancient architecture to the Renaissance. 68 black-and-white plates. 336pp. 8⅜ × 11¼. 25239-6 Pa. $14.95

MISS MACKENZIE, Anthony Trollope. Minor masterpieces by Victorian master unmasks many truths about life in 19th-century England. First inexpensive edition in years. 392pp. 5⅜ × 8½. 25201-9 Pa. $8.95

THE RIME OF THE ANCIENT MARINER, Gustave Doré, Samuel Taylor Coleridge. Dramatic engravings considered by many to be his greatest work. The terrifying space of the open sea, the storms and whirlpools of an unknown ocean, the ice of Antarctica, more—all rendered in a powerful, chilling manner. Full text. 38 plates. 77pp. 9¼ × 12. 22305-1 Pa. $4.95

THE EXPEDITIONS OF ZEBULON MONTGOMERY PIKE, Zebulon Montgomery Pike. Fascinating first-hand accounts (1805-6) of exploration of Mississippi River, Indian wars, capture by Spanish dragoons, much more. 1,088pp. 5⅜ × 8½. 25254-X, 25255-8 Pa. Two-vol. set $25.90

A CONCISE HISTORY OF PHOTOGRAPHY: Third Revised Edition, Helmut Gernsheim. Best one-volume history—camera obscura, photochemistry, daguerreotypes, evolution of cameras, film, more. Also artistic aspects—landscape, portraits, fine art, etc. 281 black-and-white photographs. 26 in color. 176pp. 8⅜ × 11¼. 25128-4 Pa. $13.95

THE DORÉ BIBLE ILLUSTRATIONS, Gustave Doré. 241 detailed plates from the Bible: the Creation scenes, Adam and Eve, Flood, Babylon, battle sequences, life of Jesus, etc. Each plate is accompanied by the verses from the King James version of the Bible. 241pp. 9 × 12. 23004-X Pa. $9.95

WANDERINGS IN WEST AFRICA, Richard F. Burton. Great Victorian scholar/adventurer's invaluable descriptions of African tribal rituals, fetishism, culture, art, much more. Fascinating 19th-century account. 624pp. 5⅜ × 8½. 26890-X Pa. $12.95

FLATLAND, E. A. Abbott. Intriguing and enormously popular science-fiction classic explores the complexities of trying to survive as a two-dimensional being in a three-dimensional world. Amusingly illustrated by the author. 16 illustrations. 103pp. 5⅜ × 8½. 20001-9 Pa. $2.50

THE HISTORY OF THE LEWIS AND CLARK EXPEDITION, Meriwether Lewis and William Clark, edited by Elliott Coues. Classic edition of Lewis and Clark's day-by-day journals that later became the basis for U.S. claims to Oregon and the West. Accurate and invaluable geographical, botanical, biological, meteorological and anthropological material. Total of 1,508pp. 5⅜ × 8½.
21268-8, 21269-6, 21270-X Pa. Three-vol. set $26.85

LANGUAGE, TRUTH AND LOGIC, Alfred J. Ayer. Famous, clear introduction to Vienna, Cambridge schools of Logical Positivism. Role of philosophy, elimination of metaphysics, nature of analysis, etc. 160pp. 5⅜ × 8½. (Available in U.S. and Canada only) 20010-8 Pa. $3.95

MATHEMATICS FOR THE NONMATHEMATICIAN, Morris Kline. Detailed, college-level treatment of mathematics in cultural and historical context, with numerous exercises. For liberal arts students. Preface. Recommended Reading Lists. Tables. Index. Numerous black-and-white figures. xvi + 641pp. 5⅜ × 8½.
24823-2 Pa. $11.95

HANDBOOK OF PICTORIAL SYMBOLS, Rudolph Modley. 3,250 signs and symbols, many systems in full; official or heavy commercial use. Arranged by subject. Most in Pictorial Archive series. 143pp. 8⅛ × 11. 23357-X Pa. $6.95

INCIDENTS OF TRAVEL IN YUCATAN, John L. Stephens. Classic (1843) exploration of jungles of Yucatan, looking for evidences of Maya civilization. Travel adventures, Mexican and Indian culture, etc. Total of 669pp. 5⅜ × 8½.
20926-1, 20927-X Pa., Two-vol. set $11.90

SIR HARRY HOTSPUR OF HUMBLETHWAITE, Anthony Trollope. Incisive, unconventional psychological study of a conflict between a wealthy baronet, his idealistic daughter, and their scapegrace cousin. The 1870 novel in its first inexpensive edition in years. 250pp. 5⅜ × 8½. 24953-0 Pa. $6.95

LASERS AND HOLOGRAPHY, Winston E. Kock. Sound introduction to burgeoning field, expanded (1981) for second edition. Wave patterns, coherence, lasers, diffraction, zone plates, properties of holograms, recent advances. 84 illustrations. 160pp. 5⅜ × 8¼. (Except in United Kingdom) 24041-X Pa. $3.95

INTRODUCTION TO ARTIFICIAL INTELLIGENCE: SECOND, ENLARGED EDITION, Philip C. Jackson, Jr. Comprehensive survey of artificial intelligence—the study of how machines (computers) can be made to act intelligently. Includes introductory and advanced material. Extensive notes updating the main text. 132 black-and-white illustrations. 512pp. 5⅜ × 8½. 24864-X Pa. $8.95

HISTORY OF INDIAN AND INDONESIAN ART, Ananda K. Coomaraswamy. Over 400 illustrations illuminate classic study of Indian art from earliest Harappa finds to early 20th century. Provides philosophical, religious and social insights. 304pp. 6⅝ × 9⅝. 25005-9 Pa. $11.95

THE GOLEM, Gustav Meyrink. Most famous supernatural novel in modern European literature, set in Ghetto of Old Prague around 1890. Compelling story of mystical experiences, strange transformations, profound terror. 13 black-and-white illustrations. 224pp. 5⅜ × 8½. (Available in U.S. only) 25025-3 Pa. $6.95

PICTORIAL ENCYCLOPEDIA OF HISTORIC ARCHITECTURAL PLANS, DETAILS AND ELEMENTS: With 1,880 Line Drawings of Arches, Domes, Doorways, Facades, Gables, Windows, etc., John Theodore Haneman. Sourcebook of inspiration for architects, designers, others. Bibliography. Captions. 141pp. 9 × 12. 24605-1 Pa. $7.95

BENCHLEY LOST AND FOUND, Robert Benchley. Finest humor from early 30's, about pet peeves, child psychologists, post office and others. Mostly unavailable elsewhere. 73 illustrations by Peter Arno and others. 183pp. 5⅜ × 8½. 22410-4 Pa. $4.95

ERTÉ GRAPHICS, Erté. Collection of striking color graphics: *Seasons, Alphabet, Numerals, Aces* and *Precious Stones*. 50 plates, including 4 on covers. 48pp. 9⅜ × 12¼. 23580-7 Pa. $7.95

THE JOURNAL OF HENRY D. THOREAU, edited by Bradford Torrey, F. H. Allen. Complete reprinting of 14 volumes, 1837–61, over two million words; the sourcebooks for *Walden*, etc. Definitive. All original sketches, plus 75 photographs. 1,804pp. 8½ × 12¼. 20312-3, 20313-1 Cloth., Two-vol. set $125.00

CASTLES: THEIR CONSTRUCTION AND HISTORY, Sidney Toy. Traces castle development from ancient roots. Nearly 200 photographs and drawings illustrate moats, keeps, baileys, many other features. Caernarvon, Dover Castles, Hadrian's Wall, Tower of London, dozens more. 256pp. 5⅜ × 8¼. 24898-4 Pa. $6.95